T0140109

Studies in Computational Intelligence

Volume 675

Series editor

Janusz Kacprzyk, Polish Academy of Sciences, Warsaw, Poland
e-mail: kacprzyk@ibspan.waw.pl

About this Series

The series "Studies in Computational Intelligence" (SCI) publishes new developments and advances in the various areas of computational intelligence—quickly and with a high quality. The intent is to cover the theory, applications, and design methods of computational intelligence, as embedded in the fields of engineering, computer science, physics and life sciences, as well as the methodologies behind them. The series contains monographs, lecture notes and edited volumes in computational intelligence spanning the areas of neural networks, connectionist systems, genetic algorithms, evolutionary computation, artificial intelligence, cellular automata, self-organizing systems, soft computing, fuzzy systems, and hybrid intelligent systems. Of particular value to both the contributors and the readership are the short publication timeframe and the worldwide distribution, which enable both wide and rapid dissemination of research output.

More information about this series at http://www.springer.com/series/7092

Honghai Liu · Zhaojie Ju
Xiaofei Ji · Chee Seng Chan
Mehdi Khoury

Human Motion Sensing and Recognition

A Fuzzy Qualitative Approach

Honghai Liu
School of Computing
University of Portsmouth
Portsmouth
UK

Zhaojie Ju
School of Computing
University of Portsmouth
Portsmouth
UK

Xiaofei Ji
School of Automation
Shenyang Aerospace University
Shenyang
China

Chee Seng Chan
Faculty of Computer Science and Information Technology
University of Malaya
Kuala Lumpur
Malaysia

Mehdi Khoury
Mathematics and Physical Sciences
University of Exeter
Exeter
UK

ISSN 1860-949X ISSN 1860-9503 (electronic)
Studies in Computational Intelligence
ISBN 978-3-662-57152-1 ISBN 978-3-662-53692-6 (eBook)
DOI 10.1007/978-3-662-53692-6

Printed on acid-free paper

This Springer imprint is published by Springer Nature
The registered company is Springer-Verlag GmbH Germany
The registered company address is: Heidelberger Platz 3, 14197 Berlin, Germany

Preface

Human motion sensing and recognition are currently one of the most important and active research topics because of the strongly growing interests driven by a wide spectrum of promising applications in many areas such as surveillance and security, human computer interactions/interfaces, human skill transfer, human robot corporation, animation and gaming, sport analysis, *etc*. With recent advances in sensing techniques, various motion capture devices have been developed and applied to the above applications, including high-resolution cameras, depth cameras, marker-based motion capture devices, data gloves, biosignal sensors, *etc*. The major challenges for recognising human motion stem mainly from first the dynamic environment where there exist some unpredicted lighting conditions, concurrent body structures or motion occlusions, second from multi-modal data which requires refined sensor integration and fusion algorithms, third from uncertain sensory information with inaccuracy, noise or fault, and finally from the limited response time required in real-time systems. Therefore, the sensing and recognition systems need to not only extract the most useful sensory information from single or multiple sources, but also avoid heavy computational cost.

This book aims to introduce readers to the latest progresses made on human motion sensing and recognition, including 2D and 3D vision-based human motion analysis, hand motion analysis and view-invariant motion recognition, from the perspective of Fuzzy Qualitative (FQ) techniques, which provide practical solutions to the above challenges by properly balancing the system's effectiveness and efficiency. Chapters in this book are deliberately arranged to cover both the theories of the fuzzy approaches and their applications in motion analysis. The book is primarily intended to be used as a valuable reference for related researchers and engineers, who are seeking practical solutions to human motion sensing and recognition. It can also be useful to practitioners for advanced knowledge, and university undergraduates and postgraduates with interest in fuzzy techniques and their applications in motion recognition.

Twelve chapters are carefully selected and arranged from our recent research work. Research background is discussed in Chap. 1. Chapters 2, 3, 5 and 6 focus on the important theory concept development of fuzzy techniques, including FQ trigonometry, FQ robot kinematic, Fuzzy Empirical Copula (FEC) and Fuzzy Gaussian Mixture Models (FGMMs), while Chaps. 4, 7, 8 and 9 explore their applications in human motion analysis. Chapters 10, 11 and 12 are self-contained ones providing practical solutions particularly to 3D and view-invariant motion recognition using novel approaches.

Chapter 1 gives a general introduction of the human motion sensing techniques, including vision-based sensing, wearable sensing and multimodal sensing, and recent recognition methods, such as probabilistic graphical models, support vector machines, neural networks and fuzzy approaches, and then highlights the motivation of this book with discussion on the current challenges in the human motion analysis.

Chapter 2 presents a novel FQ representation of conventional trigonometry to effectively bridge the gap between symbolic cognitive functions and numerical sensing and control tasks in physical systems and intelligent robotics. Fuzzy logic and qualitative reasoning techniques are employed to convert conventional trigonometric functions, rules and their extensions to triangles in Euclidean space into their counterparts in FQ coordinates. This proposed approach provides a promising representation transformation interface to easily communicate between the numeric world and qualitative world in general trigonometry-related physical systems from an artificial intelligence perspective.

Chapter 3 extends the FQ trigonometry to FQ robot kinematics by replacing the trigonometry role in robot kinematics. FQ robot kinematics have been explored and discussed in terms of FQ transformation, position and velocity of a serial kinematics. An aggregation operator is proposed to extract robot behaviour giving prominence to the impact of FQ Robot Kinematics to intelligent robotics. The proposed two versions of FQ approaches have been integrated into XTRIG MATLAB toolbox with a case study on a PUMA robot.

Chapter 4 combines the FQ robot kinematics with human motion tracking and recognition algorithms to analyse vision-based human motion analysis, especially to recognise human motions. A data quantisation process is proposed with the aim of saving computational cost and Qualitative Normalised Template (QNT) is developed by adapting the FQ robot kinematics to represent human motions. Comparative experimental results have been reported to support the effectiveness of the proposed method.

Chapter 5 presents a novel fuzzy approach, FEC, to reduce the computational cost of empirical copula, which is a non-parametric algorithm to estimate the dependence structure of high-dimensional arbitrarily distributed data, with a new version of Fuzzy Clustering by Local Approximation of Memberships (FLAME). Two case studies demonstrate the efficiency of the proposed method.

Chapter 6 introduces a fuzzy version of Gaussian Mixture Models (GMMs) with a fast convergence speed by using a degree of fuzziness onto the dissimilarity function based on distances or probabilities. The FGMMs outperform traditional GMMs because they not only suit for nonlinear datasets but also saves half of computational cost.

Chapter 7 proposes a unified fuzzy framework of a set of recognition algorithms: Time Clustering (TC), FGMMs and FEC, from numerical clustering to data dependency structure in the context of optimally real-time human hand motion recognition. TC is a fuzzy time modelling approach based on fuzzy clustering and Takagi-Sugeno (TS) modelling with numerical value as output; FGMMs efficiently extract abstract Gaussian pattern to represent components of hand gestures; FEC utilises the dependence structure among the finger joint angles to recognise motion types. Various tests have been reported to evaluate this framework's performance.

Chapter 8 develops a generalised framework to study and analyse multimodal data of human hand motions, with modules of sensor integration, signal preprocessing, correlation study of sensory information and motion identification. Fuzzy approaches have been utilised, *e.g.* FEC is used to reveal significant relationships among the sensory data; and FGMMs are employed to identify different hand grasps and manipulations based on the forearm electromyography (EMG) signals.

Chapter 9 shows a new approach to extract human hand gesture features in real-time from RGB-D images by the earth mover's distance and Lasso algorithms. A modified finger earth mover's distance algorithm is employed to locate the palm image and then a Lasso algorithm is proposed to effectively extract the fingertip feature from a hand contour curve.

Chapter 10 focuses on the problem of recognising occluded 3D human motion assisted by the recognition context. Fuzzy Quantile Generation (FQG) is proposed by using metrics derived from the probabilistic quantile function to represent uncertainties via a fuzzy membership function. Time-dependent and context-aware rules are then generated to smooth the qualitative outputs represented by fuzzy membership functions. Feature selection and reconstruction are also discussed to deal with the motion occlusion.

Chapter 11 improves 3D human motion recognition by combining a local spatiotemporal feature with a global Positional Distribution Information (PDI) of interest points. For each interest point, 3D Scale-Invariant Feature Transform (SIFT) descriptors are extracted and then integrated with the computed PDI. The combined feature is evaluated and compared on the public KTH dataset using the Support Vector Machine (SVM) recognition algorithm.

Chapter 12 addresses the issue of view-invariant action recognition by presenting a multi-view space Hidden Markov Models (HMMs). A view-insensitive feature is proposed by combining the bag-of-words of interest point with the amplitude histogram of optical flow. Based on the above combined feature, HMMs are then used to recognise unknown viewpoint by analysing the derived similarity in each sub-view space. Experimental results on multi-view action dataset IXMAS are discussed.

Portsmouth, UK Honghai Liu
Portsmouth, UK Zhaojie Ju
Shenyang, China Xiaofei Ji
Kuala Lumpur, Malaysia Chee Seng Chan
Exeter, UK Mehdi Khoury

Contents

Acronyms

AcaGMMs	Active Curve Axis Gaussian Mixture Models
CCA	Curvilinear Component Analysis
CRF	Conditional Random Fields
CSO	Cluster Supporting Object
DSP	Digital Signal Processor
EM	Expectation–Maximization
EMG	Electromyography
FA	Factor Analysis
FCMs	Fuzzy C-means
FEC	Fuzzy Empirical Copula
FGMMs	Fuzzy Gaussian Mixture Models
FHMM	Fuzzy Hidden Markov Model
FLAME	Fuzzy Clustering by Local Approximation of Memberships
FPF	False Positive Fraction
FQ	Fuzzy Qualitative
FQG	Fuzzy Quantile Generation
FQT	Fuzzy Qualitative Trigonometry
FSM	Finite State Machine
GMMs	Gaussian Mixture Models
GMR	Gaussian Mixture Regression
HMMs	Hidden Markov Models
ICA	Independent Component Analysis
KL	Kullaback–Leibler
KNN	K-Nearest Neighbours
MEMM	Maximum Entropy Markov Models
MEI	Motion Energy Image
MHI	Motion History Image
MR	Midpoint-Radius
NP	Normal Points
PC	Principal Curves

PCA	Principal Component Analysis
PDI	Positional Distribution Information
PMK	Pyramid Match Kernel
PP	Projection Pursuit
QNT	Qualitative Normalised Template
RMS	Root Mean Square
ROC	Receiver Operating Characteristic
RVM	Relevance Vector Machines
SIFT	Scale-Invariant Feature Transform
SMC	Sequential Monte Carlo
SOM	Self-Organizing Maps
SSM	Self-Similarity Matrix
SVM	Support Vector Machine
TC	Time Clustering
TDNN	Time Delay Neural Network
TPF	True Positive Fraction
TS	Takagi–Sugeno
TSK	Takagi–Sugeno–Kang

Chapter 1
Introduction

Human motion analysis mainly involves body sensing, which employs sensors to extract static or dynamic information of the human body while performing behaviours including gestures, actions and activities, and behaviour recognition, which uses software tools to automatically interpret and understand these behaviours to help the computer make further actions, for example, alarming or decision making. Human motion analysis has been receiving considerable attention for more than twenty years from computer vision researchers in applications such as surveillance and security, image storage and retrieval, human-computer interaction, etc. Recent advances in the sensing techniques greatly broaden the interest to other research areas, such as robotics, healthcare, computer games, animation and film, etc. This chapter firstly gives a brief introduction of the human motion sensing techniques and recent recognition approaches, and then highlights the motivation of this book with discussion on the current challenges of the human motion analysis.

1.1 Human Motion Sensing Techniques

Human motion sensing systems are expected to consistently capture spatio-temporal information that dynamically represents the movements of a human body or a part of it. This useful information extracted from the sensing data can be on position/location, speed/acceleration, joint angles, contact pressure, vibration, heart rate, muscle contraction, etc., subject to types of the employed sensors. The sensing techniques for human motions are generally categorised into three types: vision-based sensing, wearable sensing and multimodal sensing. Figure 1.1 shows a classification of recent popular sensor techniques that will be discussed in this session.

H. Liu et al., *Human Motion Sensing and Recognition*,
Studies in Computational Intelligence 675, DOI 10.1007/978-3-662-53692-6_1

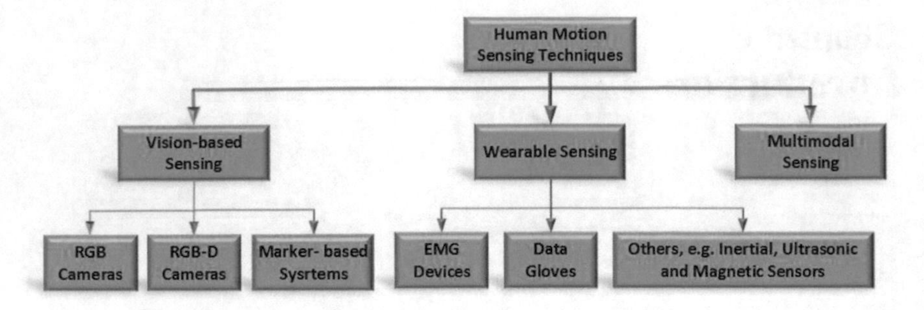

Fig. 1.1 Human motion sensing techniques

1.1.1 Vision-Based Sensing

Cameras are normally used to improve the accuracy in position/location estimation in the vision-based sensing systems, which provide a natural way of measuring human motions without any physical restrictions on the human body. The vision-based sensing systems can be mainly classified into Red-Green-Blue (RGB) cameras, Red-Green-Blue-Depth (RGB-D) cameras, and marker-based sensing systems.

1.1.1.1 RGB Cameras

Cameras have been used extensively to track human motions for decades and are still one of the most popular devices in motion sensing [1–4]. As a low cost device, a RGB camera can easily achieve millions of pixels with above twenty frames per second, which provide a rich source and a high accuracy in detecting human motions. The size of the camera is becoming much smaller, and it has been integrated in most personal mobile phones, computers and smart devices. It has been widely used in surveillance applications, such as the human action recognition [5, 6]. Since the human body has high dynamics and the occlusion often occurs due to the perspective projection, it is challenging to automatically extract human movements from the monocular RGB images or videos. To alleviate the above problem, multiple RGB cameras, which are separated by certain angles, are employed to observe three-dimensional (3D) human motions from different directions with a higher accuracy. For example, as shown in Fig. 1.2, two estimated skeletons in different views from two pre-set monocular cameras were matched and reconstructed into a 3D skeleton for human behaviour recognition [7]. However, it requires these cameras to be pre-calibrated and mounted at fixed locations, which limits the sensing area with a higher hardware expense.

A stereo camera, which integrates two or more lenses with a separate image sensor or film frame for each lens, is used to capture three-dimensional images as a single device or to improve the performance of the motion recognition [8–11]. For example, a single stereo camera was used to recognise human actions [12], where any actions taken from different viewpoints can be recognised independently of the viewpoints

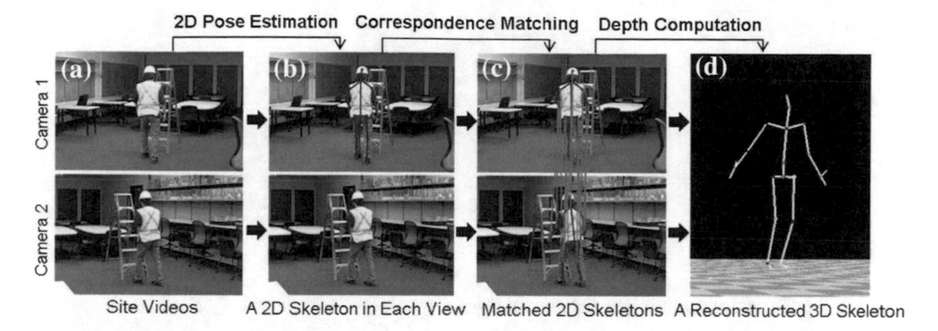

Fig. 1.2 Two monocular cameras were used for human behaviour recognition [7]

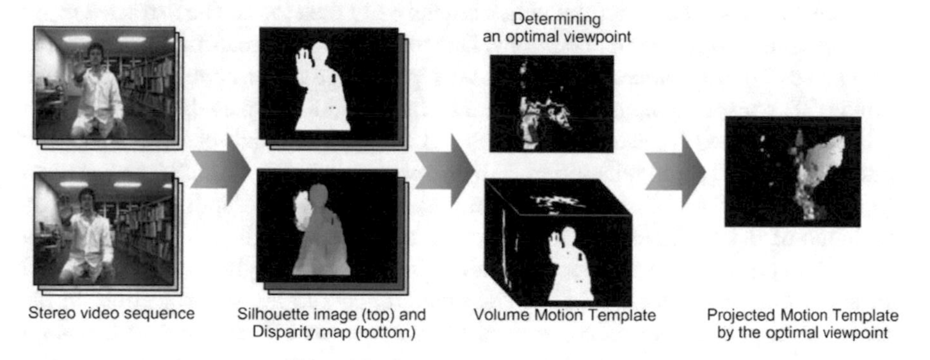

Fig. 1.3 An example of employing a single stereo camera for human action recognition [12]

and their fused 3D motion template can then be projected into an optimal 2D plane to increase the recognition rate, as shown in Fig. 1.3. The stereo camera has fixed lens angles and internal pre-calibration, which gives the camera freedom of moving, but the angle between two lenses is usually too small to cover the occlusion area in human motions. Additionally, the manifold RGB images or videos in the multiple-camera system and stereo camera dramatically increase the computational cost and create a burden for the real-time processing.

1.1.1.2 RGB-D Cameras

Depth cameras often go with names such as: ranging camera, flash lidar, time-of-flight (ToF) camera, and RGB-D camera. The underlying vision-based sensing mechanisms of the depth measurement are equally varied: structured light, ToF, triangulation via multiple camera views, and laser scan, as shown in Fig. 1.4. Triangulation is to calculate the depth information based on multiple sensing sources, for example, using two cameras in a stereo camera as discussed above. Structured light and ToF are the two most popular approaches, and they capture both traditional RGB images and depth information for each pixel at each frame. The additional depth image is

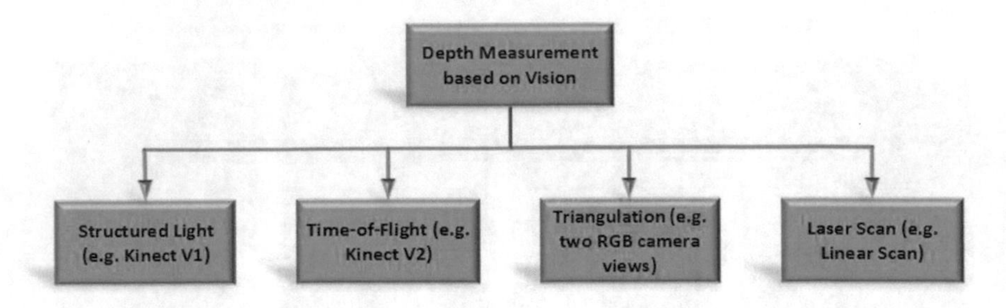

Fig. 1.4 Depth measurement techniques via light waves

converted into a point cloud, from which human body skeleton can be extracted using geometrical and topological operations. Due to its stable and real-time performance and low cost, depth cameras have been very popular in many applications such as gaming, 3D construction, augmented reality, human motion analysis, etc.

Kinect, designed by Microsoft, is the most typical of depth cameras. The first version, Kinect V1, was released in 2010 in cooperation with PrimeSense based on the sensing principle of structured light, delivering reliable depth images with a resolution of 640×480 at 30 Hz. Using structured light imaging, Kinect projects a known light pattern into the 3D scene, which is viewed by the light detector integrated in the Kinect. The distortion of the light pattern, caused by the projection on the surface of objects in the scene, is used to compute the 3D structure of the point cloud. To improve the sensing accuracy, the second version of Kinect, Kinect V2, also known as Kinect One, was released in 2013 with a higher specification than the first version, for example, the resolution went up to 1920×1080 and the skeleton joints increased to 26 from 20. The Kinect V2 is based on the ToF sensing mechanism, which measures the time that light emitted by an illumination unit requires to travel to an object and back to the sensor array, using the continuous wave intensity modulation approach. Figure 1.5 shows a comparison of the two Kinect versions in human motion sensing and their detailed and deep evaluation with pros and cons can be found in [13].

In recent years, extensive work in human motion analysis, such as the human body movements, hand gestures and facial motions, uses RGB-D information as input, and their contributions are well summarised in recent survey papers [14–16]. Some examples of the RGB-D sensor applications are shown in Fig. 1.6. Although the depth cameras have apparent advantages, it is still difficult to capture human motions via a single depth camera in cluttered indoor environments, due to the noise, body occlusion and missing data [17]. To solve this problem, multiple RGB-D cameras have be simultaneously used to track the human motions and to analyse the actions of a group of people [18].

(a) The Kinect Version 1

(b) The RGB image with 640 × 480 pixels from Kinect V1.

(c) The depth image with 320 × 240 pixels from Kinect V1.

(d) The Kinect Version 2

(e) The RGB image with 1920 × 1080 pixels from Kinect V2.

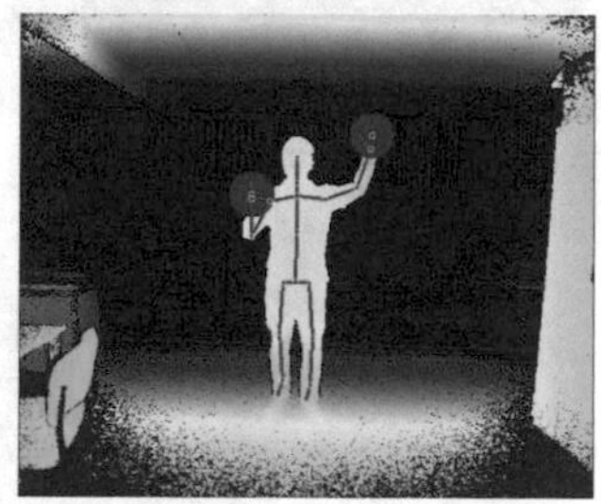

(f) The depth image with 512 × 424 pixels from Kinect V2.

Fig. 1.5 Kinect Version 1 versus Kinect Version 2 in RGB and depth images

1.1.1.3 Marker-Based Sensing Systems

Visual marker-based sensing is a technique where cameras are applied to only track the markers placed on the human body, instead of capturing the RGB/RGB-D images and then extracting joint positions/locations. As a highly articulated structure, the human body skeleton can generate twists and rotations at high degrees of freedom, some of which are hard to estimate based on the analysis of the pixel images. Since the markers are specially made to be tracked by the cameras that cover different view angles in the environment, the marker-based sensing can provide direct, reliable, accurate and fast joint positions/locations, even in clustered scenes with varied lighting conditions, object occlusions, and insufficient or incomplete scene knowledge. It is often used as a 'golden standard' or 'ground truth' in human motion analysis compared with other visual sensing techniques, for example, as a tool to assess the concurrent validity of Kinect V1 for motion analysis [25–27].

The commercial marker-based motion capture systems, e.g. Vicon systems, are widely used in clinical gait analysis, animation and computer games, industry measurement, and more recently in robot monitoring and humanoid robot research [28–33], and some of their examples are shown in Fig. 1.7. In [34], the Vicon MX motion capture system was used for obtaining the positional information of human motions and robot postures to evaluate their collaborative manipulation performance, and in [30], a motion capture system based on Vicon Blade 2 was used to capture the interactions between the human and animal, such as during equine movements. However,

(a) Real-time posture reconstruction when playing a basketball

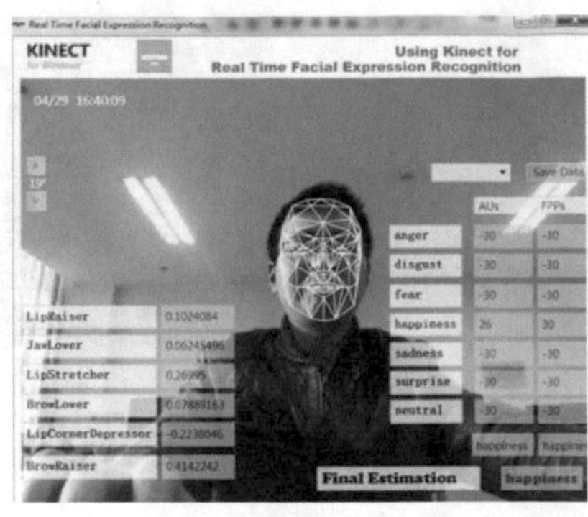

(b) Emotion recognition via facial expressions

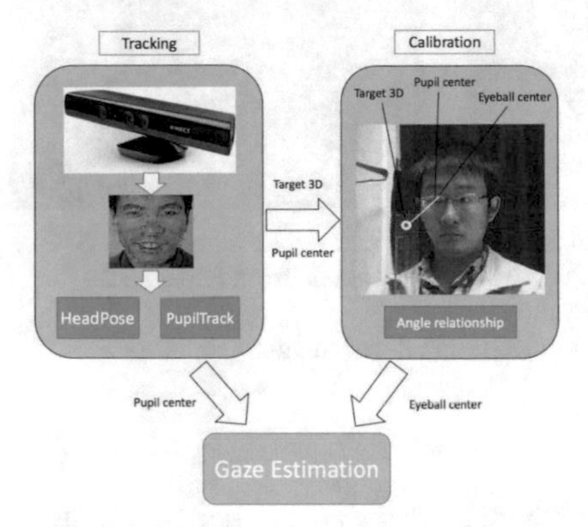

(c) Eye-model-based gaze estimation

(d) Human pose estimation from single depth images

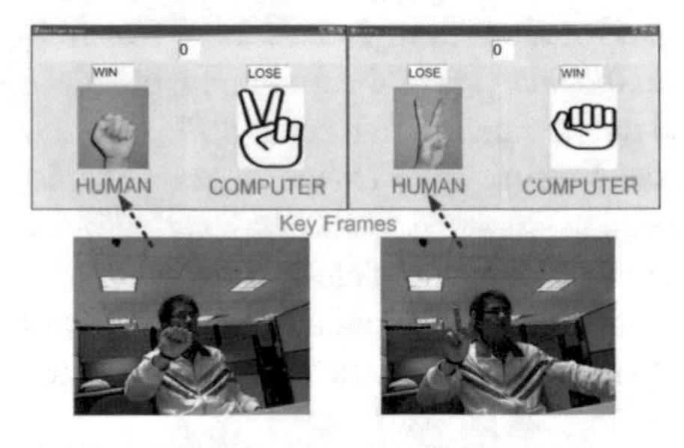

(e) Part-based hand gesture recognition

(f) Tracking people group in a crowded pedestrian area

Fig. 1.6 Application examples using RGB-D sensors [19–24]

the requirements of the constrained measurement space, special-purpose cameras, inconvenient markers or suits, and the high equipment expense restrict the use of the marker-based sensing system to specialised clinics and laboratories [35].

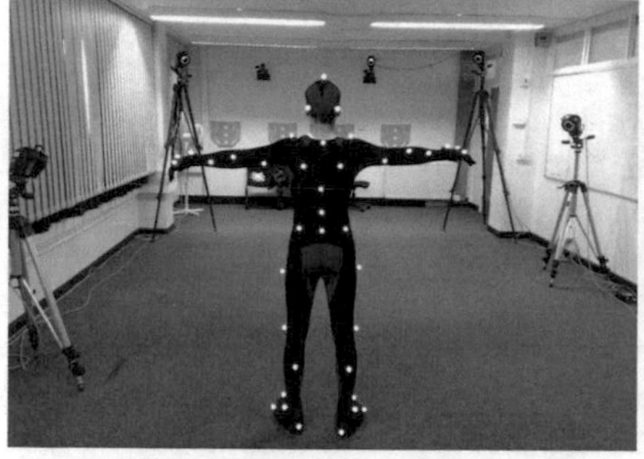

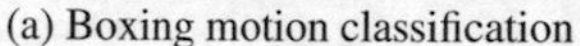

(a) Boxing motion classification

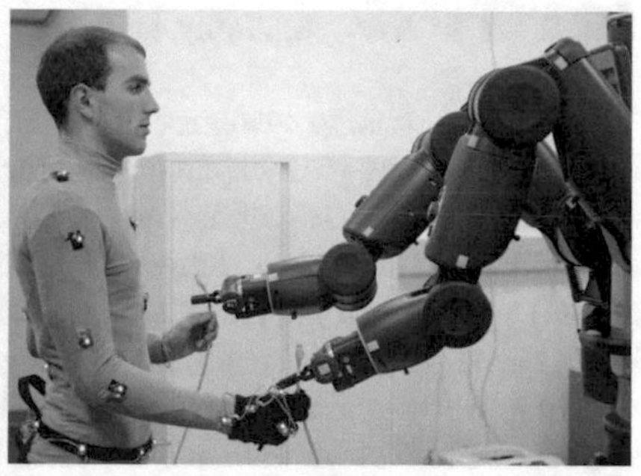

(b) Human-robot collaborative manipulation

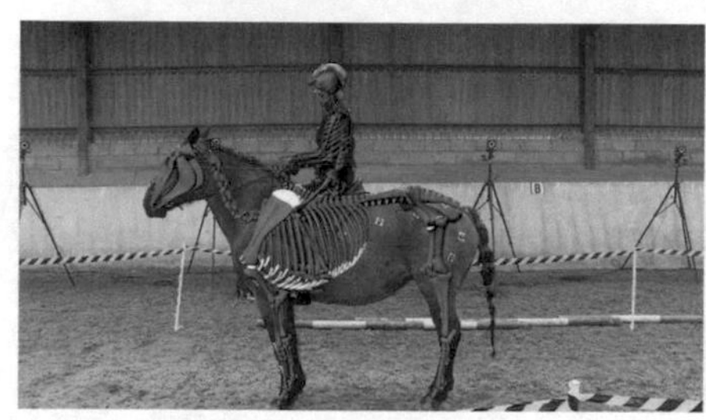

(c) Human animal interaction

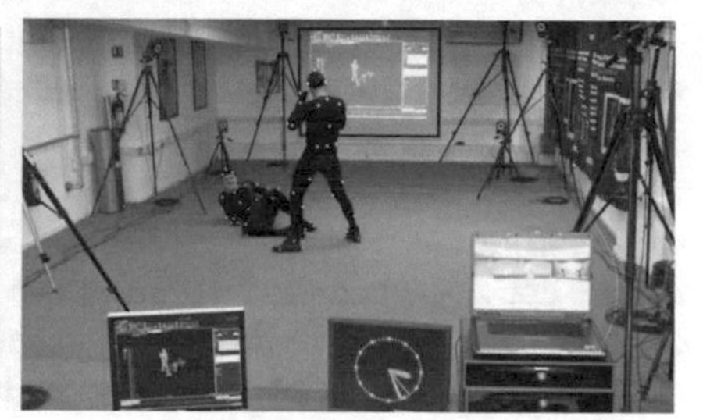

(d) Human fighting analysis

Fig. 1.7 Applications using marker-based motion capture systems [28–30]

1.1.2 Wearable Sensing

Advances in recent technologies, such as telecommunications, microelectronics, sensor manufacturing and data analysis techniques, have opened up new possibilities for using wearable sensing technology to monitor human body status, especially in the digital health ecosystem to achieve a range of health outcomes. Embedded with miniature circuits, micro-controller functions, front-end amplification, wireless data transmission, and large capacity battery, powerful and long-lasting wearable sensors can now be made small enough for people to carry and deploy in digital health monitoring systems. These wearable sensors can be integrated into various personal accessories, such as garments, necklaces, hats, gloves, wristbands, socks, shoes, eyeglasses and other devices such as wristwatches, headphones and smartphones. The sensing techniques are mainly based on inertial, magnetic, ultrasonic, electromyography, and other similar sensing techniques. Unlike the vision-based sensing techniques, the wearable sensing does not suffer from the 'line-of-sight' problem, which cannot be effectively dealt with in a home or cluttered based environment.

1.1.2.1 Electromyography Sensors

Vision-based sensing systems can get sufficient information of human joint angles or locations, but they are incapable of capturing the force involved in the human motions. One intuitive way of getting force information is to employ the force sensors such as FlexiForce sensors and Finger TPS. An alternative is to use the electromyography (EMG) signal from the muscle to estimate the applied force. With the developments in the bio-signal exploration, EMG has gained more and more interests especially in studying the human motion and manipulation. Figure 1.8 shows the Trigno wireless systems from Delsys, Inc. and the Arm-band100 from Elonxi Ltd. Table 1.1 shows configurations of some popular EMG capturing systems currently available in the market [36, 37].

EMG is a technique for evaluating and recording the activation signal of muscles, which is the electrical signal that causes muscles to contract. It is a complicated signal controlled by the nervous system, dependent on the anatomical and physiological properties of muscles. It includes surface EMG (sEMG) and needle EMG (nEMG)/intramuscular EMG (iEMG). sEMG is a non-invasive approach to measure muscle activity where surface electrodes are placed on the skin overlying a muscle or group of muscles. Because of the advantages of non-invasion and easy operation, sEMG has been used intensively in the human movement studies, for example, to control human-assisting manipulator by sEMG signals [38–41] and human body vibration analysis [42, 43], and some examples are shown in Fig. 1.9. Not only

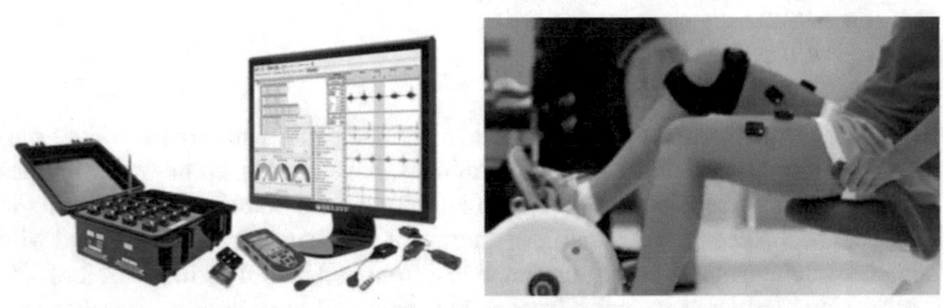

(a) Trigno wireless systems

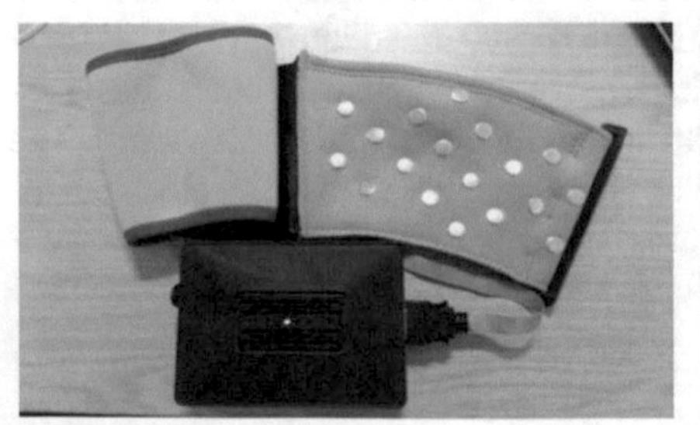

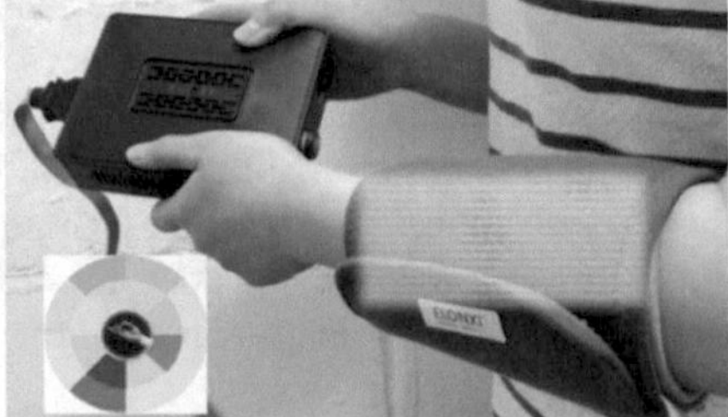

(b) Elonxi Arm-band100

Fig. 1.8 Wireless EMG capturing systems

Table 1.1 EMG capturing systems available in the market

Device	Elonxi Armband100	Biometrics MWX8	Delsys Trigno	Thalmic MyoArmband
Number of channels	16	8	16	8
Wearable	Yes	Yes	No	Yes
Electrode type	Dry	Dry	Dry	Dry
Input configuration	Differential	Differential	Differential	Differential
Bandwidth	20–500 Hz	0–1 k Hz	0–500 Hz	Unknown
Sampling rate	1000 Hz	20k Hz	2000 Hz	200 Hz
ADC Resolution	12 bit	12 bit	16 bit	8 bit
Communication interface	USB/Bluetooth	Bluetooth	USB	Bluetooth
Gain	3000	4098	909	Unknown

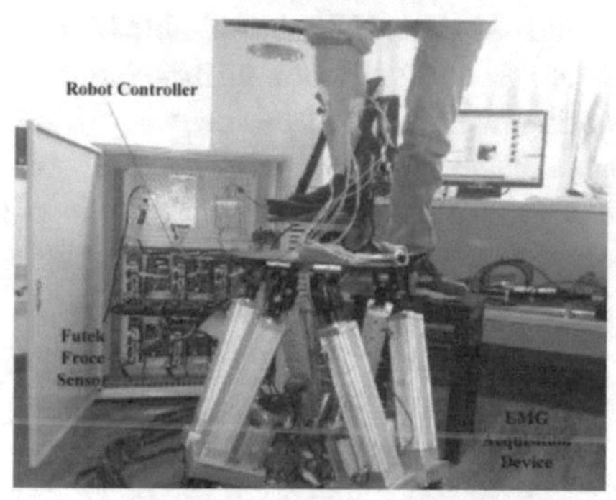

(a) Active interaction control of a lower limb rehabilitation robot

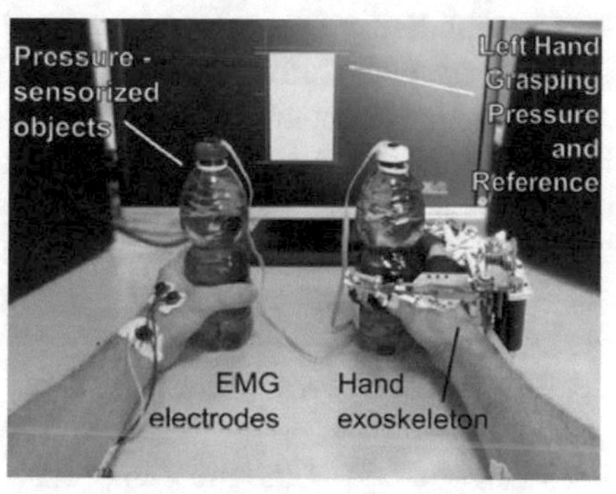

(b) An EMG-Controlled Robotic Hand Exoskeleton for Bilateral Rehabilitation

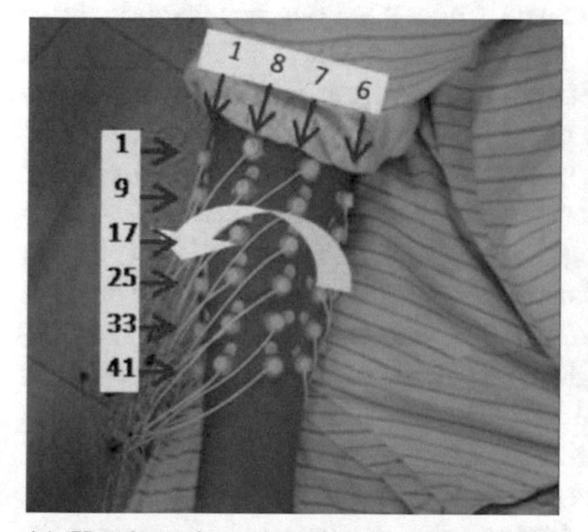

(c) Hand motion classification using high-density EMG

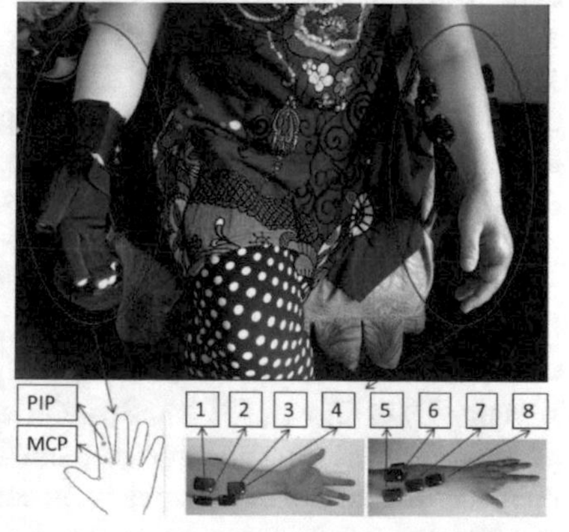

(d) Continuous estimation of finger joint angles

Fig. 1.9 Application examples using EMG sensors [46–49]

angles but also hand forces can be estimated through sEMG analysis, for example, to accurately estimate muscle forces, angle-based EMG amplitude calibration and parallel cascade identification modelling are combined for EMG-based force estimation in dynamic contractions [44]. Conventional sEMG uses one bipolar electrode pair over each muscle, while high density-surface EMG employs multiple (more than two) electrodes closely spaced overlying a restricted area of the skin. With high density-surface EMG, information on muscle-fibre conduction velocity can be used to supplement the information at the muscle-fibre level [45].

nEMG/iEMG is another EMG technique inserting a needle electrode through the skin into the muscle tissue, and it is predominantly used to evaluate motor unit function in clinical neurophysiology. nEMG can provide focal recordings from deep muscles and independent signals relatively free of crosstalk. Due to the improvement of the reliable and implantable electrodes, the use of nEMG for human movement studies has been more explored. Ernest et al. proved that a selective iEMG recording is representative of the applied grasping force and can potentially be suitable for proportional control of prosthetic devices [50]. More recently, it is demonstrated in both able-bodied and amputee subjects that probability-weighted regression using iEMG is a promising method for extracting human muscle movements to simultaneous control wrist and hand DOFs [51]. Though iEMG may provide strengthened muscle signals, its invasion and additional clinical issuc have greatly limited its applications.

1.1.2.2 Data Gloves

Data glove is one of the most important input devices for analysing the hand configuration, measuring the movements of the wearer's fingers. Various sensor technologies, e.g. magnetic tracking devices and inertial tracking devices, are used to capture physical data such as bending of fingers, finger positions, acceleration, force and even haptic feedback. The first data glove, the Sayre Glove, was developed in 1977 by Thomas de Fanti and Daniel Sandin. It used light based sensors, which are flexible tubes with a light source at one end and a photocell at the other. As the fingers with tubes were bent, the amount of light that hit the photocells varied, thus providing a measure of finger flexion. After that, numerical gloves have been proposed using different sensor technologies over the last three decades. The most popular sensors used in data gloves include piezoresistive sensor such as P5 Glove [52] and CyberGlove [53, 54], fiber optic sensor such as SpaceGlove [55] and 5DT Glove, hall-effect sensor such as Humanglove [56], and magnetic sensors such as 3d Imaging Data Glove [57] and StrinGlove [58]. They translate the fingers inflection or abduction movements precisely into electric signals at a fast speed. Laura et al. presented a comprehensive review on data glove systems and their applications [59]. Here the focus is on the more recent data gloves available in the market, shown in the Fig. 1.10. These gloves employ more sensors, including bend sensors, touch sensors and accelerometers, to measure not only finger angles but also finger abductions, pressure applied and hand tilt. They are predominant in the competitive market due to their high precision at a high speed, and their parameters are listed in Table 1.2.

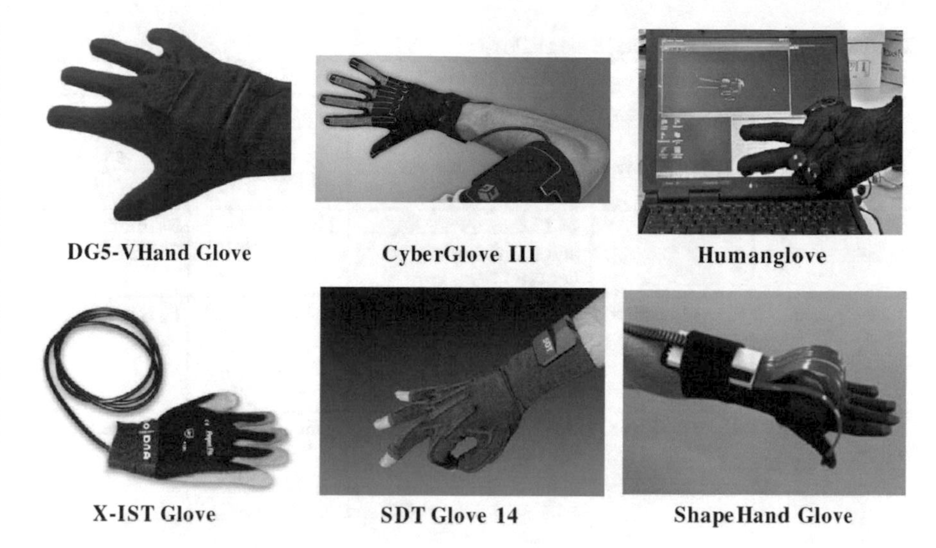

Fig. 1.10 Data gloves on the market

In addition, their considerate accuracy and easy control have gained a number of interests.

Due to the improvements of the sensor technology, glove base system is more used in human hand motion analysis [60–62]. CyberGlove is used to gather training samples from human in-hand manipulation for teaching the required behaviour [63]. Not only could the fingers' angles be recorded in the human hand skill models, but also the finger tip haptic information and the positions can be integrated in the models through data gloves [64]. Ma et al. designed a haptic glove, which is a lightweight, portable, and self-contained mechatronic system, and it fits on a bare hand and provides haptic force feedback to each finger of the hand without constraining their movements [65]. However, before using the data gloves, a time-consuming calibration phase is needed in order to account for differences in hand size and proportion when mapping from raw sensor readings to joint angles of the user's hand. How an optimal calibration can be achieved is still an unsettled question. Due to cross-couplings between the sensors, more complex forms of calibration are necessary to achieve a satisfactory fidelity.

1.1.2.3 Other Wearable Sensors

Many other wearable sensing techniques have been applied in human movement capturing, e.g. inertial sensors, magnetic sensors, ultrasonic sensors, etc. Inertial sensors, such as accelerometers and gyroscopes, have been widely used in navigation and augmented reality modelling, due to its easy use and cost efficiency in human motion detection with high sensitivity and large capturing area [66]. The angular

Table 1.2 Data gloves available in the market

Device	Technology	Sensors and locations	Precision	Speed
DG5-VHand	Accelerometer and piezo-resistive	6 (a 3 axes accelerometer in wrist; one bend sensor per finger)	10 bit	25 Hz
5DT glove 14	Fiber optic	14 (2 sensors per finger and abduction sensors between fingers)	8 bit	Minimum 75 Hz
X-IST glove	Piezo-resistive, pressure sensor and accelerometer	14 (Sensor selection: 4 bend sensor lenghts, 2 pressure sensor sizes, 1 two-axis accelerometer)	10 bit	60 Hz
CyberGlove III	Piezo-resistive	22 (three flexion sensors per finger, four abduction sensors, a palm-arch sensor, and sensors to measure flexion and abduction)	8 bit	Minimum 100 Hz
Humanglove	Hall-effect sensors	20/22 (three sensors per finger and the dbduction sensors between fingers)	0.4 deg	50 Hz
Shapehand glove	Bend sensors	40 (flexions and adductions of wrist, fingers and thumb)	n.a.	100 Hz maximum

velocities of the human joints are measured by gyroscopes and their positions are usually obtained by double integration of the translational accelerations measured by the accelerometers. For example, shoulder motions can be assessed using inertial sensors placed on the upper arm and forearm [67], shown in Fig. 1.11a. A significant problem with these measurements is that inaccuracies inherent in the measurements, caused by the fluctuation of offsets and noise, quickly accumulate and rapidly degrade their precision [68]. Hence, how to reduce and get rid of this accuracy drift in the inertial sensor based systems is the key in current research.

Magnetic tracking systems are based on the local magnetic field, via the relative flux of three orthogonal coils on the transmitter and receiver, to measure the position and orientation in 6 DoFs. The receivers, the tracking sensors, are placed on the subject to capture its spatial relationship relative to the magnetic transmitter. The magnetic tracking system does not suffer from drift errors over time as the inertial sensors, it outputs 6-DoF position and orientation in real time without post-processing, and it is less expensive than vision-based tracking systems. It has been extensively used in virtual reality [72]. Figure 1.11b shows that inertial and magnetic sensors were used to analyse human walking motions [69]. Two sensing suits with inertial sensors and magnetic sensors are shown in Fig. 1.11c. However, since the magnetic tracking system is directly related to the physical characteristics of magnetic fields,

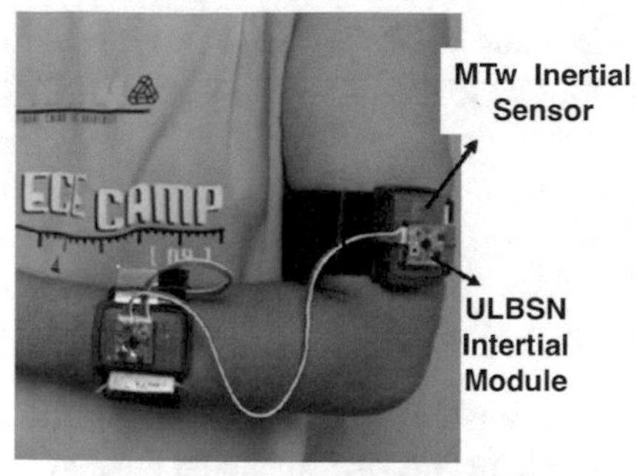

(a) Shoulder motion assessment using inertial sensors [67].

(b) Human walking analysis using inertial and magnetic sensors [69].

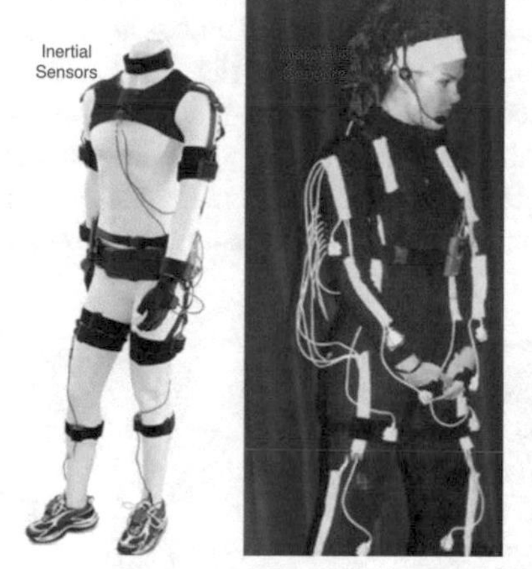

(c) Motion sensing suits with inertial sensors and magnetic sensors[70].

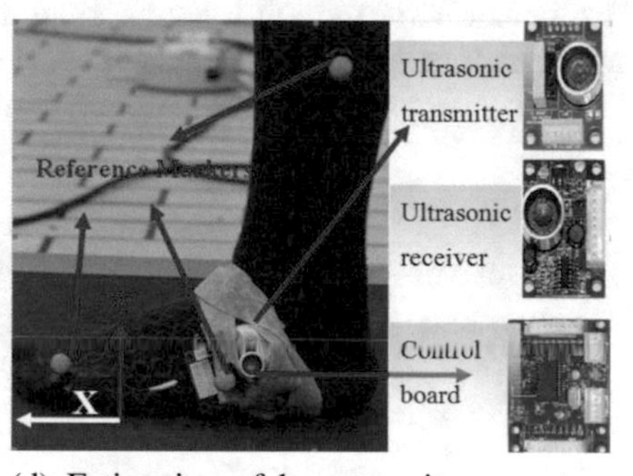

(d) Estimation of human gait parameters using a ultrasonic motion analysis system [71].

Fig. 1.11 Application examples using inertial sensors, magnetic sensors and ultrasonic sensors [67, 69–71]

the magnetic filed decreases in power rapidly as the distance from the generating source increases and it is susceptible to magnetic and electrical interference within the measurement volume [73].

Ultrasonic sensors measure positions through time-of-flight, triangulation, or phase coherence. Ultrasonic tracking systems use sonic emitters to emit ultrasound, sonic discs to reflect ultrasound and multiple sensors to time the return pulses. The positions and orientations of gestures are computed based on the propagation, reflection, speed of time and triangulation. For example, human gait parameters can be effectively assessed during human walking using a ultrasonic motion analysis system [71], shown in Fig. 1.11d. Since they have the advantages of being free to magnetic obstacles and illumination conditions, ultrasonic tracking systems have been employed in the human motion sensing systems, often combined with other sensing techniques to provide supplement motion information [74, 75]. However, they are still suffering the low resolution, low sampling rate and being lack of precision in their outputs [4].

1.1.3 Multimodal Sensing

Multimodal sensing, using two or more types of sensors at the same time, is often employed to achieve more comprehensive and reliable information about the human motions. For example, Daniel et al. integrated magnetic measurements and inertial sensors for fully ambulatory position and orientation tracking of human motions. The magnetic system provides 6 DOF measurements at a relatively low update rate while the inertial sensors track the changes in the position and orientation between the magnetic updates [76], and its sensor locations are shown in Fig. 1.12a. Hassen et al. presented the design of a viable quaternion-based complementary observer for human body motion tracking using inertial/magnetic sensor modules containing orthogonally mounted triads of accelerometers, gyroscopes, and magnetometers [73]. To overcome the drift problem of the inertial sensors, ultrasonic sensors and

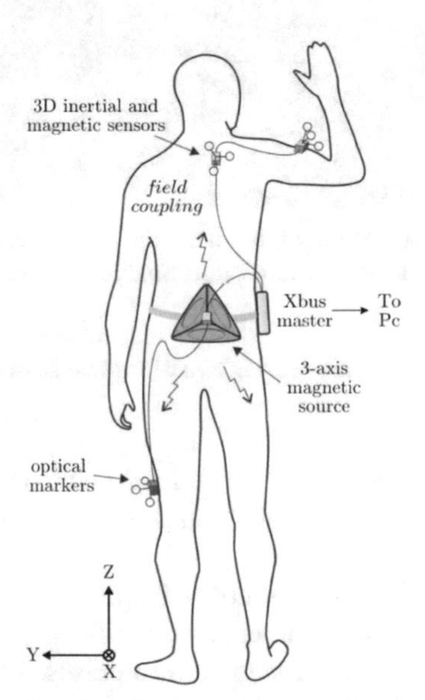

(a) Body-mounted sensing system consisting of a three-axis magnetic dipole-source and three-axis magnetic and inertial sensors

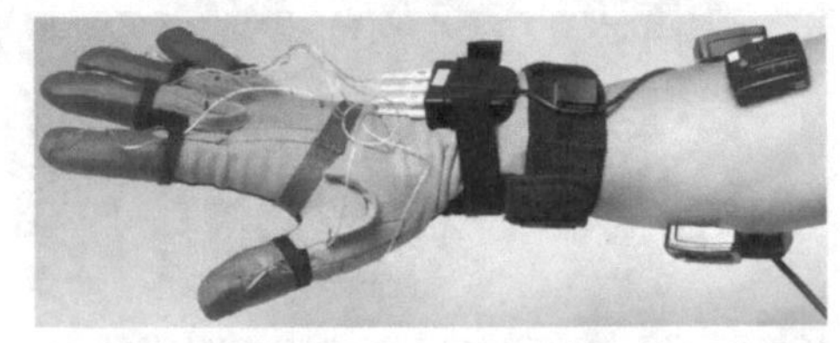

(b) Human hand motion analysis with a data glove, force sensors and EMG sensors

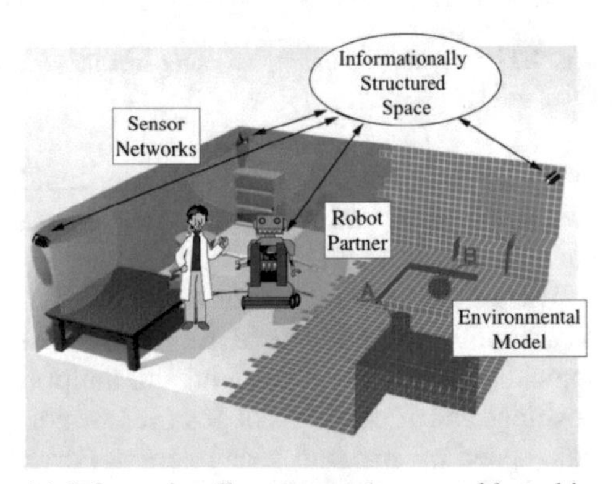

(c) Informationally structured space with multiple sensors

Fig. 1.12 Multimodal sensing applications [76–78]

magnetometers were used to calibrate the drift of accelerometers in position measurement, and the accelerometer complemented the low sampling rate of the ultrasonic sensor and worked independently when the ultrasonic signals were temporarily not available [74]. In order to analyse human hand motions with multimodal information, three types of sensors, including the data glove, force sensor and EMG sensor, were integrated to simultaneously capture the finger angle trajectories, the hand contact forces, and the forearm muscle signals [77], shown in Fig. 1.12b. Ramakant et al. investigated sign language recognition using the information captured from the data glove and a camera [61]. Three wearable magnetic angular rate and gravity sensors were deployed to calculate instantaneous orientation, velocity, and position of the human lower limbs, and Vicon was also integrated to validate the magnetic system [27]. In Fig. 1.12c, an indoor informationally structured space was formed to monitor the elderly people using multiple Kinects and wireless optical oscillo-sensors that are used to estimate the human status on a bed [78]. Though they can capture more useful data than single-modal sensing systems, the multimodal sensing systems have more complex architectures, for example, multi-system calibration and communication, data synchronisation and information fusion, and it takes much more computational cost to analyse the sensed multimodal data.

1.2 Recognition Approaches

Human motions are expressive and meaningful, involving physical movements of the fingers, hands, arms, legs, head, face, and other body movements with the intent to convey meaningful information or to interact with the environment [79]. Motion recognition is to automatically identify ongoing human motions using sensory information, e.g. a sequence of image frames from a camera. There are different tools for motion recognition, ranging from machine learning and pattern recognition, computer vision and image processing, connectionist systems, etc. [80]. There are different ways to categorise those recognition approaches, for example: template-based and state-space approaches based on the way to match motions [81]; single-layer and multi-layer approaches based on the recognition architecture [82]; vision-based, marker-based and EMG-based methods based on the sensor types [83]; recognition methods based on different body parts [84]. Among these approaches, the most popular methods are probabilistic graphical models, support vector machines, neural networks and fuzzy approaches, which will be discussed in the following part.

1.2.1 Probabilistic Graphical Models

Probabilistic graphical models, which span methods such as Bayesian networks and Markov random fields, use a graph-based representation as the basis to efficiently encode and manipulate probability distributions over high-dimensional spaces [85].

They express jointly distributed probability distributions on a set of probabilistic variables, which are consistent with a set of graph relations in that values of nodes depend directly only on immediately connected nodes. These methods are capable to reason coherently from limited and noisy observations and have been used in an enormous range of application domains, which include: fault diagnosis, image understanding, speech and motion recognition, natural language processing, robot navigation, and many more [85].

As one of the most popular methods in probabilistic graphical models, Hidden Markov Models (HMMs) and their derivatives have been successfully used in human motion recognition. Markov chain is defined as the restriction on the finite-state automaton that a state can have only one transition arc with a given output; the restriction that makes Markov chains deterministic. A HMM can be considered as a generalisation of a Markov chain without this Markov-chain restriction, operating on two layers: hidden states and observed states, shown in Fig. 1.13. The probabilistic links between the hidden and observed states can be equivalent to the likelihood that a particular hidden state will create the observed state. By training HMMs, it learns to weigh the importance of motion information for detection and classification, and determines the correct number of states for each motion to maximise the performance. Lina et al. proposed a reliable and low-cost human fall prediction and detection method using tri-axial accelerometer, where the acceleration time series, extracted from the human fall courses but before the collision, were used to train HMMs so as to build a random process mathematical model [86]. In [87], the 3D motion scale invariant feature transform for the description of the depth and human motion information was combined with HMMs to reduce the data noise and improve the accuracy of human behaviour recognition. Natraj and Stephen proposed a non-parametric hierarchical HMM structure to model activities with actions at the top level and factorised skeleton and object poses at the bottom level, with the advantages of automatic inference of the number of states from data, information sharing and semi-supervised learning [88]. HMMs and their derivatives have the advantages of incorporation of prior knowledge, the ability of being combined into larger HMMs, and mathematical analysis of the results and processes. On the other hand, HMMs are computationally expensive and require large amount of training data. Structure

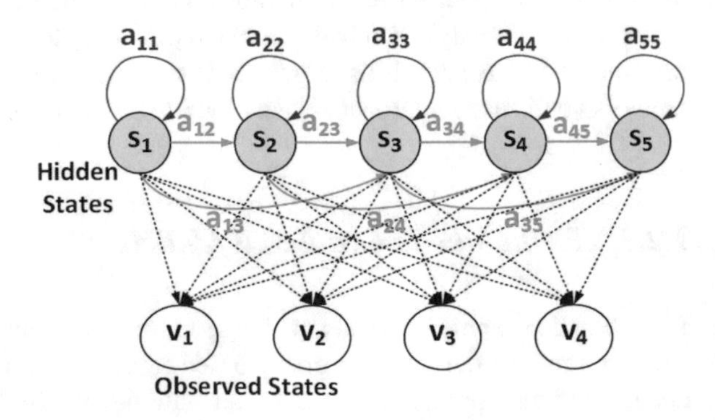

Fig. 1.13 A HMM with five hidden states and four observed states [83]

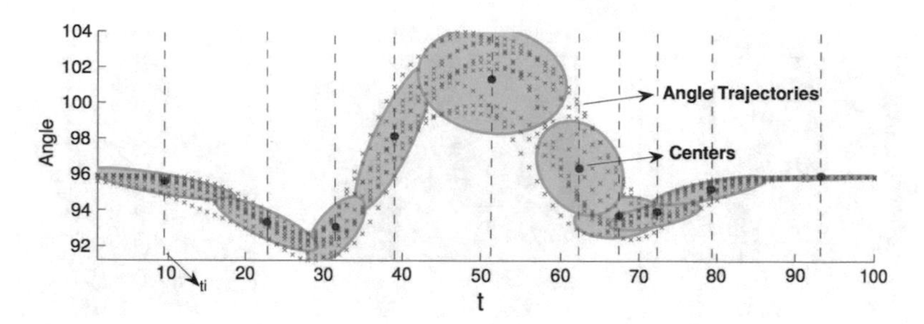

Fig. 1.14 An example of GMMs for human grasping recognition [93]

complexity is not always good, because it may lead to over-fitting problem. Moreover, HMMs may converge to local maxima instead of the truly optimal parameter set for a given training set.

Another popular probabilistic approach is Gaussian Mixture Models (GMMs), which can be regarded as a hybrid between a parametric and nonparametric density model. They are well suited for multivariate densities capable of representing arbitrary densities and used in a biometric system, such as vocal tract-related spectral features in a recognition system [129]. They assume that all the data points are generated from a mixture of a finite number of Gaussian distributions with unknown parameters, and try to describe the characteristic of these data points with a mixture of Gaussian distributions. They draw confidence ellipsoids for multivariate models and compute the Bayesian Information Criterion to assess the number of clusters in the data, as shown in Fig. 1.14. GMMs and their derivatives have been employed widely in machine learning, pattern recognition, image processing, robotics, etc., especially in motion recognition [89], for example: Yingli et al. proposed a hierarchical filtered motion method based on GMMs to recognise actions in crowded videos by the use of motion history image as basic representations of motions because of its robustness and efficiency [90]; Di and Ling showed that better action recognition using skeletal features can be achieved by replacing GMMs by deep neural networks that contain many layers of features to predict probability distributions over states of HMMs [91]; Ruikun and Dmitry designed an approach consisting of a two-layer library of GMMs that can be used both for recognition and prediction of human reaching motion in industrial manipulation tasks [92]. GMMs normally have a fast convergence process, which is stable with high computation efficiency, however they normally need more components to fit the datasets with nonlinearity and their parameter estimation is based on the local optimum, which depends on the initial component states.

1.2.2 Support Vector Machines

Support Vector Machines (SVMs) are one supervised learning approach based on the concept of decision planes, which define decision boundaries and separate between

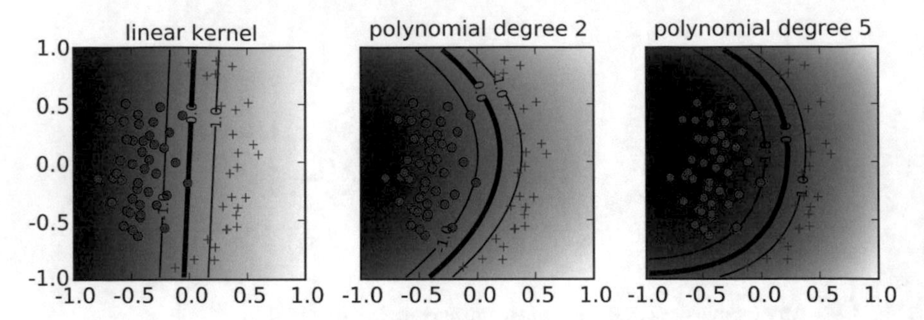

Fig. 1.15 An example of SVMs with linear and nonlinear (polynomial) kernels [94]

a set of objects having different class memberships. The decision plane maps these objects so that the examples of the separate classes are divided by a clear boundary that is as wide as possible. New objects are then located into the same space and recognised to belong to a class based on which area they fall in, as shown in Fig. 1.15. SVMs are effective in high dimensional space and compatible with different kernel functions specified for the decision function, with common kernels well provided and freedom to specify custom kernels.

Extensive researches have used SVMs and their derivatives to meet the requirements in terms of recognition accuracy and robustness for human motion recognition [95]. Huimin et al. proposed a new system framework for recognising multiple human activities from video sequences, using a SVM multi-class classifier with binary tree architecture [96]. Each SVM in the multi-class classifier is trained separately to achieve its best classification performance by choosing proper features before they are aggregated. Salah et al. presented two sets of features obtained from RGB-D data acquired using a Kinect sensor for human activity recognition [97]. These features were fused via the SimpleMKL algorithm and multiclass-SVM using two kernel functions in order to enhance the classification performance for similar activities. Though SVMs are powerful in terms of the performance, selection of the kernel function parameters is difficult in real applications, the kernel models can be quite sensitive to over-fitting the model selection, and they need high algorithmic complexity and extensive memory in large-scale tasks [98, 99].

1.2.3 Artificial Neural Networks and Deep Learning

Artificial Neural Networks (ANNs) are inspired by biological neural networks, such as the animal brain, and used to estimate or approximate functions that can depend on a large number of inputs and are generally unknown. They uses the node as its fundamental unit, the links as its associated weight, and activation function such as the step, sign, sigmoid functions as the transfer function, normally with three layers that are input layer, hidden layer and output layer as shown in Fig. 1.16a. They can be

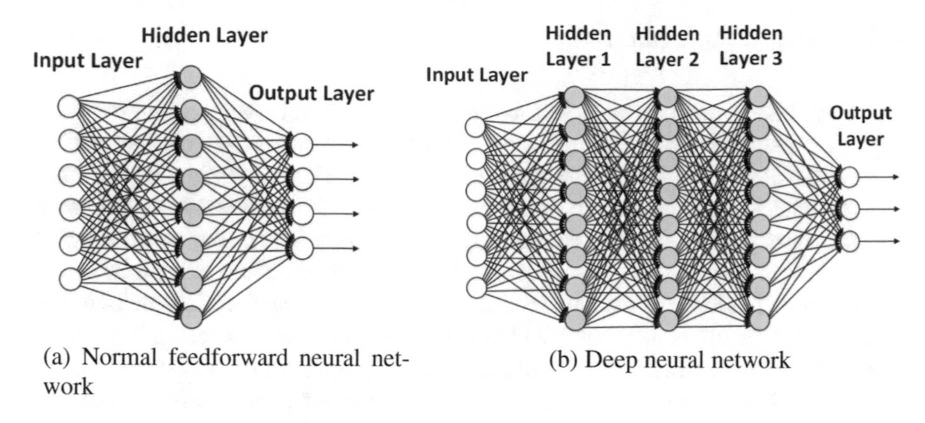

(a) Normal feedforward neural network

(b) Deep neural network

Fig. 1.16 Architectures of normal neural network and deep neural network

configured for a specific application, such as pattern recognition or data classification, through a learning process. They can handle very complex situations which are too difficult to model using traditional approaches such as the inferential statistics or programming logic [100]. Hand gesture recognition based on shape analysis was achieved by utilising a neural network-based approach with a unique multi-layer perception and a back propagation learning algorithm [101]. An ensemble of Radial Basis Function (RBF) neural networks were used to learn the affective/stylistic features for human motion recognition, following time warping and principal component analysis, so that the system can learn the components and classify stylistic motion sequences into distinct affective and stylistic classes [102].

To model more complex and high-level abstractions or structures from the training data, deep neural architectures have been proposed, following the recent achievement of ANN by Hinton and others [103]. A deep neural architecture has multiple hidden layers of units between the input and output layers, as shown in Fig. 1.16b. The extra layers enable composition of features from lower layers, giving the potential of modelling complex data with fewer units than a similarly performing shallow network [104]. Various deep learning architectures such as deep neural networks (DNN), convolutional deep neural networks, deep belief networks and recurrent neural networks have been applied to fields like computer vision and automatic speech/motion recognition where they won numerous contests in pattern recognition and representation learning [105, 106]. Di et al. proposed an approach to learn high-level spatio-temporal representations using deep neural networks suited to the input modality, with a Gaussian-Bernouilli Deep Belief Network to handle skeletal dynamics and a 3D Convolutional Neural Network to manage and fuse batches of depth and RGB images [107]. It showed that DNN is capable of capturing the full context of each body joint and it is substantially simpler to formulate than methods based on graphical models since there is no need to explicitly design feature representations and detectors for parts and no need to explicitly design a model topology and interactions between joints [108].

1.2.4 *Fuzzy Approaches*

The first version of fuzzy sets, known as type-1 fuzzy sets, whose elements have degrees of membership, was introduced by Lotfi A. Zadeh and Dieter Klaua in 1965, as an extension of the traditional notion of set [109, 110]. Type-2 or type-n fuzzy sets and systems generalise the traditional fuzzy sets and systems to handle more uncertainty about the value of the membership function [111]. In the past decades, the fuzzy sets and systems have been widely applied and demonstrated to be one of the most effective approaches in different areas, such as decision-making, pattern recognition and control theory [112–115]. In human motion recognition, fuzzy approaches, such as fuzzy clustering, hybrid fuzzy approaches and fuzzy qualitative approaches, have shown good performance in terms of both effectiveness and efficiency.

1.2.4.1 Fuzzy Clustering

Distinct clusters are employed in the hard clustering approaches, where each data point only belongs to exactly one cluster. For example, the method of K-means clustering partitions data points into clusters in which each point belongs to the cluster with the nearest mean, serving as a prototype of the cluster. Fuzzy clustering utilises a membership function to enable each data point potentially belong to multiple clusters with different membership values. One of the most widely used fuzzy clustering methods is the Fuzzy C-Means (FCM) algorithm, which adds membership values to represent the degrees of truth that one point belongs to one cluster, and the fuzzifier to determine the level of cluster fuzziness. The solution of membership functions and fuzzifier works better for human motion analysis in some challenging environments such as the complex backgrounds, dynamic lighting conditions, and the deformable body shapes with real-time computational speeds [116, 117]. Juan et al. proposed a hand gesture recognition system, where FCM was utilised to handle feature-weighted clustering problem with smaller training sets and short training times [116]. Palm et al. proposed a fuzzy time clustering method to recognise human grasps, which models time dependent trajectories using fuzzy modelling [118]. It showed this fuzzy method can fast represent the dynamic hand grasps by a small number of local linear models and few parameters, and it can also be used as a nonlinear filtering of noisy trajectories and a simple interpolation between data samples. To improve the time clustering method, the degree of membership of the model was extended with both time instance weights and different learning motion weights in [119]. The extended degree of membership enabled the method to be more comprehensive to learn repeated motions from the same subject or similar gestures from various subjects.

1.2.4.2 Fuzzy Inference System

The Fuzzy Inference System (FIS) is a fuzzy decision making system that utilises the fuzzy set theory to map inputs that are features into outputs that are classes in the case of fuzzy classification. It has the capability to model the uncertainty involved in the data and compromise the vague, noisy, missing, and ill-defined data. There are three main components in the type-1 FIS: fuzzification, inference, and defuzzification, as shown in Fig. 1.17a. The fuzzification component maps the crisp input data from a set of sensors, which normally is the features or attributes, to the membership functions to generate the fuzzy input sets with linguistic support [120]. Based on the fuzzy set inputs, the fuzzy rules component infers the fuzzy output sets, which are finally defuzzified into the crisp outputs. To improve the performance of the type-1 FIS, type-2 FIS utilises type-2 fuzzy sets with the membership function with uncertainty to handle higher uncertainty levels presenting in the real world dynamic environments [121]. Therefore, the input data is first fuzzified into type-2 input fuzzy sets, and then go through the inference process with similar rules as the type-1 FIS. Before the defuzzification component which generates crisp outputs, the type-2 output fuzzy sets must be reduced from type-2 to type-1 output fuzzy sets. This is conducted by using a type-reducer, which is unique to type-2 FIS, as shown in Fig. 1.17b.

The FIS can be efficiently used to distinguish the human motion patterns and recognise the human motions with its capability of modelling the uncertainty and the fusion of different features in the classification process. Mehdi and Liu presented

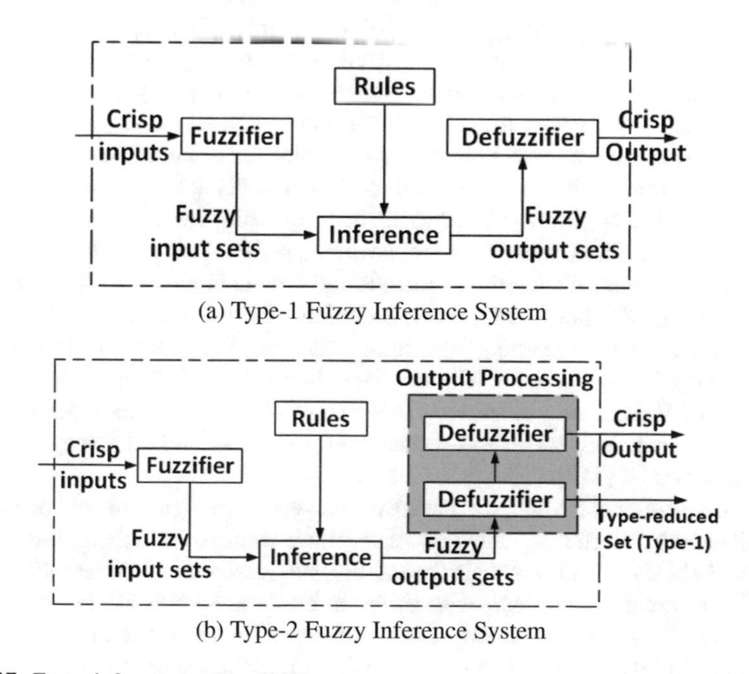

Fig. 1.17 Fuzzy inference systems [122]

a novel FIS for recognising occluded 3D human motion assisted by the recognition context [123]. Uncertainties were wrapped into a fuzzy membership function via metrics derived from the probabilistic quantile function, and time-dependent and context-aware rules were learnt via a genetic programming to smooth the qualitative outputs represented by fuzzy membership functions. In [124], four fuzzy spatio-temporal relations that relate an entity to an object of the environment were proposed to identify the human motions in real time. The semantic interpretation of the entitys trajectory was created to consider the uncertainties of the activities structures. Similarly, the spatio-temporal features, which are the silhouette slices and the movement speed from human silhouette in video sequences, were also investigated and used as the inputs in the FIS [125]. In this work, the membership functions were automatically learned by FCM and good performance was demonstrated compared to the state-of-the-art approaches. Those FIS based approaches all aimed to handle high uncertainty levels and the complexities involved in both the spatial and temporal features for realtime human motion recognition in the real world applications.

1.2.4.3 Hybrid Fuzzy Approaches

In the fuzzy systems, once the parameters, such as the interval values of the membership function and the threshold value for the inference step, are properly chosen, they can achieve superior performance in human motion recognition. However, since the above fuzzy approaches have no capability to learn from training data, the optimisation of these parameters is usually difficult in real applications, and it normally needs intervention from experts or people with extensive experiments, which might be very subjective, expensive in time cost and error prone [126]. In addition, some traditional machine learning methods assign one data point only to one cluster. To solve the above problems, researchers try to combine fuzzy techniques with machine learning algorithms, which can be used to automatically learn or tune those parameters in specified applications. Those methods are called hybrid fuzzy approaches, and they are expected to possess the all advantages from both aspects, i.e. the ability to learn from machine learning methods and simple implementation in realtime systems from fuzzy approaches. Fuzzy techniques have been integrated into most of the popular machine learning algorithms, such as Fuzzy Hidden Markov Models (FHMMs), Fuzzy Gaussian Mixture Models (FGMMs), Fuzzy Support Vector Machines (FSVMs), Fuzzy Neural Networks (FNNs), etc., which have demonstrated to have satisfying performance in terms of effectiveness and efficiency in human motion recognition [113].

In the conventional HMMs, each observation vector is assigned to only one cluster with full degree of certainty, which in consequence would reduce their classification rate. The FHMMs extend HMMs in the training stage, where each observation vector has different membership degrees to different clusters, by considering the inverse of the distance as the membership degree of the observation vector to the cluster [127]. FHMMs were applied to recognise human actions with good accuracy for the similar actions such as walk and run. Magdi et al. described a generalisation of

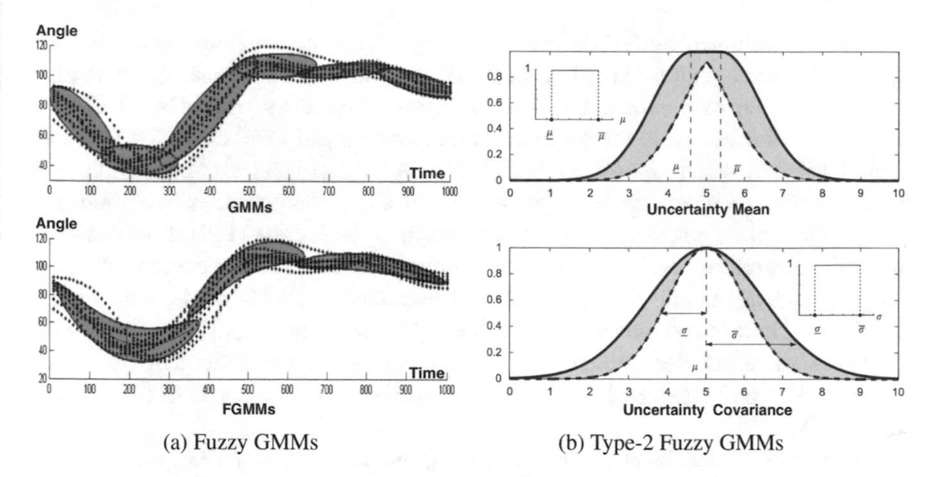

(a) Fuzzy GMMs (b) Type-2 Fuzzy GMMs

Fig. 1.18 Fuzzy gaussian mixture models in type 1 and type 2 [132, 134]

classical HMMs using fuzzy measures and fuzzy integrals resulting in a fuzzy hidden Markov modelling framework [128]. This fuzzy model has the advantage that the non-stationary behaviour is achieved naturally and dynamically as a byproduct of the nonlinear aggregation of information using the fuzzy integral. It is also capable of dealing with varied lengths of the observation sequences and learning with limited training data. To demonstrate its performance, this fuzzy framework was applied to recognise the handwritten word recognition [128]. Moreover, it was further adapted to identify dynamic facial expressions with a reduction in training time, and improved flexibility for use as a valuable part of a larger multimodal system for natural human-device interaction [129].

FGMMs combine a generalised GMMs with FCM in the mathematical modelling by introducing the degree of fuzziness on the dissimilarity function based on distances or probabilities, with a non-linearity and a faster convergence process [130–132]. They not only possess the non-linearity to fit datasets with curve manifolds but also have a much faster convergence process saving more than half computational cost than GMMs', and one example of comparison is shown in Fig. 1.18a. They achieved the best performance in recognising various human manipulation tasks compared to HMMs, GMMs and SVM with limited training datasets [119, 133]. They were also applied to identify the EMG-based hand motions with a better performance than the popular statistical methods [77]. Another fuzzy version of GMMs is the Type-2 FGMMs, which use interval likelihoods to deal with the insufficient or noise data [134]. The footprint of uncertainty in type-2 FGMMs is used to deal with the uncertain mean vector or uncertain covariance matrix in the GMMs, as shown in Fig. 1.18b.

FSVMs employ fuzzy membership functions in traditional support vector machines, solving inseparable area problems when two-class problems expanding to multi-class problems faced by traditional SVMs [135]. In FSVMs, the decision functions determined by SVMs are used to generate membership functions with hyper-polyhedral regions [136]. Using the membership functions, the existence of unclassifiable regions and multi-label regions caused by multi-class SVMs is resolved. To further solve the problem of undefined multi-label classes, a new version of FSVMs was proposed, in which for each multi-label class, a region with the associated membership function is defined and a data point is classified into a multi-label class whose membership function is the largest [137]. It can resolve unclassifiable regions and classification to undefined label set that occur in the conventional one-against-all SVM. Yan et al. constructed a FSVMs model to recognise EMG-based human motions with about 5% higher accuracy than Back Propagation (BP) network [138]. Kan applied FSVMs to recognise human actions in Weizmann action dataset and received high correct recognition rate compared to the conventional SVMs [139].

FNNs, also called Neuro-Fuzzy Systems, are learning machines that find the parameters of fuzzy systems, including the fuzzy sets and fuzzy rules by utilising approximation techniques from neural networks. A FNN was designed to recognise human body postures with the ability to detect the emergencies caused by the accidental falls or when a person remains in the lying posture for a period of time [140]. Based on the FNNs, Annamaria et al. developed a CircularFuzzy Neural Network (CFNN) with interval arithmetic to convert the coordinate model of the given hand to an fuzzy hand-posture model in the system of gesture recognition [141]. The CFNN has an improved topology with a circular network, as shown in Fig. 1.19a for the topology of FNN and in Fig. 1.19b for the topology for CFNN. In [142], a fuzzy rule-based reasoning system, combining the decision tree, neural networks and FIS, was proposed as a classifier, which adopts statistical information from a set of audiovisual features as its crisp input values and produces semantic concepts corresponding to the occurred events. A set of fuzzy rules, generated by traversing

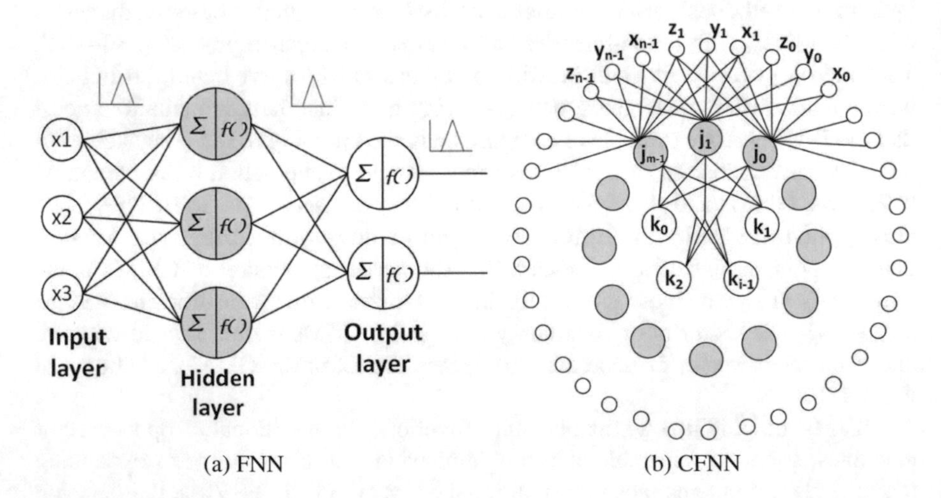

(a) FNN (b) CFNN

Fig. 1.19 Topologies of FNN and CFNN [141]

each path from root toward leaf nodes of constructed decision tree, are utilised in a fuzzy inference engine to perform decision-making process and predict the occurred events in the input video.

1.2.4.4 Fuzzy Qualitative Approaches

Fuzzy qualitative reasoning (FQR) is a form of approximate reasoning that can be defined as the fusion between the fuzzy set theory and the qualitative reasoning [143]. Fuzzy reasoning, the main component in the fuzzy set theory, provides a model free way to handle mathematical and engineering systems by utilising powerful reasoning strategies through conditional statements. It is also an effective tool to naturally deal with the uncertainties in significantly noisy environments. However, one of its main problems is that its knowledge is primarily shallow and the questions over the computational overhead associated with handling grades of membership of discrete fuzzy sets must be taken into account if deeper reasoning is to be carried out. On the other hand, qualitative reasoning can reason about the behaviour of physical systems, without precise quantitative information. It operates at the conceptual modelling level, being symbolic, and maintains the mathematical structures of the problem, without requiring the exact numerical information about the constitutive relations of the model. Qualitative reasoning can make use of multiple ontologies, explicitly represent causality, enable the construction of more sophisticated models from simpler constituents by means of compositional modelling, and infer the global behaviour of a system from a description of its structure [144]. FQR that combines the two concepts captures the strengths of both these approaches: on the one hand, higher levels of precision compared to the FIS can be achieved; on the other hand, the integrated system has an improved interpretability and robustness of nonlinear system identification from input-output data. So FQR enables promising performance of reasoning tasks, at appropriate levels in applications that are particularly problematic from the modelling point of view for either on its own [145].

FQR has been employed to model the kinematic chain that has been demonstrated to be an effective approach to represent the human activities, alleviating the hard boundary or the crisp values of the ordinary measurement space. For example, Liu et al. proposed the Fuzzy Qualitative Trigonometry (FQT), where the ordinary Cartesian space and the unit circle were substituted with the combination of membership functions yielding the fuzzy qualitative coordinate and the fuzzy qualitative unit circle [146, 147]. Furthermore, the FQT was extended to robot kinematics with the goal of solving the intelligent connection problem for physical systems, in particular robotic systems [144, 148]. Motivated by the above approaches, a Qualitative Normalised Template (QNT) based on the FQT was proposed by Chan et al. for human action recognition [149, 150]. A data quantisation process was employed to model the uncertainties during the kinematic chain tracking process for the activity representation, handling the tradeoffs between the tracking precision and the computational cost. The QNT was constructed according to the fuzzy qualitative robot kinematics framework [144, 148]. An empirical comparison with the conventional

HMM and FHMM using both the KTH and the Weizmannn datasets has shown the effectiveness of the proposed solution [149]. More recently, Chern et al. incorporated the Poisson solution and the Fuzzy Quantity Space from the FQT to generate a new human model, which is invariant to different human anatomy (e.g. size, height, and pose) and is position standardised [151]. This work has the potential to be integrated into the view specific action recognition framework as a prerequisite.

1.3 Challenges and Motivation

Current techniques in sensing human motions provide a rich source, including both the detailed dynamic motions and the environment, for motion recognition in many applications. Different sensors can also simultaneously provide different characteristics of the human motions due to the fast capturing speed of the advanced hardware. However, how to utilising the sensed data to analyse and recognise human motions is still challenging. Firstly, the dynamic environment where the people perform activities varies a lot in terms of lighting conditions, backgrounds and surroundings, and the motions performed by the people also include both small and big movements, such as sign language and playing tennis. These varying factors must be carefully considered or modelled in the design of the analyse methods. Secondly, the multimodal data from different sensors may contain redundant information, which must be removed by dedicated fusing algorithms to increase the system's performances. Thirdly, the captured data is facing high levels of uncertainties that can affect the performances of such applications, such as the inaccuracy and noise inherited in the sensors or sensing systems. Fourthly, the training samples are limited or insufficient in many applications, which require the method to have a fast learning process. Last but more importantly, the extensive data, for example from the high-frequency capturing devices, would cost huge computational time, which would greatly limit applications in real-time systems. Thus, not only is one ideal system expected to have satisfying performance of the motion recognition, but it also needs to be computationally inexpensive.

On the other hand, the non-fuzzy approaches, such as the probabilistic graphical models and neural networks, have demonstrated great effectiveness in recognising human motions in many applications, as discussed above. Some of them are cost expensive, especially in the training stage, and their good performances may not be guaranteed when there are insufficient training samples. Fortunately, fuzzy techniques, especially the hybrid fuzzy approaches and fuzzy qualitative approaches, provide a promising way to handle the trade-off between human motion recognition precision and its computational efficiency. For example, the hybrid fuzzy approaches possess advantages from both machine learning techniques, which have a proper mathematical mechanism to guarantee the system performance, and the fuzzy techniques, which provide a practical solution of simple implementation in realtime systems; the fuzzy qualitative approaches also offer an effective way to represent human motion with reduced computational complexity and also bridge the semantic gap that exists between the sensed data (signal) and the human observer (symbol). Therefore, the rest of the book will introduce recent progresses on human motion sensing and recognition, such as human motion recognition in 2D and 3D, hand motion analysis with contact sensors, and vision-based view-invariant motion recognition, by using fuzzy approaches, with a focus on the fuzzy qualitative techniques and hybrid fuzzy approaches.

References

1. Muhammad Hassan, Tasweer Ahmad, Nudrat Liaqat, Ali Farooq, Syed Asghar Ali, and Syed Rizwan Hassan. A review on human actions recognition using vision based techniques. *Journal of Image and Graphics*, 2(1):28–32, 2014.
2. Tracey KM Lee, Mohammed Belkhatir, and Saeid Sanei. A comprehensive review of past and present vision-based techniques for gait recognition. *Multimedia Tools and Applications*, 72(3):2833–2869, 2014.
3. Haitham Hasan and Sameem Abdul-Kareem. Human–computer interaction using vision-based hand gesture recognition systems: a survey. *Neural Computing and Applications*, 25(2):251–261, 2014.
4. Siddharth S Rautaray and Anupam Agrawal. Vision based hand gesture recognition for human computer interaction: a survey. *Artificial Intelligence Review*, 43(1):1–54, 2015.
5. Jianfang Dou and Jianxun Li. Robust human action recognition based on spatio-temporal descriptors and motion temporal templates. *Optik-International Journal for Light and Electron Optics*, 125(7):1891–1896, 2014.
6. Basura Fernando, Efstratios Gavves, Jose M Oramas, Amir Ghodrati, and Tinne Tuytelaars. Modeling video evolution for action recognition. In *Proceedings of the IEEE Conference on Computer Vision and Pattern Recognition*, pages 5378–5387, 2015.
7. SangUk Han and SangHyun Lee. A vision-based motion capture and recognition framework for behavior-based safety management. *Automation in Construction*, 35:131–141, 2013.
8. Pedram Azad, Tamim Asfour, and Rudiger Dillmann. Robust real-time stereo-based markerless human motion capture. In *Humanoids 2008-8th IEEE-RAS International Conference on Humanoid Robots*, pages 700–707. IEEE, 2008.
9. Rustam Rakhimov Igorevich, Pusik Park, Jongchan Choi, and Dugki Min. Two hand gesture recognition using stereo camera. *International Journal of Computer and Electrical Engineering*, 5(1):69, 2013.
10. Jake K Aggarwal and Lu Xia. Human activity recognition from 3d data: A review. *Pattern Recognition Letters*, 48:70–80, 2014.
11. Kui Liu and Nasser Kehtarnavaz. Real-time robust vision-based hand gesture recognition using stereo images. *Journal of Real-Time Image Processing*, 11(1):201–209, 2016.
12. Myung-Cheol Roh, Ho-Keun Shin, and Seong-Whan Lee. View-independent human action recognition with volume motion template on single stereo camera. *Pattern Recognition Letters*, 31(7):639–647, 2010.
13. Hamed Sarbolandi, Damien Lefloch, and Andreas Kolb. Kinect range sensing: Structured-light versus time-of-flight kinect. *Computer Vision and Image Understanding*, 139:1–20, 2015.
14. Jungong Han, Ling Shao, Dong Xu, and Jamie Shotton. Enhanced computer vision with microsoft kinect sensor: A review. *Cybernetics, IEEE Transactions on*, 43(5):1318–1334, 2013.
15. Lulu Chen, Hong Wei, and James Ferryman. A survey of human motion analysis using depth imagery. *Pattern Recognition Letters*, 34(15):1995–2006, 2013.
16. Roanna Lun and Wenbing Zhao. A survey of applications and human motion recognition with microsoft kinect. *International Journal of Pattern Recognition and Artificial Intelligence*, 29(05):1555008, 2015.

17. Yue Gao, You Yang, Yi Zhen, and Qionghai Dai. Depth error elimination for rgb-d cameras. *ACM Transactions on Intelligent Systems and Technology*, 6(2):13, 2015.
18. Zhenbao Liu, Jinxin Huang, Junwei Han, Shuhui Bu, and Jianfeng Lv. (in press) human motion tracking by multiple rgbd cameras. *IEEE Transactions on Circuits and Systems for Video Technology*, 2016.
19. Hubert PH Shum, Edmond SL Ho, Yang Jiang, and Shu Takagi. Real-time posture reconstruction for microsoft kinect. *IEEE transactions on cybernetics*, 43(5):1357–1369, 2013.
20. Qi-rong Mao, Xin-yu Pan, Yong-zhao Zhan, and Xiang-jun Shen. Using kinect for real-time emotion recognition via facial expressions. *Frontiers of Information Technology & Electronic Engineering*, 16(4):272–282, 2015.
21. Jianfeng Li and Shigang Li. Eye-model-based gaze estimation by rgb-d camera. In *Proceedings of the IEEE Conference on Computer Vision and Pattern Recognition Workshops*, pages 592–596, 2014.
22. Jamie Shotton, Ross Girshick, Andrew Fitzgibbon, Toby Sharp, Mat Cook, Mark Finocchio, Richard Moore, Pushmeet Kohli, Antonio Criminisi, Alex Kipman, et al. Efficient human pose estimation from single depth images. *IEEE Transactions on Pattern Analysis and Machine Intelligence*, 35(12):2821–2840, 2013.
23. Zhou Ren, Junsong Yuan, Jingjing Meng, and Zhengyou Zhang. Robust part-based hand gesture recognition using kinect sensor. *IEEE transactions on multimedia*, 15(5):1110–1120, 2013.
24. Timm Linder and Kai O Arras. Multi-model hypothesis tracking of groups of people in rgb-d data. In *International Conference on Information Fusion*, pages 1–7. IEEE, 2014.
25. Alexandra Pfister, Alexandre M West, Shaw Bronner, and Jack Adam Noah. Comparative abilities of microsoft kinect and vicon 3d motion capture for gait analysis. *Journal of medical engineering & technology*, 38(5):274–280, 2014.
26. Bruno Bonnechere, Bart Jansen, P Salvia, H Bouzahouene, L Omelina, Fedor Moiseev, Victor Sholukha, Jan Cornelis, Marcel Rooze, and S Van Sint Jan. Validity and reliability of the kinect within functional assessment activities: comparison with standard stereophotogrammetry. *Gait & posture*, 39(1):593–598, 2014.
27. Sen Qiu, Zhelong Wang, Hongyu Zhao, and Huosheng Hu. Using distributed wearable sensors to measure and evaluate human lower limb motions. *IEEE Transactions on Instrumentation and Measurement*, 65(4):939–950, 2016.
28. Mehdi Khoury and Honghai Liu. Boxing motions classification through combining fuzzy gaussian inference with a context-aware rule-based system. In *IEEE International Conference on Fuzzy Systems*, pages 842–847. IEEE, 2009.
29. Aaron M Bestick, Samuel A Burden, Giorgia Willits, Nikhil Naikal, S Shankar Sastry, and Ruzena Bajcsy. Personalized kinematics for human-robot collaborative manipulation. In *IEEE/RSJ International Conference on Intelligent Robots and Systems*, pages 1037–1044. IEEE, 2015.
30. Karl Abson and Ian Palmer. Motion capture: capturing interaction between human and animal. *The Visual Computer*, 31(3):341–353, 2015.
31. Federico L Moro, Nikos G Tsagarakis, and Darwin G Caldwell. Walking in the resonance with the coman robot with trajectories based on human kinematic motion primitives (kmps). *Autonomous Robots*, 36(4):331–347, 2014.
32. Catalin Ionescu, Dragos Papava, Vlad Olaru, and Cristian Sminchisescu. Human3. 6m: Large scale datasets and predictive methods for 3d human sensing in natural environments. *IEEE transactions on pattern analysis and machine intelligence*, 36(7):1325–1339, 2014.
33. Dong K Noh, Nam G Lee, and Joshua H You. A novel spinal kinematic analysis using x-ray imaging and vicon motion analysis: A case study. *Bio-medical materials and engineering*, 24(1):593–598, 2014.
34. Weihua Sheng, Anand Thobbi, and Ye Gu. An integrated framework for human–robot collaborative manipulation. *IEEE transactions on cybernetics*, 45(10):2030–2041, 2015.
35. Lu Bai, Matthew G Pepper, Yong Yan, Sarah K Spurgeon, and Mohamed Sakel. Application of low cost inertial sensors to human motion analysis. In *IEEE International Conference on Instrumentation and Measurement Technology*, pages 1280–1285. IEEE, 2012.

36. Yinfeng Fang, Xiangyang Zhu, and Honghai Liu. Development of a surface emg acquisition system with novel electrodes configuration and signal representation. In *International Conference on Intelligent Robotics and Applications*, pages 405–414. Springer, 2013.
37. Yinfeng Fang, Honghai Liu, Gongfa Li, and Xiangyang Zhu. A multichannel surface emg system for hand motion recognition. *International Journal of Humanoid Robotics*, 12(02):1550011, 2015.
38. Zhaojie Ju, Gaoxiang Ouyang, Marzena Wilamowska-Korsak, and Honghai Liu. Surface emg based hand manipulation identification via nonlinear feature extraction and classification. *IEEE Sensors Journal*, 13(9):3302–3311, 2013.
39. Gaoxiang Ouyang, Zhaojie Ju, and Honghai Liu. Changes in emg–emg coherence during hand grasp movements. *International Journal of Humanoid Robotics*, 11(01):1450002, 2014.
40. Gaoxiang Ouyang, Xiangyang Zhu, Zhaojie Ju, and Honghai Liu. Dynamical characteristics of surface emg signals of hand grasps via recurrence plot. *IEEE journal of biomedical and health informatics*, 18(1):257–265, 2014.
41. Mark Ison and Panagiotis Artemiadis. Proportional myoelectric control of robots: muscle synergy development drives performance enhancement, retainment, and generalization. *IEEE Transactions on Robotics*, 31(2):259–268, 2015.
42. Karin Lienhard, Aline Cabasson, Olivier Meste, and Serge S Colson. Semg during whole-body vibration contains motion artifacts and reflex activity. *Journal of sports science & medicine*, 14(1):54, 2015.
43. Yinfeng Fang, Nalinda Hettiarachchi, Dalin Zhou, and Honghai Liu. Multi-modal sensing techniques for interfacing hand prostheses: a review. *IEEE Sensors Journal*, 15(11):6065–6076, 2015.
44. Javad Hashemi, Evelyn Morin, Parvin Mousavi, and Keyvan Hashtrudi-Zaad. Enhanced dynamic emg-force estimation through calibration and pci modeling. *IEEE Transactions on Neural Systems and Rehabilitation Engineering*, 23(1):41–50, 2015.
45. B Afsharipour, K Ullah, and R Merletti. Amplitude indicators and spatial aliasing in high density surface electromyography recordings. *Biomedical Signal Processing and Control*, 22:170–179, 2015.
46. Wei Meng, Quan Liu, Zude Zhou, and Qingsong Ai. Active interaction control applied to a lower limb rehabilitation robot by using emg recognition and impedance model. *Industrial Robot: An International Journal*, 41(5):465–479, 2014.
47. Daniele Leonardis, Michele Barsotti, Claudio Loconsole, Massimiliano Solazzi, Marco Troncossi, Claudio Mazzotti, Vincenzo Parenti Castelli, Caterina Procopio, Giuseppe Lamola, Carmelo Chisari, et al. An emg-controlled robotic hand exoskeleton for bilateral rehabilitation. *IEEE transactions on haptics*, 8(2):140–151, 2015.
48. Yanjuan Geng, Xiufeng Zhang, Yuan-Ting Zhang, and Guanglin Li. A novel channel selection method for multiple motion classification using high-density electromyography. *Biomedical engineering online*, 13(1):1–16, 2014.
49. Lizhi Pan, Dingguo Zhang, Jianwei Liu, Xinjun Sheng, and Xiangyang Zhu. Continuous estimation of finger joint angles under different static wrist motions from surface emg signals. *Biomedical Signal Processing and Control*, 14:265–271, 2014.
50. Ernest Nlandu Kamavuako, Dario Farina, Ken Yoshida, and Winnie Jensen. Relationship between grasping force and features of single-channel intramuscular emg signals. *Journal of Neuroscience Methods*, 185(1):143–150, 2009.
51. Lauren Hart Smith. *Use of Intramuscular Electromyography for the Simultaneous Control of Multiple Degrees of Freedom in Upper-Limb Myoelectric Prostheses*. PhD thesis, Northwestern University, 2015.
52. DJ Sturman, D. Zeltzer, and P. Medialab. A survey of glove-based input. *IEEE Computer Graphics and Applications*, 14(1):30–39, 1994.
53. G. Burdea and P. Coiffet. Virtual reality technology. *Presence: Teleoperators & Virtual Environments*, 12(6):663–664, 2003.
54. Masatake Sato, Vladimir Savchenko, and Ryutarou Ohbuchi. 3d freeform design: interactive shape deformations by the use of cyberglove. In *International Conference on Cyberworlds*, pages 147–154. IEEE, 2004.

55. D.J. Sturman. *Whole-hand input*. PhD thesis, Massachusetts Institute of Technology, 1992.
56. L. Dipietro, A.M. Sabatini, and P. Dario. Evaluation of an instrumented glove for hand-movement acquisition. *Journal of rehabilitation research and development*, 40(2):179–190, 2003.
57. Y. Su, CR Allen, D. Geng, D. Burn, U. Brechany, GD Bell, and R. Rowland. 3-D motion system ("data-gloves"): application for Parkinson's disease. *IEEE Transactions on Instrumentation and Measurement*, 52(3):662–674, 2003.
58. T. Kuroda, Y. Yabata, A. Goto, H. Ikuta, and M. Murakami. Consumer price data-glove for sign language recognition. *International Conference on Disability, Virtual Reality & Assoc. Tech*, 2004.
59. L. Dipietro, A.M. Sabatini, and P. Dario. A Survey of Glove-Based Systems and Their Applications. *IEEE Transactions on Systems, Man and Cybernetics - Part C: Applications and Reviews*, 38(4):461–482, 2008.
60. J Bukhari, Maryam Rehman, Saman Ishtiaq Malik, Awais M Kamboh, and Ahmad Salman. American sign language translation through sensory glove; signspeak. *Int. J. u-and e-Service, Science and Technology*, 8, 2015.
61. Noor-e-Karishma Shaik, Lathasree Veerapalli, et al. Sign language recognition through fusion of 5dt data glove and camera based information. In *IEEE International Advance Computing Conference*, pages 639–643. IEEE, 2015.
62. An Nayyar and Vikram Puri. Data glove: Internet of things (iot) based smart wearable gadget. *British Journal of Mathematics & Computer Science*, 15(5), 2016.
63. Satoshi Funabashi, Alexander Schmitz, Takashi Sato, Sophon Somlor, and Shigeki Sugano. Robust in-hand manipulation of variously sized and shaped objects. In *IEEE/RSJ International Conference on Intelligent Robots and Systems*, pages 257–263. IEEE, 2015.
64. M.G. Ceruti, V.V. Dinh, N.X. Tran, H. Van Phan, L.R.T. Duffy, T.A. Ton, G. Leonard, E. Medina, O. Amezcua, S. Fugate, et al. Wireless communication glove apparatus for motion tracking, gesture recognition, data transmission, and reception in extreme environments. In *Proceedings of the 2009 ACM symposium on Applied Computing*, pages 172–176. ACM, 2009.
65. Zhou Ma and Pinhas Ben-Tzvi. Rml glovean exoskeleton glove mechanism with haptics feedback. *Mechatronics, IEEE/ASME Transactions on*, 20(2):641–652, 2015.
66. Huiyu Zhou and Huosheng Hu. Human motion tracking for rehabilitationa survey. *Biomedical Signal Processing and Control*, 3(1):1–18, 2008.
67. Yu-Liang Hsu, Jeen-Shing Wang, Yu-Ching Lin, Shu-Min Chen, Yu-Ju Tsai, Cheng-Ling Chu, and Che-Wei Chang. A wearable inertial-sensing-based body sensor network for shoulder range of motion assessment. In *International Conference On Orange Technologies*, pages 328–331. IEEE, 2013.
68. Mahmoud El-Gohary and James McNames. Human joint angle estimation with inertial sensors and validation with a robot arm. *IEEE Transactions on Biomedical Engineering*, 62(7):1759–1767, 2015.
69. Xiaoping Yun, James Calusdian, Eric R Bachmann, and Robert B McGhee. Estimation of human foot motion during normal walking using inertial and magnetic sensor measurements. *IEEE Transactions on Instrumentation and Measurement*, 61(7):2059–2072, 2012.
70. Mehdi Khoury. *A fuzzy probabilistic inference methodology for constrained 3D human motion classification*. PhD thesis, University of Portsmouth, 2010.
71. Yongbin Qi, Cheong Boon Soh, Erry Gunawan, Kay-Soon Low, and Rijil Thomas. Estimation of spatial-temporal gait parameters using a low-cost ultrasonic motion analysis system. *Sensors*, 14(8):15434–15457, 2014.
72. Sangwoo Cho, Jeonghun Ku, Yun Kyung Cho, In Young Kim, Youn Joo Kang, Dong Pyo Jang, and Sun I Kim. Development of virtual reality proprioceptive rehabilitation system for stroke patients. *Computer methods and programs in biomedicine*, 113(1):258–265, 2014.
73. Hassen Fourati, Noureddine Manamanni, Lissan Afilal, and Yves Handrich. Complementary observer for body segments motion capturing by inertial and magnetic sensors. *IEEE/ASME Transactions on Mechatronics*, 19(1):149–157, 2014.

74. He Zhao and Zheyao Wang. Motion measurement using inertial sensors, ultrasonic sensors, and magnetometers with extended kalman filter for data fusion. *Sensors Journal, IEEE*, 12(5):943–953, 2012.
75. Roland Cheng, Wendi Heinzelman, Melissa Sturge-Apple, and Zeljko Ignjatovic. A motion-tracking ultrasonic sensor array for behavioral monitoring. *Sensors Journal, IEEE*, 12(3):707–712, 2012.
76. Daniel Roetenberg, Per J Slycke, and Peter H Veltink. Ambulatory position and orientation tracking fusing magnetic and inertial sensing. *IEEE Transactions on Biomedical Engineering*, 54(5):883–890, 2007.
77. Zhaojie Ju and Honghai Liu. Human hand motion analysis with multisensory information. *IEEE/ASME Transactions on Mechatronics*, 19(2):456–466, 2014.
78. Dalai Tang, Bakhtiar Yusuf, János Botzheim, Naoyuki Kubota, and Chee Seng Chan. A novel multimodal communication framework using robot partner for aging population. *Expert Systems with Applications*, 42(9):4540–4555, 2015.
79. S. Mitra and T. Acharya. Gesture Recognition: A Survey. *IEEE Transactions on systems, man and cybernetics. Part C, Applications and reviews*, 37(3):311–324, 2007.
80. Paulo Vinicius Koerich Borges, Nicola Conci, and Andrea Cavallaro. Video-based human behavior understanding: a survey. *IEEE Transactions on Circuits and Systems for Video Technology*, 23(11):1993–2008, 2013.
81. X. Ji and H. Liu. Advances in View-Invariant Human Motion Analysis: A Review. *IEEE Transactions on Systems, Man and Cybernetics Part C*, 40(1):13–24, 2010.
82. Jake K Aggarwal and Michael S Ryoo. Human activity analysis: A review. *ACM Computing Surveys (CSUR)*, 43(3):16, 2011.
83. H. Liu. Exploring human hand capabilities into embedded multifingered object manipulation. *IEEE Transactions on Industrial Informatics*, 7(3):389–398, 2011.
84. T.B. Moeslund, A. Hilton, and V. Krüger. A survey of advances in vision-based human motion capture and analysis. *Computer Vision and Image Understanding*, 104(2–3):90–126, 2006.
85. Daphne Koller and Nir Friedman. *Probabilistic graphical models: principles and techniques*. MIT press, 2009
86. Lina Tong, Quanjun Song, Yunjian Ge, and Ming Liu. Hmm-based human fall detection and prediction method using tri-axial accelerometer. *Sensors Journal, IEEE*, 13(5):1849–1856, 2013.
87. Yuexin Wu, Zhe Jia, Yue Ming, Juanjuan Sun, and Liujuan Cao. Human behavior recognition based on 3d features and hidden markov models. *Signal, Image and Video Processing*, pages 1–8, 2015.
88. Natraj Raman and SJ Maybank. Activity recognition using a supervised non-parametric hierarchical hmm. *Neurocomputing*, 199:163–177, 2016.
89. Thanh Minh Nguyen and QM Jonathan Wu. Fast and robust spatially constrained gaussian mixture model for image segmentation. *IEEE transactions on circuits and systems for video technology*, 23(4):621–635, 2013.
90. YingLi Tian, Liangliang Cao, Zicheng Liu, and Zhengyou Zhang. Hierarchical filtered motion for action recognition in crowded videos. *IEEE Transactions on Systems, Man, and Cybernetics, Part C: Applications and Reviews*, 42(3):313–323, 2012.
91. Di Wu and Ling Shao. Leveraging hierarchical parametric networks for skeletal joints based action segmentation and recognition. In *Proceedings of the IEEE Conference on Computer Vision and Pattern Recognition*, pages 724–731, 2014.
92. Ruikun Luo and Dmitry Berenson. A framework for unsupervised online human reaching motion recognition and early prediction. In *IEEE/RSJ International Conference on Intelligent Robots and Systems*, pages 2426–2433. IEEE, 2015.
93. Z. Ju, H. Liu, X. Zhu, and Y. Xiong. Dynamic Grasp Recognition Using Time Clustering, Gaussian Mixture Models and Hidden Markov Models. *Journal of Advanced Robotics*, 23:1359–1371, 2009.
94. Asa Ben-Hur and Jason Weston. A users guide to support vector machines. *Data mining techniques for the life sciences*, pages 223–239, 2010.

95. Chen Chen, Roozbeh Jafari, and Nasser Kehtarnavaz. Improving human action recognition using fusion of depth camera and inertial sensors. *IEEE Transactions on Human-Machine Systems*, 45(1):51–61, 2015.
96. Huimin Qian, Yaobin Mao, Wenbo Xiang, and Zhiquan Wang. Recognition of human activities using svm multi-class classifier. *Pattern Recognition Letters*, 31(2):100–111, 2010.
97. Salah Althloothi, Mohammad H Mahoor, Xiao Zhang, and Richard M Voyles. Human activity recognition using multi-features and multiple kernel learning. *Pattern recognition*, 47(5):1800–1812, 2014.
98. Johan AK Suykens. *Advances in learning theory: methods, models, and applications*, volume 190. IOS Press, 2003.
99. Gavin C Cawley and Nicola LC Talbot. On over-fitting in model selection and subsequent selection bias in performance evaluation. *The Journal of Machine Learning Research*, 11:2079–2107, 2010.
100. Anders Krogh, Jesper Vedelsby, et al. Neural network ensembles, cross validation, and active learning. *Advances in neural information processing systems*, 7:231–238, 1995.
101. Haitham Hasan and S Abdul-Kareem. Static hand gesture recognition using neural networks. *Artificial Intelligence Review*, 41(2):147–181, 2014.
102. S Ali Etemad and Ali Arya. Classification and translation of style and affect in human motion using rbf neural networks. *Neurocomputing*, 129:585–595, 2014.
103. Geoffrey E Hinton, Simon Osindero, and Yee-Whye Teh. A fast learning algorithm for deep belief nets. *Neural computation*, 18(7):1527–1554, 2006.
104. Yoshua Bengio. Learning deep architectures for ai. *Foundations and trends® in Machine Learning*, 2(1):1–127, 2009.
105. Di Wu and Ling Shao. Deep dynamic neural networks for gesture segmentation and recognition. In *Workshop at the European Conference on Computer Vision*, pages 552–571. Springer, 2014.
106. Martin Längkvist, Lars Karlsson, and Amy Loutfi. A review of unsupervised feature learning and deep learning for time-series modeling. *Pattern Recognition Letters*, 42:11–24, 2014.
107. D. Wu, L. Pigou, P. J. Kindermans, N. LE, L. Shao, J. Dambre, and J. M. Odobez. Deep dynamic neural networks for multimodal gesture segmentation and recognition. *IEEE Transactions on Pattern Analysis and Machine Intelligence*, PP(99):1–1, 2016.
108. Alexander Toshev and Christian Szegedy. Deeppose: Human pose estimation via deep neural networks. In *Proceedings of the IEEE Conference on Computer Vision and Pattern Recognition*, pages 1653–1660, 2014.
109. L. Zadeh. Fuzzy sets. *Information and Control*, 8:338–353, 1965.
110. Siegfried Gottwald. An early approach toward graded identity and graded membership in set theory. *Fuzzy Sets and Systems*, 161(18):2369–2379, 2010.
111. L.A. Zadeh. The concept of a linguistic variable and its applications to approximate reasoning - i. *Information Sciences*, 8:199–249, 1975.
112. Patricia Melin and Oscar Castillo. A review on type-2 fuzzy logic applications in clustering, classification and pattern recognition. *Applied soft computing*, 21:568–577, 2014.
113. Chern Hong Lim, Ekta Vats, and Chee Seng Chan. Fuzzy human motion analysis: A review. *Pattern Recognition*, 48(5):1773–1796, 2015.
114. MRH Mohd Adnan, Arezoo Sarkheyli, Azlan Mohd Zain, and Habibollah Haron. Fuzzy logic for modeling machining process: a review. *Artificial Intelligence Review*, 43(3):345–379, 2015.
115. Cengiz Kahraman, Sezi Cevik Onar, and Basar Oztaysi. Fuzzy multicriteria decision-making: a literature review. *International Journal of Computational Intelligence Systems*, 8(4):637–666, 2015.
116. Juan P Wachs, Helman Stern, and Yael Edan. Cluster labeling and parameter estimation for the automated setup of a hand-gesture recognition system. *IEEE Transactions on Systems, Man, and Cybernetics-Part A: Systems and Humans*, 35(6):932–944, 2005.
117. R. Verma and A. Dev. Vision based hand gesture recognition using finite state machines and fuzzy logic. In *International Conference on Ultra Modern Telecommunications Workshops*, pages 1–6, 12–14 2009.

118. R. Palm, B. Iliev, and B. Kadmiry. Recognition of human grasps by time-clustering and fuzzy modeling. *Robotics and Autonomous Systems*, 57(5):484–495, 2009.
119. Z. Ju and H. Liu. A unified fuzzy framework for human hand motion recognition. *IEEE Transactions on Fuzzy Systems*, 19(5):901–913, 2011.
120. L. A. Zadeh. Fuzzy logic. *Computer*, 21(4):83–93, April 1988.
121. Qilian Liang and Jerry M Mendel. Interval type-2 fuzzy logic systems: theory and design. *IEEE Transactions on Fuzzy systems*, 8(5):535–550, 2000.
122. Jerry M Mendel, Robert I John, and Feilong Liu. Interval type-2 fuzzy logic systems made simple. *IEEE Transactions on Fuzzy Systems*, 14(6):808–821, 2006.
123. Mehdi Khoury and Honghai Liu. Extending evolutionary fuzzy quantile inference to classify partially occluded human motions. In *IEEE International Conference on Fuzzy Systems*, pages 1–8. IEEE, 2010.
124. Jean-Marie Le Yaouanc and Jean-Philippe Poli. A fuzzy spatio-temporal-based approach for activity recognition. In *International Conference on Conceptual Modeling*, pages 314–323. Springer, 2012.
125. Bo Yao, Hani Hagras, Mohammed J Alhaddad, and Daniyal Alghazzawi. A fuzzy logic-based system for the automation of human behavior recognition using machine vision in intelligent environments. *Soft Computing*, 19(2):499–506, 2015.
126. Fida El Baf, Thierry Bouwmans, and Bertrand Vachon. A fuzzy approach for background subtraction. In *2008 15th IEEE International Conference on Image Processing*, pages 2648–2651. IEEE, 2008.
127. Kourosh Mozafari, Nasrollah Moghadam Charkari, Hamidreza Shayegh Boroujeni, and Mohammad Behrouzifar. A novel fuzzy hmm approach for human action recognition in video. In *Knowledge Technology*, pages 184–193. Springer, 2012.
128. Magdi A Mohamed and Paul Gader. Generalized hidden markov models. i. theoretical frameworks. *IEEE Transactions on fuzzy systems*, 8(1):67–81, 2000.
129. Ben W Miners and Otman A Basir. Dynamic facial expression recognition using fuzzy hidden markov models. In *2005 IEEE International Conference on Systems, Man and Cybernetics*, volume 2, pages 1417–1422. IEEE, 2005.
130. Zhaojie Ju and Honghai Liu. Applying fuzzy em algorithm with a fast convergence to gmms. In *IEEE International Conference on Fuzzy Systems*, pages 1–6. IEEE, 2010.
131. Zhaojie Ju and Honghai Liu. Hand motion recognition via fuzzy active curve axis gaussian mixture models: A comparative study. In *IEEE International Conference on Fuzzy Systems*, pages 699–705. IEEE, 2011.
132. Z. Ju and H. Liu. Fuzzy gaussian mixture models. *Pattern Recognition*, 45(3):1146–1158, 2012.
133. Zhaojie Ju. *A fuzzy framework for human hand motion recognition*. PhD thesis, University of Portsmouth, 2010.
134. J. Zeng, L. Xie, and Z.Q. Liu. Type-2 fuzzy Gaussian mixture models. *Pattern Recognition*, 41(12):3636–3643, 2008.
135. Chun-Fu Lin and Sheng-De Wang. Fuzzy support vector machines. *IEEE Transactions on neural networks*, 13(2):464–471, 2002.
136. Daisuke Tsujinishi and Shigeo Abe. Fuzzy least squares support vector machines for multi-class problems. *Neural Networks*, 16(5):785–792, 2003.
137. Shigeo Abe. Fuzzy support vector machines for multilabel classification. *Pattern Recognition*, 48(6):2110–2117, 2015.
138. Zhiguo Yan, Zhizhong Wang, and Hongbo Xie. Joint application of rough set-based feature reduction and fuzzy ls-svm classifier in motion classification. *Medical & biological engineering & computing*, 46(6):519–527, 2008.
139. Kan Li. Human action recognition based on fuzzy support vector machines. In *Computational Intelligence and Design (ISCID), 2012 Fifth International Symposium on*, volume 1, pages 45–48. IEEE, 2012.
140. Chia-Feng Juang and Chia-Ming Chang. Human body posture classification by a neural fuzzy network and home care system application. *IEEE Transactions on Systems, Man, and Cybernetics-part A: Systems and Humans*, 37(6):984–994, 2007.

141. Annamária R Várkonyi-Kóczy and Balázs Tusor. Human–computer interaction for smart environment applications using fuzzy hand posture and gesture models. *IEEE Transactions on Instrumentation and Measurement*, 60(5):1505–1514, 2011.
142. Monireh-Sadat Hosseini and Amir-Masoud Eftekhari-Moghadam. Fuzzy rule-based reasoning approach for event detection and annotation of broadcast soccer video. *Applied Soft Computing*, 13(2):846–866, 2013.
143. Qiang Shen and Roy Leitch. Fuzzy qualitative simulation. *IEEE Transactions on Systems, Man, and Cybernetics*, 23(4):1038–1061, 1993.
144. H. Liu, D.J. Brown, and G.M. Coghill. Fuzzy qualitative robot kinematics. *IEEE Transactions on Fuzzy Systems*, 16(3):802–822, 2008.
145. Chee Seng Chan, George M Coghill, and Honghai Liu. Recent advances in fuzzy qualitative reasoning. *International Journal of Uncertainty, Fuzziness and Knowledge-Based Systems*, 19(03):417–422, 2011.
146. H. Liu and G.M. Coghill. Fuzzy qualitative trigonometry. *Proc. IEEE International Conference on Systems, Man and Cybernetics, Hawaii, USA.*, 2005.
147. Honghai Liu, George M Coghill, and Dave P Barnes. Fuzzy qualitative trigonometry. *International Journal of Approximate Reasoning*, 51(1):71–88, 2009.
148. H. Liu and D.J. Brown. An extension to fuzzy qualitative trigonometry and its application to robot kinematics. In *Proceedings of the IEEE International Conference on Fuzzy Systems*, pages 1111–1118, 2006.
149. C. S. Chan and H. Liu. Fuzzy Qualitative Human Motion Analysis. *IEEE Transactions on Fuzzy Systems*, 17(4):851–862, 2009.
150. Chern Hong Lim and Chee Seng Chan. A fuzzy qualitative approach for scene classification. In *IEEE International Conference on Fuzzy Systems*, pages 1–8. IEEE, 2012.
151. Chern Hong Lim and Chee Seng Chan. Fuzzy qualitative human model for viewpoint identification. *Neural Computing and Applications*, 27(4):845–856, 2016.

Chapter 2
Fuzzy Qualitative Trigonometry

2.1 Introduction

Trigonometry is a branch of mathematics that deals with the relationships between the sides and angles of triangles and with the properties and application of trigonometric functions of angles. It began as the computational component of geometry in the second century BC and plays a crucial role in domains such as mathematics and engineering. In order to bridge the gap between qualitative and quantitative descriptions of physical systems, we propose a fuzzy qualitative representation of trigonometry (FQT), which provides theoretical foundations for the representation of trigonometric properties. It is often desirable and sometimes necessary to reason about the behaviour of a system on the basis of incomplete or sparse information. The methods of model-based technology provide a means of doing this [1, 2]. The initial approaches to model-based reasoning were seminal but focused on symbolic qualitative reasoning (QR) only, providing a means whereby the global picture of how a system might behave could be generated using only the sign of the magnitude and direction of change of the system variables. This makes qualitative reasoning complementary to quantitative simulation. However, quantitative and qualitative simulation form the two ends of a spectrum; and semi-quantitative methods were developed to fill the gap. For the most part these were interval reasoners bolted on to existing qualitative reasoning systems (e.g. [3]), Blackwell did pioneering work on spatial reasoning on robots [4]; however, one exception to this was fuzzy qualitative reasoning which integrated the strengths of approximate reasoning with those of qualitative reasoning to form a more coherent semiquantitative approach than their predecessors [5, 6]. Model-based technology methods have been successfully applied to a number of tasks in the process domain. However, while some effort has been expended on developing qualitative kinematic models, the results have been limited [7–10]. The basic requirement for progressing in this domain is the development of a qualitative version of the trigonometric rules. Buckley and Eslami [11] proposed the definition of fuzzy trigonometry from the fuzzy perspective without consideration of the geometric meaning of trigonometry. Some progress has been made in this direction

H. Liu et al., *Human Motion Sensing and Recognition*,
Studies in Computational Intelligence 675, DOI 10.1007/978-3-662-53692-6_2

by Liu [12], but as with other applications of qualitative reasoning, the flexibility gained in variable precision by integrating fuzzy and qualitative approaches is no less important in the kinematics domain. In this chapter an extension of the rules of trigonometry to the fuzzy qualitative case is presented, which will serve as the basis for fuzzy qualitative reasoning about the behaviour and possible diagnosis of kinematic robot devices.

Fuzzy qualitative reasoning combines the advantages of fuzzy reasoning and qualitative reasoning techniques. Research into the integration of fuzzy reasoning and qualitative reasoning has been carried out in both theory and application in the past two decades [5, 6, 13–18]. The use of fuzzy reasoning methods are becoming more and more popular in intelligent systems [19, 20], especially hybrid methods and their applications integrating with evolutionary computing [21–25], decision trees [26, 27], neural networks [28–32], data mining [13], and so on [33–37]. Qualitative reasoning is reviewed in [38–41]. The integration of fuzzy reasoning and qualitative reasoning (i.e., fuzzy qualitative reasoning) provides an opportunity to explore research (e.g., spatial reasoning) with both the advantages of fuzzy reasoning and qualitative reasoning. Some of fuzzy qualitative reasoning contributions can be found in [5, 6, 14–16]. Shen and Leitch [5] use a fuzzy quantity space (i.e., normalized fuzzy partition), which allows for a more detailed description of the values of the variables. Such an approach relies on the extension principle and approximation principle in order to express the results of calculations in terms of the fuzzy sets of the fuzzy quantity space.

Fuzzy reasoning has been significantly developed and has attracted much attention and exploitation from industry and research communities in the past four decades. Fuzzy reasoning is good at communicating with sensing and control level subsystems by means of fuzzification and defuzzification methods. It has powerful reasoning strategies utilising compiled knowledge through conditional statements so as to easily handle mathematical and engineering systems in model free manner. Fuzzy reasoning also provides a means of handling uncertainty in a natural way making it robust in significantly noisy environments. However, the fact that its knowledge is primarily shallow, and the questions over the computational overhead associated with handling grades of membership of discrete fuzzy sets must be taken into account if multi-step reasoning is to be carried out. On the other hand, Qualitative and Model-based Reasoning has been successfully deployed in many applications such are autonomous spacecraft support [42], Systems Biology [43] and qualitative systems identification [44]. It has the advantage of operating at the conceptual modelling level, reasoning symbolically with models which retain the mathematical structure of the problem rather than the input/output representation of rule bases (fuzzy or otherwise). These models are incomplete in the sense that, being symbolic, they do not contain, or require, exact parameter information in order to operate. Qualitative reasoning can make use of multiple ontologies, can explicitly represent causality, enable the construction of more sophisticated models from simpler constituents by means of compositional modelling, and infer the global behaviour of a system from a description of its structure [38, 41]. These features can, when combined with fuzzy values and operators, compensate for the lack of ability in fuzzy reasoning alone to deal with that kind of inference about complex systems. The computational cause-effect relations contained in qualitative models facilitates analysing and explaining the behaviour of a structural model. Based on a scenario generated from fuzzy reasoning's fuzzification process, fuzzy qualitative reasoning may be able to build a behavioural model automatically, and use this model to generate a behaviour description, acceptable by symbolic systems, either by abstraction and qualitative simulation or as a comprehensive representation of all possible behaviours utilising linguistic fuzzy values. Liu, Coghill and Brown had attempted two completely different approaches [45–47] based on fuzzy qualitative trigonometry [48]. Research reported in [45] proposed a normalised based qualitative representation from cognition perspective, it converts both numeric and sub-symbolic data into a normalisation reference where transfer of different types of data is carried out, the method was not implemented into spatial robots due to its costly computational cost and complex spatial relation though it was applied to planar robots. On the other hand, conventional robotics had been adapted with the fuzzy qualitative trigonometry, not only did it implement feasible for spatial robots but also it shows the promising potential for intelligent robotics [46]. A full account of fuzzy qualitative trigonometry is presented in this chapter with the goal of solving the intelligent connection problem (also known as symbol grounding problem) for physical systems, in particular robotic systems. This problem is one of the key issues in AI robotics [49] and relates to a wide range of research areas such as computer vision [50]. This chapter is organised as follows. Section 2.2 presents fuzzy qualitative (FQ) trigonometry. Section 2.3 derives FQ trigonometric functions. Section 2.4 proposes the derivative extension to FQ trigonometry. Section 2.5 addresses discussions and conclusions.

2.2 Fuzzy Qualitative Trigonometry

The membership distribution of a normal convex fuzzy number can be approximated by a 4-tuple fuzzy number representation. The representation of the 4-tuple fuzzy numbers is a better qualitative representation for trigonometry, because this representation has high resolution and good compositionality [5, 51]. The degree of resolution can be adjusted by the choice of fuzzy numbers. Such a representation provides the flexibility to carry out computation based on real numbers, intervals, triangular numbers and trapezoidal fuzzy intervals, which comprise nearly all of the computing elements in fuzzy qualitative reasoning. This representation has the ability to combine representations for different aspects of a phenomenon or system to create a representation of the phenomenon or systems as a whole. The computation of fuzzy numbers is based on its arithmetic operations. A four-tuple fuzzy number representation, namely $[a,\ b,\ \tau\ \beta]$) with the condition $a <= b$ and $a \times b >= 0$, can be used to approximate the membership distribution of a normal convex fuzzy number. The four tuple set, $[a, b, \tau, \beta]$, can be defined by the following membership function, $\mu_a(x)$,

$$\mu_A(x) = \begin{cases} 0 & x < a - \tau \\ \tau^{-1}(x - a + \tau) & x \in [a - \tau, a] \\ 1 & x \in [a, b] \\ \beta^{-1}(b + \beta - x) & x \in [b, b + \beta] \\ 0 & x > b + \beta \end{cases} \quad (2.1)$$

Note that if $a = b$, Eq. 2.1 is a triangular fuzzy number with its membership as $\mu_A(x) = [\tau^{-1}(x - a + \tau), \beta^{-1}(a + \beta - x)]$, where $x \in [a - \tau, a + \beta]$. Besides, if $a = b$ and $\tau = \beta = 0$, it turns the 4-tuple fuzzy number into a real number. For instance [1, 1, 0, 0] stands for the real number 1. In addition, the arithmetic operations on the 4-tuple parametric representation of fuzzy numbers are given in Appendix A, where $<_0$ is the partial order $<_\alpha$ when $\alpha = 0$, which was used in chapter [5]. The partial order has been given in a general form, i.e., $\{\mathbb{O}_\alpha \,|(\mathbb{O} \in \{>, <, =, \leq, \geq, \in, \subset, \supset, \subseteq, \supseteq\})$. For instance, partial order $<_\alpha$ is defined such that for m, n $(m \neq n)$, we say m is α-less than n, $m <_\alpha n$, iff $a < b$, $a \in m_\alpha, b \in n_\alpha$ with m_α and n_α being the α-cuts of m and n, respectively. It is believed that fuzzy arithmetic is still a developing research topic, usage of the above-mentioned fuzzy operators is constrained by the fact that an assignment of α value is required in a calculation cycle.

FQ trigonometry actually is the FQ description of conventional trigonometry. The unit circle of conventional trigonometry has been modified by the introduction of quantity spaces for orientation and translation. Conventional trigonometric functions (e.g. a sine function) have been abstracted to obtain FQ versions of these operations. For a good survey on FQ reasoning please see [48]. The quantity space for every variable in the system is a finite and convex discretisation of the real number line. FQ quantity space Q consists of an orientation component Q^a and a translation component Q^d. It can be described as,

$$\begin{aligned} Q^a &= [QS_a(\theta_1), \ldots, QS_a(\theta_i), \ldots QS_a(\theta_m)] \\ Q^d &= \left[QS_d(l_1), \ldots, QS_d(l_j), \ldots QS_d(l_n)\right] \end{aligned} \quad (2.2)$$

where $QS_a(\theta_i)$ denotes the state of an angle θ_i, $QS_d(l_j)$ denotes the state of a distance l_j, m and n are the number of the elements of the two components. The position measurement of $P(QS_a(\theta_i), QS_d(l_j))$ determined by both the characteristics of the fuzzy membership functions of $QS_a(\theta_i)$ and $QS_d(l_j)$. The geometric meaning of FQ trigonometry is demonstrated in a proposed FQ unit circle, in which the motion is described by an orientation component and a translation component. A FQ unit circle is nothing but a conventional trigonometric circle whose axes are replaced by unit quantity spaces. The position state of a FQ point is defined by the projections of the point into FQ axes. For instance, let the X axis projection of a FQ point P be X_p, and the Y axis projection as Y_p, its FQ position can be described as $P = (X_p, Y_p)$. It means that a point position in FQ coordinates is described by fuzzy sets, hence, fuzzy arithmetic and its characteristics can be applied to calculation of FQ states in FQ coordinates. In accordance with Cartesian coordinates convention, the position, $P_o(X_o, Y_o) = ([0, 0, 0, 0], [0, 0, 0, 0])$, is the FQ origin, where the axes intersect. Counterclockwise is the positive orientation, the number of qualitative orientation

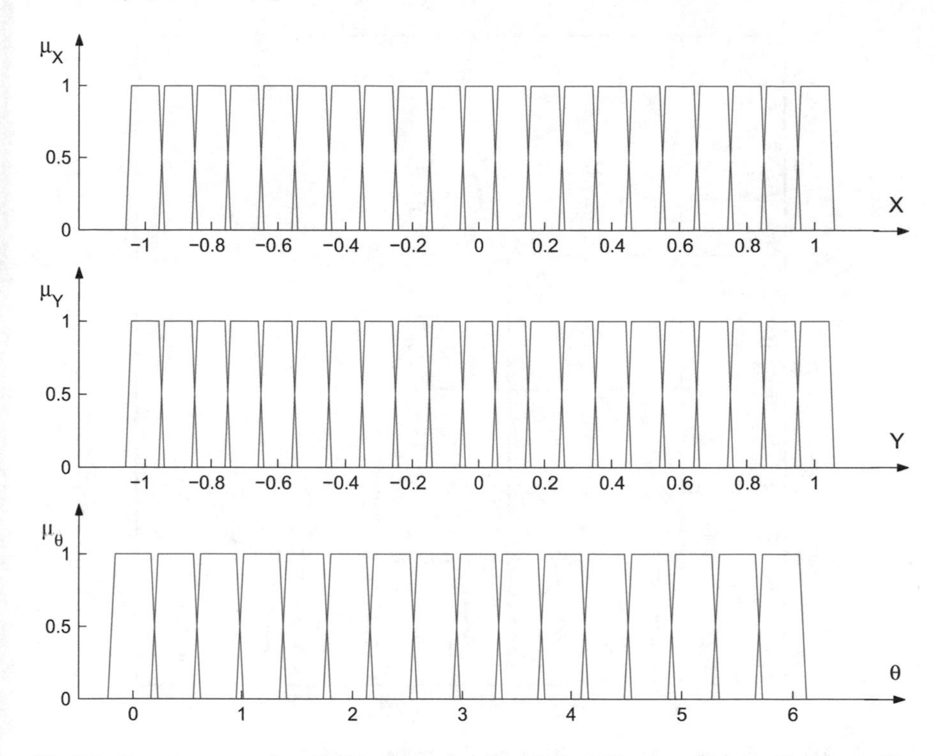

Fig. 2.1 Quantity spaces described by fuzzy membership functions (p = 17, q_x = 23 and q_y = 23)

states in a full circle starting from the X axis is denoted by p, the number of qualitative translation states is denoted by q_{axis}, (i.e. q_x, q_y). Though different functions can be employed to generate fuzzy numbers of a quantity space, two conditions must hold. First, fuzzy numbers in each component must be origin symmetric. Second, special real numbers (e.g., 0, $\pi/2$, π, $3\pi/2$ in an orientation, and -1, 0, -1 in a translation) are the centres of some certain fuzzy numbers in their corresponding quantity space. For instance, a right angle in conventional trigonometry is corresponded to fuzzy number $QS_a(p/4+1)$, whose centre is equal to $\pi/2$. Please note that, for a 4-tuple fuzzy number (e.g., $[a,\ b,\ \tau,\ \beta]$), its centre is defined by $((a+b)/2)\times 2\pi$. Note that the fuzzy numbers in the orientation quantity space need to be multiplied by 2π before they are applied to the arithmetic in Appendix A.

A MATLAB toolbox named XTRIG has been developed by [48] to implement the proposed FQ trigonometry. XTRIG has a function for the generation of 4-tuple fuzzy numbers, say, a fuzzy number, $[a,\ b,\ \tau,\ \beta]$, the function is $b-a=\kappa_0\tau$ with the condition $\tau=\beta$. κ_0 is a threshold parameter to define the shape of fuzzy numbers. For instance, let the number of the quantity space of an orientation p be 17 for a FQ trigonometric circle, those of a translation, q_x and q_y, be 23 each, and set κ_0 as 5, XTRIG generates the corresponding quantity spaces of the orientation Q^a and translation Q^d, which are shown in Fig. 2.1. It shows that the fuzzy-number generation

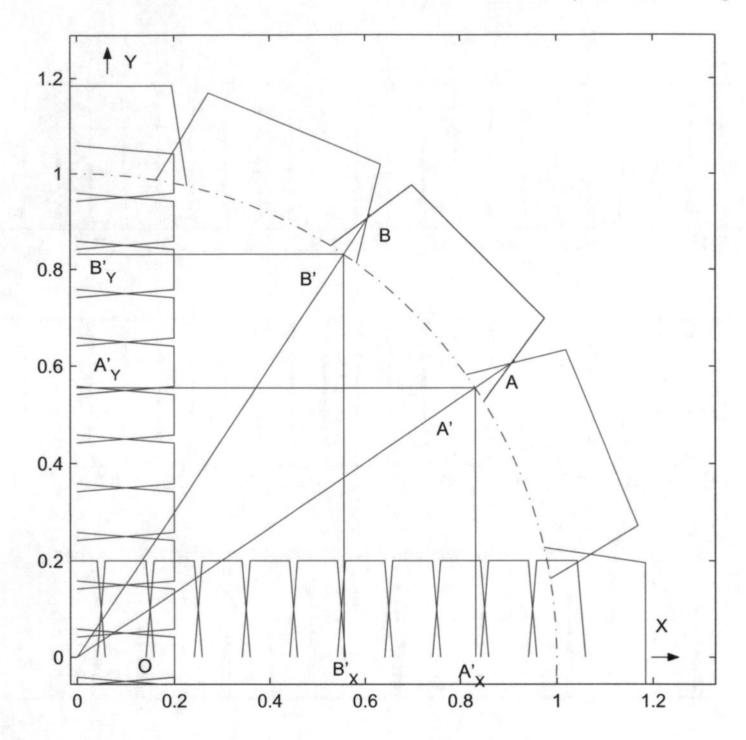

Fig. 2.2 The relation between qualitative translation and orientation

function distributes the 16 fuzzy numbers in the translation component in the range [0 1), and two sets of the 21 fuzzy numbers in the range [−1 1]. Figure 2.2 shows the conversion from a fuzzy qualitative angle to its FQ position, where the qualitative angle is the third orientation state in the orientation quantity space Q^a (i.e., [0.099 0.151 0.0104 0.0104]). The conversion is carried out as follows. Firstly, we calculate the positions of crossing points A, B between the fuzzy number (i.e., the third orientation angle) and its adjacent fuzzy numbers; Secondly, we calculate crossing point positions A', B' between the unit circle curve and the two line segments OA and OB; Thirdly, we obtain two line segments $A'_x B'_x$ and $A'_y B'_y$ by projecting points A', B' into X and Y coordinates, respectively. Finally we generate two sets of fuzzy numbers which totally or partially are in the range $A'_x B'_x$ and $A'_y B'_y$. The set of fuzzy numbers are within $A'_x B'_x$ is the FQ position on X axis of the orientation angle (i.e., $QS_a(3)_X$); the other set within $A'_y B'_y$ is for Y axis projection (i.e., $QS_a(3)_Y$). The FQ position is,

$$P_{QS_a(3)} = (P_{QS_a(3)_X},\ P_{QS_a(3)_Y}) \tag{2.3a}$$

where

$$P_{QS_a(3)_X} =_0 \begin{bmatrix} 0.5583\ 0.6417\ 0.0167\ 0.0167 \\ 0.6583\ 0.7417\ 0.0167\ 0.0167 \\ 0.7583\ 0.8417\ 0.0167\ 0.0167 \end{bmatrix} \quad (2.3b)$$

$$P_{QS_a(3)_Y} =_0 \begin{bmatrix} 0.5583\ 0.6417\ 0.0167\ 0.0167 \\ 0.6583\ 0.7417\ 0.0167\ 0.0167 \\ 0.7583\ 0.8417\ 0.0167\ 0.0167 \end{bmatrix} \quad (2.3c)$$

The FQ values of the third orientation angle in the X and Y axes are denoted by a set of three 4-tuple fuzzy numbers and a set of five fuzzy numbers generated from its translation quantity space when its α cut is defaulted as zero. The conversion methodology can also be applied to the conversion from fuzzy numbers in its translation quantity space to those in its orientation quantity space.

2.3 Fuzzy Qualitative Trigonometric Functions

FQ trigonometric functions are derived from conventional trigonometric functions based on the extension principle, which allows the extension of classical mathematical operators to the fuzzy domain. It means that an arithmetic operation performed between n fuzzy sets will yield a fuzzy set of the same form. Each trigonometric function is derived and illustrated using FQ trigonometry and the quantity spaces of the FQ coordinates. For simplicity, the symbols of quantitative trigonometric functions are used to describe their counterparts in FQ trigonometry but with qualitative variables instead (e.g., $QS_i(j)$, where $i \in \{a, d\}$ and $j \in \{p, q\}$). In order to give a clear explanation of FQ trigonometric functions, let us consider the example for the third FQ angle in Fig. 2.2 as a general case of the ith orientation angle (i.e., $QS_a(i)$). $P_{QS_a(i)}$ stands for the FQ position of the intersection of $QS_a(i)$ to a unit circle curve, and its projected FQ positions on the axes of X and Y are $QS_d(|A'_xB'_x|)$ and $QS_d(|A'_yB'_y|)$. For now we are ready to define FQ trigonometric functions,

$$\sin\left(QS_a\left(i\right)\right) =_\alpha \frac{QS_d\left(|A'_yB'_y|\right)}{[1\,1\,0\,0]} =_\alpha QS_d\left(|A'_yB'_y|\right) \quad (2.4a)$$

$$\cos\left(QS_a\left(i\right)\right) =_\alpha \frac{QS_d\left(|A'_xB'_x|\right)}{[1\,1\,0\,0]} =_\alpha QS_d\left(|A'_xB'_x|\right) \quad (2.4b)$$

$$\sec\left(QS_a\left(i\right)\right) =_\alpha \frac{[1\,1\,0\,0]}{QS_d\left(|A'_xB'_x|\right)} =_\alpha \frac{[1\,1\,0\,0]}{\cos\left(QS_a\left(i\right)\right)} \quad (2.4c)$$

$$\csc\left(QS_a\left(i\right)\right) =_\alpha \frac{\left[1\,1\,0\,0\right]}{QS_d\left(|A'_yB'_y|\right)} =_\alpha \frac{\left[1\,1\,0\,0\right]}{\sin\left(QS_a\left(i\right)\right)} \tag{2.4d}$$

$$\arcsin\left(QS_d\left(|A'_yB'_y|\right)\right) =_\alpha QS_a\left(i\right) \tag{2.4e}$$

$$\arccos\left(QS_d\left(|A'_xB'_x|\right)\right) =_\alpha QS_a\left(i\right) \tag{2.4f}$$

where $sin(QS_a(i))$, $cos(QS_a(i))$, $sec(QS_a(i))$ and $csc(QS_a(i))$ FQ functions are straightforward as shown in Fig. 2.2. The ith fuzzy qualitative position's arcsine and arccosine functions shown in Eqs. 2.4e and 2.4f, i.e., $arcsin(QS_d(i))$ and $arccos(QS_d(i))$, are the inverse of its sine and cosine functions given in Eqs. 2.4a and 2.4b. Refer to Fig. 2.2, if points B'_y and A'_y are the crossing points of the ith fuzzy number and its adjacent fuzzy numbers in Y coordinate, the fuzzy numbers in the orientation quantity space within arc $\widehat{A'B'}$ are its corresponding qualitative angle. FQ arcsine and arccosine functions, as their quantitative counterparts have, are constrained that $\arcsin\left(QS_d\left(i\right)\right) \subseteq_\alpha \left[-\frac{p}{4}-1, \frac{p}{4}+1\right]$, and $\arccos\left(QS_d\left(i\right)\right) \subseteq_\alpha \left[1, \frac{p}{2}+2\right]$. Further, the tangent of $QS_a\left(i\right)$, written as $tan(QS_a\left(i\right))$, is defined as the ratio of the opposite to the adjacent sides, $QS_d\left(|A'_yB'_y|\right)/QS_d\left(|A'_xB'_x|\right)$. It can clearly see that the tangent of $QS_a\left(i\right)$ is equal to the sine of $QS_a\left(i\right)$ divided by the cosine of $QS_a\left(i\right)$. The same can apply to cotangent function, then we obtain

$$\tan\left(QS_a\left(i\right)\right) =_\alpha \frac{QS_d\left(|A'_yB'_y|\right)}{QS_d\left(|A'_xB'_x|\right)} =_\alpha \frac{\sin\left(QS_a\left(i\right)\right)}{\cos\left(QS_a\left(i\right)\right)} \tag{2.5a}$$

$$\cot\left(QS_a\left(i\right)\right) =_\alpha \frac{QS_d\left(|A'_xB'_x|\right)}{QS_d\left(|A'_yB'_y|\right)} =_\alpha \frac{\cos\left(QS_a\left(i\right)\right)}{\sin\left(QS_a\left(i\right)\right)} \tag{2.5b}$$

where

$$\cos^2\left(QS_a(i)\right) + \sin^2\left(QS_a(i)\right) =_\alpha QS_d(p) \tag{2.5c}$$

With the setting of the example in Fig. 2.2, we obtain the following results,

$$cos(QS_a(114)) =_0 \begin{bmatrix} 0.7583\ 0.8417\ 0.0167\ 0.0167 \\ 0.8583\ 0.9417\ 0.0167\ 0.0167 \\ 0.9583\ 1.0417\ 0.0167\ 0.0167 \end{bmatrix}$$

$$sec(QS_a(14)) =_0 \begin{bmatrix} 4.1379\ 6.3158\ 0.2670\ 0.7430 \\ 2.9268\ 3.8710\ 0.1361\ 0.2670 \\ 2.2642\ 2.7907\ 0.0823\ 0.1361 \\ 1.8462\ 2.1818\ 0.0551\ 0.0823 \\ 1.5584\ 1.7910\ 0.0395\ 0.0551 \end{bmatrix}$$

$$tan(QS_a(3)) =_0 \begin{bmatrix} 0.6634\ 0.8462\ 0.0323\ 0.0415 \\ 0.8876\ 1.1266\ 0.0415\ 0.0552 \\ 1.1818\ 1.5075\ 0.0552\ 0.0772 \end{bmatrix}$$

$$cot(QS_a(3)) =_0 \begin{bmatrix} 1.1818\ 1.5075\ 0.0552\ 0.0783 \\ 0.8876\ 1.1266\ 0.0415\ 0.0561 \\ 0.6634\ 0.8462\ 0.0323\ 0.0423 \end{bmatrix}$$

$$\arcsin(QS_d(19)) =_0 \begin{bmatrix} 0.0990\ 0.1510\ 0.0104\ 0.0104 \\ 0.1615\ 0.2135\ 0.0104\ 0.0104 \end{bmatrix}$$

$$\arccos(QS_d(20)) =_0 \begin{bmatrix} 0.0365\ 0.0885\ 0.0104\ 0.0104 \end{bmatrix}$$

The example above indicates two characteristics of FQ trigonometry. First, FQ trigonometric functions have period characteristic, e.g., $cos(QS_a(114)) = cos(QS_a(2 + 16 \times 7))$. Secondly, FQ tangent and cotangent functions are calculated based on FQ sine, cosine functions and the fundamental trigonometric identity. The use of the latter ensures the elimination of those fuzzy numbers that are the results generated by FQ sine and cosine functions but they lose their geometric meaning.

2.4 Derivatives of Fuzzy Qualitative Trigonometry

Combining FQ trigonometry with the extension principle [52] and the principle of convex and normal fuzzy sets' mapping [53], we have the lemmas below for the derivatives of FQ trigonometric functions.

Lemma I *If $y = d^n(x)$ is a n order derivative function, then with A being a fuzzy set in quantity space X ($x \in \bigcup FQT_\alpha(Q_X)$), $d^n()$ maps from A to a fuzzy set B in quantity space Y ($y \in \bigcup FQT_\alpha(Q_Y)$) such that*

$$\mu_B(y) = \sup_{x \in d^{-n}(y)} \mu_A(x) \tag{2.6}$$

where $FQT_\alpha()$ is one of FQT functions, α stands for α-cut and $Q_x, Q_y \subset Q$.

Fuzzy numbers herein are fuzzy sets of the real line $\mathbb{R}$ with a normal, fuzzy convex and continuous membership function of bounded support. In order to demonstrate the proposed extension, we have implemented it into XTRIG MATLATB toolbox. Recall the example used in Fig. 2.1. For first-order derivative, Lemma I can be rewritten as,

$$\mu_B(y) = \sup_{x \in d^{-1}(y)} \mu_A(x) \tag{2.7}$$

For instance, first order derivative of $cos(QS_a(3))$ can be calculated as,

$$\begin{aligned} d^1\left(\cos\left(QS_a\left(3\right)\right)\right) &=_\alpha -\sin\left(QS_a\left(3\right)\right) \\ &=_0 \begin{bmatrix} -0.7119 & -0.7966 & 0.0169 & 0.0169 \\ -0.8136 & -0.8983 & 0.0169 & 0.0169 \\ -0.9153 & -1.000 & 0.0169 & 0 \end{bmatrix} \end{aligned} \tag{2.8}$$

With the definition of the support of a fuzzy set, $supp(A)$, Lemma I can be further developed as a lemma below for the mapping of a fuzzy set's support.

Lemma II *With the convention in Lemma I, the following is obtained,*

$$d^n(supp(A)) = supp(B). \tag{2.9}$$

Furthermore, applying Lemma II to the example, we have the fuzzy sets' support version of FQ trigonometric first-order derivative as below,

$$\begin{aligned} & d^1\left(\sup\left(\cos\left(QS_a\left(3\right)\right)\right)\right) \\ &=_0 \sup\left(d^1\left(\cos\left(QS_a\left(3\right)\right)\right)\right) =_0 \sup\left(\sin\left(QS_a\left(3\right)\right)\right) \\ &=_0 \sup\left(\begin{bmatrix} -0.7288 & -0.8135 \\ -0.8305 & -0.9152 \\ -0.9322 & -1.000 \end{bmatrix}\right) \\ &=_0 \left[-0.7288 \; -1.000\right] \end{aligned} \tag{2.10}$$

It shows that the support of the derivative output has two representations: local support and global support. The local support consists of the support values of the fuzzy numbers of the output; the global support is the support of a qualitative state. In the XTRIG toolbox, the column dimension is used to tell the representation of a fuzzy number from that of its support.

Lemma III *Quantity spaces in a unit circle must be x-axis or y-axis symmetric, or origin symmetric.*

Lemma above is the condition that Lemmas I and II hold; actually it is also the condition for FQ trigonometry. The characteristic of x-axis or y-axis symmetry is required for a translation quantity space, and that of origin symmetry is for an orientation quantity space.

Say A is a fuzzy set, $[A]^\alpha$ is a compact subset of R and $M(A)$ is a closed interval bounded by the lower and upper possibilistic mean values $[M_*(A), M^*(A)]$ of A [54], then it leads to the following definition,

Definition I α-centre of a fuzzy set is its possibilistic mean value.

$$C\left(A_{\alpha}\right)=\frac{M^{*}\left(A_{\alpha}\right)+M_{*}\left(A_{\alpha}\right)}{2} \tag{2.11}$$

where

$$M^{*}\left(A_{\alpha}\right)=\frac{\int_{0}^{1} Pos[A\geq b(\alpha)]\times \max[A]^{\alpha} d\alpha}{\int_{0}^{1} Pos[A\leq b(\alpha)]d\alpha}$$
$$M_{*}\left(A_{\alpha}\right)=\frac{\int_{0}^{1} Pos[A\leq a(\alpha)]\times \min[A]^{\alpha} d\alpha}{\int_{0}^{1} Pos[A\leq a(\alpha)]d\alpha}$$

and Pos denotes possibility.

Definition II α power support of a fuzzy set is the proportion of its α-power to its α-height.

$$PS\left(A_{\alpha}\right)=\begin{cases}\int_{R}\mu_{A}^{\alpha}\left(x\right)dx & \alpha=1\\ \frac{\int_{R}\mu_{A}^{\alpha}(x)dx}{1-\alpha} & \alpha\neq 1\end{cases} \tag{2.12}$$

Definition III A fuzzy qualitative state $QS(A)$ in a α-level of a fuzzy set $\mu_{A}(X)$ can be described by its α-centre and scaled α power support.

$$QS\left(A\right)=_{\alpha}\left[C\left(A\right),\ \gamma PS\left(A\right)\right] \tag{2.13}$$

where $\gamma\in[0\ 1]$ is a scaling factor. Hence, we have two representations for a qualitative state, four-tuple fuzzy numbers and centre-power-support fuzzy intervals. For instance, given a four tuple fuzzy number $A=[a\ b\ \tau\ \beta]$, its centre-power-support representation can be obtained below with default value $\alpha=0$,

$$\begin{aligned}QS\left(A\right)&=_{\alpha}\left[C\left(A\right),\ \gamma PS\left(A\right)\right]\\&=_{\alpha}\left[\tfrac{a+b}{2},\ b-a+\tfrac{\tau+\beta}{2}\right]\end{aligned}$$

Let us consider $QS_{d}(3)=[0.2034\ 0.2881\ 0.01069\ 0.01069]$, its fuzzy qualitative state at α_{0} level is $[0.2458,\ 0.1016\gamma]$. That is, the state is a 1D interval range with 0.2458 as its centre, 0.1016γ as its scalable radius. A fuzzy qualitative state's linguistic meaning or precision can be controlled by the combination of the adjustment of its scaling factor and fuzzy quantity spaces. Further, arithmetic for FQ states at a α level can be defined in terms of midpoint-radius (MR) representation and Definition III, for literature on interval MR representation, please see [55, 56].

Definition IV For $QS(A) =_\alpha [C(A)\ \gamma_A PS(A)]$ $(QS(A) \in Q)$, $QS(B) =_\alpha [C(B)\ \gamma_B PS(B)]$, $(QS(B) \in Q)$, γ_A and γ_B are scaling factors, respectively, we define

$$\begin{aligned} &QS\,(A) + QS\,(B) = \\ &_\alpha \left[C\,(A) + C\,(B)\,,\ \gamma_A PS\,(A) + \gamma_B PS\,(B) \right] \end{aligned} \tag{2.14a}$$

$$\begin{aligned} &QS\,(A) - QS\,(B) = \\ &_\alpha \left[C\,(A) - C\,(B)\,,\ \gamma_A PS\,(A) + \gamma_B PS\,(B) \right] \end{aligned} \tag{2.14b}$$

$$\begin{aligned} &QS\,(A) \times QS\,(B) = \\ &_\alpha \left[C\,(A) \times C\,(B)\,,\ |\rho(C\,(A)| \times \gamma_B PS\,(B) + \right. \\ &\left. |C\,(B)|\,\gamma_A PS\,(A) + \gamma_A \gamma_B PS\,(A)\,PS\,(B)) \right] \end{aligned} \tag{2.14c}$$

where

$$\begin{cases} \rho = 1\ strong\ MR\ multiplication \\ \rho = \frac{2}{3}\ \ weak\ MR\ multiplication \end{cases}$$

and

$$\frac{I}{QS(A)} =_\alpha \left[\frac{C(B)}{D},\ \frac{\gamma_B PS(B)}{D} \right] \tag{2.14d}$$

where

$$D =_\alpha C^2\,(B) - \gamma_B^2 PS^2\,(B)\,,\ and\ QS\,(B) \neq \mathbf{0}$$

and

$$\frac{QS\,(A)}{QS\,(B)} =_\alpha QS\,(A) \times \frac{I}{QS\,(B)}\ for\ QS\,(B) \neq \mathbf{0} \tag{2.14e}$$

Due to Rump's overestimation theory in [57] that the overestimation of MR interval arithmetic compared to power set operations is uniformly bounded by a factor 1.5 in radius, we introduce an index to distinguish strong and weak multiplication in terms of interval MR representation, see Eq. 2.14c. Strong multiplication with the index as 1 includes the overestimation caused by interval MR multiplication, on the other hand, weak multiplication with the index as $2/3$ does not. Generally speaking, interval computation in MR representation is faster, especially matrix operations, than infimum-supremum representation. This overestimation characteristic provides a method of controlling tradeoffs for the selection of interval representations for interval-related applications, e.g., real-time humanoid systems. The algorithms in MR representation and its conversion on four-tuple-fuzzy-number-based computation have been implemented into the XTRIG toolbox. We use MR-based computation in the rest of this chapter, not only because its faster computation nature is suitable for robotics application, but also because MR-based representation is more flexible to control precision in data conversion among numerical, symbolic and hybrid data, thanks to the fact that fuzzy quantity space with domain background allows convert parameters among different types of data.

2.5 Conclusion

A fuzzy qualitative description of conventional trigonometry has been presented in this chapter. The unit circle of conventional trigonometry has been modified by the introduction of fuzzy partitions for orientation and translation. Conventional trigonometric functions (e.g., a sine function) and rules (e.g., the sine rule) have been abstracted to give fuzzy qualitative versions of these operations. In addition, fuzzy qualitative versions of the conventional triangle theorems are also provided in FQT. Examples have been given throughout the chapter to demonstrate FQT's ability. One of these focuses on robot kinematics and explains how contributions could be made by a new type of FQT-based kinematics to achieve the intelligent connection of low-level sensing and control tasks to high-level symbolic tasks.

The proposed fuzzy qualitative variables become linguistic variables or symbolic variables when they are modified with descriptive symbols [58–60]. Besides, methods such as aggregation operators [61] can be used to select specific symbols from a set of fuzzy qualitative states. Hence, fuzzy qualitative states in FQT can be denoted by symbolic terms which provide a promising base for behavioural description in robotics [62]. For example, Jenkins and Mataric [63] have provided a skill-level interface named behaviour vocabulary for a humanoid robot. However, the connection between the behaviour vocabulary and low-level sensing and control tasks is still uncertain. FQT is an abstraction of conventional trigonometry into the domain of fuzzy logic and qualitative reasoning. This version of trigonometry can replace the role that conventional trigonometry plays in robot kinematics; so a general robot kinematics can be derived based on FQT. The new type of kinematics handles fuzzy qualitative states that allow access to both numerical data and symbolic data. Further, a FQT-based robot kinematics could provide a transit layer for communicating variables, even variable-based cognitive functions of knowledge-level tasks and numerical sensing and motion control [49, 50]. From an implementation perspective, a robotic system with this type of kinematics can be easily fitted into a conventional fuzzy model which consists of a fuzzification unit, knowledge base, inference engine and defuzzification unit. The FQT-based reasoning could play a crucial role in an inference engine whose variables are supported by a knowledge base. The output of an inference engine is able to access knowledge-based systems, e.g., symbolic planning subsystems. The fuzzification and defuzzification units are able to provide low-level sensing and control tasks, it is obvious this role can be easily replaced by other techniques, e.g., fuzzy clustering.

Fuzzy qualitative reasoning seeks to harness the strengths of fuzzy reasoning and qualitative reasoning. It is in a position to play a crucial role in closing the gap between low-level sensing and control tasks and high-level symbolic description. Our future work will focus on the development of an intelligent architecture using the FQT in the domain of AI robotics. The architecture is composed of a low-level handler, knowledge-based handler, and inference handler. The low-level handler provides an interface to numerical data; it is composed of a fuzzification unit and defuzzification unit, in each unit different techniques can be applied such as fuzzy clustering. A knowledge-based handler not only provides a knowledge base to support the system inferences, but also facilitates human supervision (e.g., cognitive inputs). FQT is the core of the inference handler which should be able to, scalable and in parallel interface both low-level handler and knowledge-based handler.

References

1. P. Bourseau, K. Bousson, J.I. Dormoy, J.M. Evrard, F. Guerrin, L. Leyval, O. Lhomme, B. Lucas, A. Missier, J. Montmain, N. Piera, N. Rakoto-Ravalontsalama, M. Steyer, J.P. Tomasena, I. Trave-Massuyes, M. Vescovi, S. Xanthakis, and B. Yannou. qualitative reasoning: a survey of techniques and applications. *AI communications*, 8(3/4):119–193, 1995.
2. B. Kuipers. Qualitative reasoning. *MIT Press*, 1994.
3. D. Berleant. Qualitative and quantitative simulation: bridging the gap. *Artificial Intelligence Journal*, 95(2):215–255, 1997.
4. A.F. Blackwell. Spatial reasoning for robots: a qualitative approach. *Master Thesis, Victoria University of Wellington*, 1988.
5. Q. Shen and R. Leitch. Fuzzy qualitative simulation. *IEEE Transactions on Systems, Man, and Cybernetics*, 23(4):1038–1061, 1993.
6. G. M. Coghill. *Mycroft: A Framework for Constraint-Based Fuzzy Qualitative Reasoning*. PhD thesis, Heriot-Watt University, 1996.
7. K.D. Forbus, P. Nielsen, and B. Faltings. Qualitative kinematics: A framework. *Proceedings of the International Joint Conference on Artificial Intelligence*, pages 430–435, 1987.
8. P.E. Nielsen. A qualitative approach to rigid body mechanics. *University of Illinois at Urbana-Champaign, PhD thesis*, 1988.
9. B. Faltings. A symbolic approach to qualitative kinematics. *Artificial Intelligence*, 56(2–3):139–170, 1992.
10. L. Magdalena and F. Monasterio-Huelin. A fuzzy logic controller with learning through the evolution of its knowledge base. *International Journal of Approximate Reasoning*, 16(3):335–358, 1997.
11. J.J. Buckley and E. Eslami. An introduction to fuzzy logic and fuzzy sets. *Springer-Verlag*, 2002.
12. J. Liu. A method of spatial reasoning based on qualitative trigonometry. *Artificial Intelligence*, 98(1-2):137–168, 1998.
13. E. Smith and J. Eloff. Cognitive fuzzy modelling for enhanced risk assessment in a health care institution. *IEEE Intelligent Systems*, 15(2):69–75, 2000.
14. J.W.T. Lee, D.S. Yeung, and E.C.C. Tsang. Ordinal fuzzy sets. *IEEE transactions on Fuzzy Systems*, 10(6):767–778, 2002.
15. A.H. Ali, D. Dubois, and H. Prade. Qualitative reasoning based on fuzzy relative orders of magnitude. *IEEE transactions on Fuzzy Systems*, 11(1):9–23, 2003.
16. Y. Li and S. Li. A fuzzy sets theoretic approach to approximate spatial reasoning. *IEEE transactions on Fuzzy Systems*, 12(6):745–754, 2004.
17. Chee Seng Chan, George M Coghill, and Honghai Liu. Recent advances in fuzzy qualitative reasoning. *International Journal of Uncertainty, Fuzziness and Knowledge-Based Systems*, 19(03):417–422, 2011.

18. Scott Friedman and Ann Kate Lockwood. Qualitative reasoning: Everyday, pervasive, and moving forwarda report on qr-15. *AI Magazine*, 37(2):95–96, 2016.
19. R. Fuller. On fuzzy reasoning schemes. *The state of the art of information systems application in 2007, TUCS General Publications, Carlsson (eds),C.*, 16:85–112, 1999.
20. B.S. Chen, Yang Y.S., B.K. Lee, and Lee T.H. Fuzzy adaptive predictive flow control of atm network traffic. *IEEE Transactions on Fuzzy Systems*, 11(4):568–581, 2003.
21. M. Setnes and H. Roubos. Ga-fuzzy modelling and classification: complexity and performance. *IEEE Transactions on Fuzzy Systems*, 8(5):35–44, 2000.
22. E. Pedrycz and M. Reformat. Evolutionary fuzzy modelling. *IEEE Transactions on Fuzzy Systems*, 11(5):652–655, 2003.
23. A. Fernández, M. del Jesus, and F. Herrera. Hierarchical fuzzy rule based classification systems with genetic rule selection for imbalanced data-sets. *International Journal of Approximate Reasoning*, 50(3):561–577, 2009.
24. M. Antonelli, P. Ducange, B. Lazzerini, and F. Marcelloni. Learning concurrently partition granularities and rule bases of mamdani fuzzy systems in a multi-objective evolutionary framework. *International Journal of Approximate Reasoning*, 50(7):1066–1080, 2009.
25. Andrea GB Tettamanzi and Marco Tomassini. *Soft computing: integrating evolutionary, neural, and fuzzy systems*. Springer Science & Business Media, 2013.
26. E.C.C. Tsang, X.Z. Wang, and D.S. Yeung. Improving learning accuracy of fuzzy decision trees by hybrid neural networks. *IEEE transactions on Fuzzy Systems*, 8(5):601–614, 2000.
27. P. Pulkkinen and H. Koivisto. Fuzzy classifier identification using decision tree and multiobjective evolutionary algorithms. *International Journal of Approximate Reasoning*, 48(2):526–543, 2008.
28. P. Vuorimaa, T. Jukarainen, and E. Karpanoja. A neuro-fuzzy system for chemical agent detection. *IEEE transactions on Fuzzy Systems*, 3(4):415–424, 1995.
29. A. Ciaramella, R. Tagliaferri, W. Pedrycz, and A. Di Nola. Fuzzy relational neural network. *International Journal of Approximate Reasoning*, 41(2):146–163, 2006.
30. L. Hung and H. Chung. Decoupled sliding-mode with fuzzy-neural network controller for nonlinear systems. *International Journal of Approximate Reasoning*, 46(1):74–97, 2007.
31. Ronald R Yager and Lotfi A Zadeh. *An introduction to fuzzy logic applications in intelligent systems*, volume 165. Springer Science & Business Media, 2012.
32. Guillermo Bosque, Inés del Campo, and Javier Echanobe. Fuzzy systems, neural networks and neuro-fuzzy systems: a vision on their hardware implementation and platforms over two decades. *Engineering Applications of Artificial Intelligence*, 32:283–331, 2014.
33. D. Schwartz. A system for reasoning with imprecise linguistic information. *International Journal of Approximate Reasoning*, 5(5):463–488, 1991.
34. H. Kwan and Y. Cai. A fuzzy neural network and its application to pattern recognition. *IEEE Transactions on Fuzzy Systems*, 2(3):185–193, 1994.
35. M. Ghalia and P. Wang. Intelligent system to support judgmental business forecasting: the case of estimating hotel room demand. In *Computational Intelligence in Economics and Finance*, pages 59–92. Springer, 2004.
36. S.M. Bae, S.C. Park, and S.H. Ha. Fuzzy web ad selector based on web usage mining. *IEEE Intelligent Systems*, 18(6):62–69, 2003.
37. J. Casillas, F. Herrera, R. Pérez, M. del Jesus, and P. Villar. Special issue on genetic fuzzy systems and the interpretability–accuracy trade-off. *International Journal of Approximate Reasoning*, 44(1):1–3, 2007.
38. D.S. Weld and J. de Kleer (eds). Reading in qualitative reasoning about physical systems. *Morgan Kaufman, San Mateo, CA*, 1990.
39. B.C. Williams and J. de Kleer (eds). Special issue on qualitative reasoning about physical systems. *Artificial Intelligence*, 51(1–3), 1991.
40. Q. Shen and R. Leitch. Combining qualitative simulation and fuzzy sets. *Recent advances in qualitative physics*, pages 83–100, 1993.
41. B. Bredeweg and P. Struss (eds). Special issues on qualitative reasoning. *AI Magzine*, 2003.

42. P. Nayak and B. Williams. A model-based approach to reactive self-configuring systems. *AAAI-96*, pages 971–978, 1996.
43. Hidde De Jong, Jean-Luc Gouzé, Céline Hernandez, Michel Page, Tewfik Sari, and Johannes Geiselmann. Qualitative simulation of genetic regulatory networks using piecewise-linear models. *Bulletin of Mathematical Biology*, 66(2):301–340, 2004.
44. G.M. Coghill, S.M. Garrett, and R.D. King. Learning qualitative metabolic models. In *European Conference on Artificial Intelligence (ECAI'04)*, 2004.
45. H. Liu, G.M. Coghill, and D.J. Brown. Qualitative kinematics of planar robots: Intelligent connection (in press). *International Journal of Approximate Reasoning*, 2007.
46. H. Liu, D.J. Brown, and G.M. Coghill. Fuzzy qualitative robot kinematics. *IEEE Transactions on Fuzzy Systems*, 16(3):802–822, 2008.
47. H. Liu. A fuzzy qualitative framework for connecting robot qualitative and quantitative representations. *IEEE Transactions on Fuzzy Systems*, 16(6):1522–1530, 2008.
48. H. Liu and G.M. Coghill. Fuzzy qualitative trigonometry. *Proc. IEEE International Conference on Systems, Man and Cybernetics, Hawaii, USA.*, 2005.
49. M. Brady. Artificial intelligence and robotics. *Artificial Intelligence*, 26:79–121, 1985.
50. P. Dpherty, G. Lakemeyer, and A.P. Del Pobil. *Proceedings of the Fourth International Cognitive Robotics Workshop*. CogRob, 2004.
51. P.P. Bonissone and K.S. Decker. Selecting uncertainty calculi and granularity: An experiment in trading-off precision and complexity. *Uncertainty in Artificial Intelligence*, pages 217–247, 1986.
52. T.J. Ross. Fuzzy logic with engineering applications. *Jonh Wiley & Sons Ltd*, 2004.
53. J.J. Saade. Mapping convex and normal fuzzy sets. *Fuzzy sets and systems*, 81:251–256, 1996.
54. C. Carlsson and R. Fuller. On possibilistic mean value and variance of fuzzy numbers. *Fuzzy Sets and Systems*, 122:315–326, 2001.
55. A. Neumaier. Interval methods for systems of equations. *Encyclopedia of mathematics and its applications, Cambridge University Press*, 1990.
56. L.D. Petković and M.S. Petković. Inequalities in circular arithmetic: A survey. In *Recent Progress in Inequalities*, pages 325–340. Springer, 1998.
57. S.M. Rump. Fast and parallel interval arithmetic. *BIT Numerical Mathematics*, 39(3):534–554, 1999.
58. L.A. Zadeh. The concept of a linguistic variable and its applications to approximate reasoning - i. *Information Sciences*, 8:199–249, 1975.
59. L.A. Zadeh. The concept of a linguistic variable and its application to approximate reasoning-ii. *Information Sciences*, 8(4):301–357, 1975.
60. L.A. Zadeh. The concept of a linguistic variable and its application to approximate reasoning-iii. *Information Sciences*, 9(1):43–80, 1975.
61. T. Calvo, A. Kolesarova, M. Komornikova, and R. Mesiar. Aggregation operators: properties, classes and construction methods. In T. Calvo, G. Mayor, and R. Mesiar, editors, *Aggregation Operators. New Trends and Applications*, pages 3–104. Physica-Verlag, Heidelberg, New York, 2002.
62. R.C. Arkin. *Behavior-based robotics*. MIT press, 1998.
63. O.C. Jenkins and M.J. Mataric. Performance-derived behavior vocabularies: data-driven acquisition of skills from motion. *International Journal of Humanoid Robotics*, pages 237–288, 2004.

Chapter 3
Fuzzy Qualitative Robot Kinematics

3.1 Introduction

Future robotics is a multidisciplinary research area. Its central aim is to integrate traditional robotics, artificial intelligence, cognitive science and neuroscience etc. Research into robotics has traditionally emphasised low-level sensing and control tasks including sensory processing, path planning, and control. Future robotics is concerned with endowing robots and software agents with higher-level cognitive functions that enable them to reason, act and perceive in dynamic, incompletely known and unpredictable environments. The methods used should be flexible enough to combine the strengths of conventional programming with those of machine learning methods, e.g., the existing combination methods of fuzzy inference and neural networks by Kiguchi and Fukuda [1] and Ge et al. [2]. On the other hand, cognitive-function descriptions for robot behaviour and behaviour-based robotics are urgently required to confront challenges raised by future robotics. Robots can be generally categorised as manipulators, mobile robots and mobile manipulators. Due to the nature of sensors (e.g. sonar) that a mobile robot has, methods for its behaviour description have been developed in the past three decades [3–5]. However, there have been very few behaviour methods for both robotic arms and mobile manipulators reported so far. For instance, Jenkins and Mataric [6] presented performance-derived behaviour vocabularies as a methodology for automatically deriving behaviour vocabularies to serve as skill-level interfaces for autonomous humanoid robots. Although Khatib et al. [7] presented a whole-body control framework that decouples the interaction between the task and postural objectives and compensates for the dynamics in their respective space, it is difficult to use the same framework to further describe the behaviours of a body.

This chapter proposes a novel representation of robot kinematics with the goal of solving the intelligent connection problem (also known as symbol grounding problem) for physical systems, in particular robotic systems. This problem is one of the key issues in AI robotics [3, 8] and relates to a wide range of research areas such as computer vision [9]. Based on FQ trigonometry [10], we present a derivative

H. Liu et al., *Human Motion Sensing and Recognition*,
Studies in Computational Intelligence 675, DOI 10.1007/978-3-662-53692-6_3

extension to FQ trigonometry; The derivative extension enables us further to derive FQ robot kinematics. This chapter is organised as follows. Section 3.2 derives FQ robot kinematics, Sect. 3.3 proposes an aggregation operator for calculation of the presented fuzzy qualitative kinematics with adjustable computational cost, Sect. 3.4 presents a case study of a Puma robot, and conclusion and discussion are included in Sect. 3.5.

3.2 Fuzzy Qualitative Robot Kinematics

Based on FQ trigonometry in Chap. 2, we first establish definitions for FQ vectors, and then introduce FQ transformations, finally derive FQ robot kinematics. The notation of conventional robotics in Craig's book [11] has been used in this chapter.

3.2.1 Fuzzy Qualitative Transformations

Necessary definitions of FQ transformation are provided in this section.

Definition V A FQ point vector is

$$QS_d(\mathbf{p}) = QS_d(x)\, i + QS_d(y)\, j + QS_d(z)\, k,$$

where $i,\ j,\ k$ are unit vectors along $x,\ y,\ z$ FQ coordinate axes, respectively. It can be represented in homogeneous FQ coordinates as,

$$QS_d(\mathbf{p}) = \left[\, QS_d(x)\ QS_d(y)\ QS_d(z)\ \mathbf{I} \right]^T,$$

where the superscript T indicates the transpose of the row vector into a column vector and $\mathbf{I}$ is a unit vector.

Definition VI Given a fuzzy qualitative point $\mathbf{p_0}$, its rigid transformation $\mathbf{p}$ by a transformation matrix $\mathbf{H}$ is represented by the matrix product,

$$QS_d(\mathbf{p}) = \mathbf{H} \cdot QS_d(\mathbf{p_0})\,. \tag{3.1}$$

FQ transformations are used to describe the fuzzy qualitative position and orientation of objects. For a rigid body, its transformations includes translation transformations $\mathbf{T}$ and orientation transformations $\mathbf{R}$, whose descriptions are given as,

$$\mathbf{T} = Trans\left(QS_d\left(p_x\right), QS_d\left(p_y\right), QS_d\left(p_z\right)\right) = \begin{bmatrix} \mathbf{I}\ \mathbf{O}\ \mathbf{O}\ QS_d\left(p_x\right) \\ \mathbf{O}\ \mathbf{I}\ \mathbf{O}\ QS_d\left(p_y\right) \\ \mathbf{O}\ \mathbf{O}\ \mathbf{I}\ QS_d\left(p_z\right) \\ \mathbf{O}\ \mathbf{O}\ \mathbf{O}\ \ \ \mathbf{I} \end{bmatrix} \quad (3.2)$$

and

$$\mathbf{R} = Rot\left(axis, QS_a\left(\theta_{axis}\right)\right) \quad for\ axis \in \{x, y, z\}. \quad (3.3)$$

where

$$Rot\left(x, QS_a\left(\theta_x\right)\right) = \begin{bmatrix} \mathbf{I} & \mathbf{O} & \mathbf{O} & \mathbf{O} \\ \mathbf{O} & cos\left(QS_a\left(\theta_x\right)\right) & -\sin\left(QS_a\left(\theta_x\right)\right) & \mathbf{O} \\ \mathbf{O} & \sin\left(QS_a\left(\theta_x\right)\right) & cos\left(QS_a\left(\theta_x\right)\right) & \mathbf{O} \\ \mathbf{O} & \mathbf{O} & \mathbf{O} & \mathbf{I} \end{bmatrix} \quad (3.4a)$$

$$Rot\left(y, QS_a\left(\theta_y\right)\right) = \begin{bmatrix} cos\left(QS_a\left(\theta_y\right)\right) & \mathbf{O} & \sin\left(QS_a\left(\theta_y\right)\right) & \mathbf{O} \\ \mathbf{O} & \mathbf{I} & \mathbf{O} & \mathbf{O} \\ -\sin\left(QS_a\left(\theta_y\right)\right) & \mathbf{O} & cos\left(QS_a\left(\theta_y\right)\right) & \mathbf{O} \\ \mathbf{O} & \mathbf{O} & \mathbf{O} & \mathbf{I} \end{bmatrix} \quad (3.4b)$$

$$Rot\left(z, QS_a\left(\theta_z\right)\right) = \begin{bmatrix} cos\left(QS_a\left(\theta_z\right)\right) & -\sin\left(QS_a\left(\theta_z\right)\right) & \mathbf{O} & \mathbf{O} \\ \sin\left(QS_a\left(\theta_z\right)\right) & cos\left(QS_a\left(\theta_z\right)\right) & \mathbf{O} & \mathbf{O} \\ \mathbf{O} & \mathbf{O} & \mathbf{I} & \mathbf{O} \\ \mathbf{O} & \mathbf{O} & \mathbf{O} & \mathbf{I} \end{bmatrix} \quad (3.4c)$$

where the elements of FQ transformations are 4-tuple fuzzy numbers in this chapter, for simplicity, though those can be replaced by any types of fuzzy numbers. Note that the elements **0** and **1** in the MR representation are

$$\mathbf{O} = \left[\mathbf{0}\ , 0\right] \ \ and\ \mathbf{I} = \left[\mathbf{1}\ , 0\right].$$

where the length of vectors **0** and **1** is determined by the maximum length of the output of FQ trigonometric functions at the same row as they are in a transformation matrix in Eqs. 3.2 and 3.3. Due to the fact, unlike conventional trigonometry, the output of a FQ trigonometric function is a subset of a normalised quantity space instead of a unique real number; the propagation of trigonometric functions in FQ terms allows FQ transformations to transform among real numbers, fuzzy intervals and linguistic variables.

Let us consider a transformation of a FQ point vector from $\mathbf{p}_0$ to $\mathbf{p}_e$ by first rotating by the y axis with an angle $QS_a(\theta)$, then translating by a position vector $QS_d(\mathbf{t})$. For demonstration purpose, parameters of its quantity space p and q are set as 16 and 4, that is to say, 16 qualitative states are used to describe a full orientation and 4 qualitative states to a unit length. The transformation is described in Eq. 3.5a, it generates all possible FQ states for the given motion. In order to compare the proposed method with conventional space transformation, Eq. 3.5a has been rewritten as Eq. 3.5b in terms of Euclidean geometry. The simulation shows that the initial position p_0 is labelled as '△', its Euclidean transformation is labelled as '□' in Fig. 3.1; the result of the fuzzy qualitative transformation are given in two representations, one is shown by a set of centres labelled as '○' in Fig. 3.1 and the other is represented by a set of overlapped interval patches in Fig. 3.2. Figure 3.1 shows that the Euclidean position '□' is not in the normally distributed centre of the fuzzy qualitative positions '○', this caused by fuzzy qualitative arithmetic [10]. This demonstrates that fuzzy qualitative transformation has extended Euclidean transformation in terms of qualitative data propagation instead of a simple replacement of Euclidean transformation. We expect that this characteristic to play a key role in the problem of robot intelligent connection.

$$\begin{aligned}\mathbf{p}_e &= Rot\,(y,\ QS_a(\theta))\,Trans(QS_d\,(\mathbf{t}))\mathbf{p_0}\\ &= Rot\,(y,\ QS_a\,([0.79,\ 0.39]))\\ &\quad \times Trans\,([4.25,\ 0.25]\,,[11.25,\ 0.25]\,,[4.25,\ 0.25])\\ &\quad \times \left([4,\ 0]\,,[3,\ 0]\,,[5,\ 0]^T\right)\end{aligned} \tag{3.5a}$$

$$\begin{aligned}\mathbf{p}_e' &= Rot\,\left(y,\ \theta'\right)Trans(\left(\mathbf{t}'\right))\mathbf{p}_0'\\ &= Rot\,(y,\ 0.79) \times Trans\,(4.25,\ 11.25,\ 4.25)\\ &\quad \times[4,\ 3,\ 5]^T\end{aligned} \tag{3.5b}$$

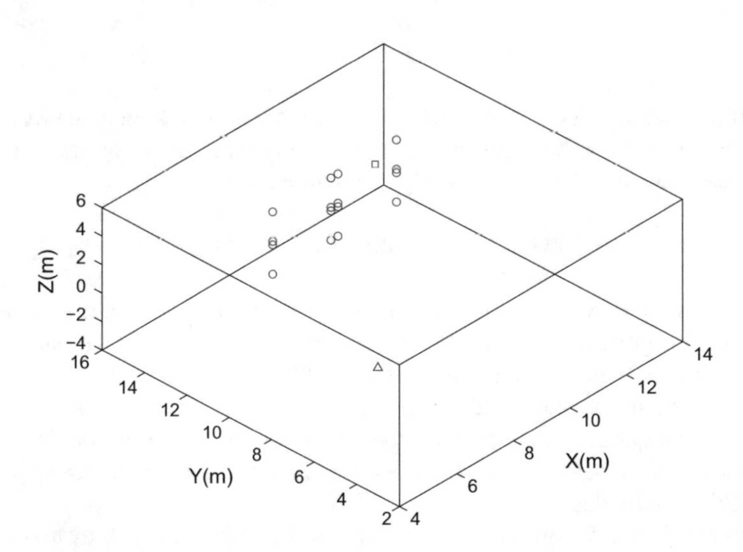

Fig. 3.1 Comparison of a numeric position's Euclidean transformation and the centres of interval patches of its fuzzy qualitative transformation

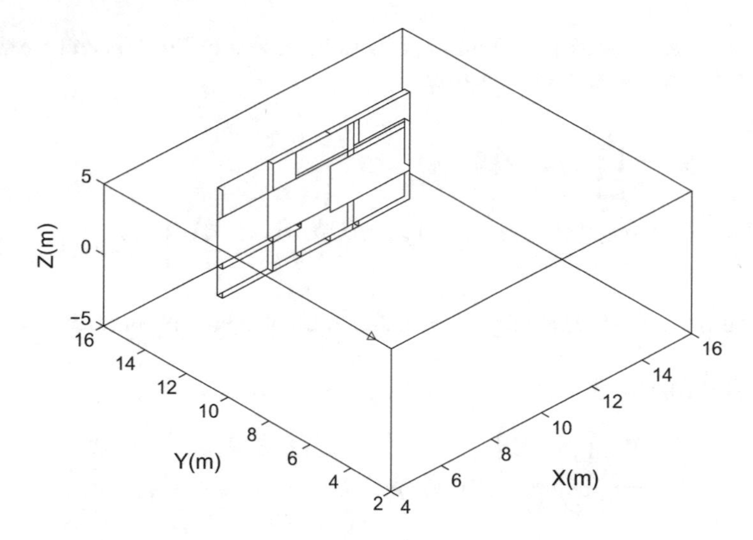

Fig. 3.2 Interval patches of the fuzzy qualitative transformation in Fig. 3.1

3.2.2 Fuzzy Qualitative Robot Kinematics

We first define a general robot kinematics in terms of FQ trigonometry, then consider its relation and advantages over conventional robotic kinematics. Denote a n-link spatial robot with its FQ home position as $QS_d(\mathbf{p_0})$, its end-effector position $QS_d(\mathbf{p})$ can be obtained by the transformations of its link components $\mathbf{H}_i$ or the transformations ${}^{i-1}_{i}\mathbf{H}_{DH}$ in D-H parameter terms, respectively,

$$QS_d(\mathbf{p}) = \prod_{i=1}^{n} \mathbf{H}_i \cdot QS_d(\mathbf{p_0}) = \prod_{i=1}^{n} {}^{i-1}_{i}\mathbf{H_{DH}} \cdot QS_d(\mathbf{p_0}) \tag{3.6}$$

where ${}^{i-1}_{i}\mathbf{H}_{DH}$ is FQ Denavit-Hartenberg kinematics structure. By replacing conventional trigonometric function with FQ trigonometry, we have the following,

$${}^{i-1}_{i}\mathbf{H}_{DH} = \begin{bmatrix} C\theta_i & -S\theta_i & \mathbf{O} & a_{i-1} \\ S\theta_i C\alpha_{i-1} & C\theta_i C\alpha_{i-1} & -S\alpha_{i-1} & -d_i S\alpha_{i-1} \\ S\theta_i S\alpha_{i-1} & C\theta_i C\alpha_{i-1} & C\alpha_{i-1} & d_i C\alpha_{i-1} \\ \mathbf{O} & \mathbf{O} & \mathbf{O} & \mathbf{I} \end{bmatrix} \tag{3.7}$$

where C stands for cos; S stands for sin. θ stands for $QS_a(\theta)$, α stands for $QS_a(\alpha)$, d stands for $QS_d(d)$, a stands for $QS_d(a)$. $QS_a(\theta_i)$ is the rotation angle from X_{i-1} to X_i measured about Z_i, joint variable from revolute joints, $QS_a(\alpha_{i-1})$ is the twist angle from Z_{i-1} to Z_i measured about X_{i-1}, $QS_d(d_i)$ is the distance from X_{i-1} to X_i measured along Z_i, joint variable for prismatic joints, $QS_d(a_{i-1})$ is the link offset, distance from Z_{i-1} to Z_i measured along X_{i-1}. X_i, Y_i, Z_i is the coordinate axes of ith coordinates system mounted on ith link segment. For a graphical representation please see Craig's book [11]. Further, robot kinematics in Eq. 3.6 can be rewritten for the end-effector of a robot as follows,

$$\begin{aligned} QS_d(\mathbf{p}_e) &= \prod_{i=1}^{n} \mathbf{H}_{DH} \cdot QS_d(\mathbf{p}_0) \\ &= \begin{bmatrix} R_{3\times 3}(QS_a(\theta_i, \alpha_{i-1})) & P_{3\times 1}(QS_a(\theta_i, a_i, d_i)) \\ O_{1\times 3} & I \end{bmatrix} \cdot QS_d(\mathbf{p}_0) \end{aligned} \tag{3.8}$$

Furthermore, the FQ velocity of the end-effector of a serial robot can be derived as,

$$\begin{aligned} QS_d^1(\mathbf{p}_e) &= QS_d(\dot{\mathbf{p}}_e) \\ &= \frac{\partial^1 \left(\prod_{i=1}^{n} {}^{i-1}_{i}\mathbf{H} \right)}{\partial QS(\mathbf{t})} \cdot QS_d(\dot{\mathbf{p}}_0) \\ &= \frac{\partial^4 \left(\prod_{i=1}^{n} \mathbf{H}_{DH} \right)}{\partial QS_a(\theta) \cdot \partial QS_a(\alpha) \cdot \partial QS_d(a) \cdot \partial QS_d(d)} \cdot QS_d(\dot{\mathbf{p}}_0) \\ &= QS(\mathbf{J}) \cdot QS_d(\dot{\mathbf{p}}_0) \end{aligned} \tag{3.9}$$

where $QS(\mathbf{J})$ is FQ Jacobian matrix which is able to transform FQ velocity between the configuration space of the orientation angles and the Cartesian space of the end-effector. Note that variable transformed herein could be numeric, symbolic or hybrid and physical constant robotic parameters, e.g., a_i and d_i, are not considered as FQ parameters. FQ Jacobian matrix $QS(\mathbf{J})$ consists of FQ trigonometric functions of $(QS_a(\theta_i, \alpha_{i-1}))$ and their FQ derivatives. FQ robot kinematics can be minished as conventional robot kinematics, for instance, when the scaling factor γ in Eq. 2.13 is set as zero and its quantity space is sufficient big, e.g., 360 elements in the MR representation. Its advantage over conventional kinematics is that it is able to handle hybrid parameters in the context of robot kinematics. FQ robot kinematics could provide a base for solving the perception-action problem which is one of the bottleneck problems leading to intelligent robots.

3.3 Quantity Vector Aggregation

Quantity vector aggregation is proposed to extract robot behaviour description of FQ states generated by FQ trigonometry and its variants. The foundation of quantity vector aggregation is quantity point vector aggregation because quantity patch vector aggregation is superset of their point vector aggregation. Both point and patch vectors can be aggregated by either point vectors or patch vectors. Understanding the geometrical motion of quantity vectors benefits the mathematical description of variable aggregation for handling and transferring different types of parameters in reasoning systems. Understanding the mathematical nature of reasoning methods such as qualitative reasoning, fuzzy reasoning and probabilistic reasoning will speed up the development of automatic specification and verification tools. It will also help to overcome limitations such as state explosion and undecidability.

For simplicity and without loss of generality, vector aggregation in a 3-dimensional quantity space is considered. Point vector aggregation can be used to describe rigid transformations of quantity point vectors. Craig's book [11] provides a thorough background in transformation terminology in Euclidean space. The general transformation of a point quantity vector $P(P_1, P_2, P_3)$ consists of translation transformations P_T and orientation transformations P_R. These transformations can be defined by either numerical vectors or quantity vectors. There is no propagation when transforming a quantity vector to another numerical location by applying real numbers to the transformation matrices because it is calculated in Euclidean space. A translation matrix of quantity vectors and an orientation matrix of either real numbers or quantity vectors are considered in FQ calculation in this section. Fuzzy quantity vector propagation in Sect. 2.4 generates complete possible states or trajectory states using the proposed quantity arithmetic with inputs described in terms of numerical, intervals or symbols. However, in order to describe the propagated states by a qualitative description, aggregation operators are introduced to select both suitable states and symbolic functions. Aggregation operators model operations such as conjunction, disjunction and averaging on intervals and fuzzy sets [12].

One of aggregation operator families is ordered weighted averaging operators $\mathbb{OWA}$ and their variants [13–16], originally studied by Yager [17]. Their general mathematical description is given below

$$\mathbb{OWA}\,(P_1, P_2, \cdots, P_n,) = \sum_{i=1}^{n} w_i P_{\sigma(i)}, \tag{3.10}$$

where σ is an ordinal sequence, $w_i \geq 0$ and $\Sigma w_i = 1$. The $\mathbb{OWA}$ operators provide a parameterised family of aggregation operators which can be used for many of the well-known operators by choosing suitable weights. It includes the average $\frac{1}{n}\sum_{i=1}^{n} P_i$, the quadratic mean $\sqrt{\frac{1}{n}\sum_{i=1}^{n} P_i^2}$, the harmonic mean $\frac{n}{\sum_{i=1}^{n}\frac{1}{P_i}}$, the geometric mean $\left(\prod_{i=1}^{n} P_i\right)^{\frac{1}{n}}$, the maximum operator and minimum operator. The $\mathbb{OWA}$ weights can be selected based on domain knowledge or application contexts where the aim is to adjust the $\mathbb{OWA}$ operator to generate suitable symbolic or qualitative functions. In addition to the method proposed in this section, there are a wide variety of other information aggregation and fusion techniques. For example, [18] could be adapted for quantity spaces and could be used to generate symbolic functions as possible to describe requested qualitative as accurately as possible. Note that It has shown that normalized quantity spaces can be used to produce grounded symbols with the aid of domain knowledge.

Quantity vector aggregation can be manipulated by the $\mathbb{OWA}$ and their variants when the dimension of the quantity vector is low or its online computation has required low priority. The issue of computational cost has been raised for quantity vector aggregation in higher dimensions, especially for real-time robotic systems. A $k-\mathbb{AGOP}$ aggregation operator is proposed to estimate the aggregation of quantity vectors of a higher dimension using a smaller k-dimensional subset. The $\mathbb{OWA}$ aggregation operators manipulate a single dataset or a data vector whereas the $k-\mathbb{AGOP}$ aggregation operator has the ability of handling multiple datasets using the proposed quantity arithmetic. Note that apart from domains where quantity space is used, calculations between datasets can be easily replaced by standard arithmetic. A k–$\mathbb{AGOP}$ aggregation operator of dimension n is a two-step mapping as follows,

$$\Gamma : \mathbb{R}^m \times \mathbb{R}^n \rightarrow \mathbb{R}^k \rightarrow \mathbb{R}$$

Function Γ has an associated weighting vector W of dimension n whose elements have the property:

$$\sum_{i=1}^{k_s} \sum_{j=1}^{k_l} w_{k_l}^{k_s} = 1, \;\; for\, w_{k_l}^{k_s} \in [0, 1]$$

The mapping is given by

$$\Gamma_k \left(P_m, P_n\right) = W_k^T \left(P_m^i \mathcal{O} P_n^j\right), \;\; for\ i,\ j \in \{k_s, k_l\} \tag{3.11}$$

where $P_m^i \mathcal{O} P_n^j$ is a k-dimensional column vector in ascending order and $\mathcal{O}$ is the quantity arithmetic operator. If the dimensions of the two vectors P_m and P_n are d_m and d_n respectively then $k_s = \min(d_m,\ d_n)$ and $k_l = \max(d_m,\ d_n)$. k_s and k_l can be obtained from the Eq. 3.12, where ceil(X) rounds the element of X to the nearest integer greater than or equal to X.

$$\begin{aligned} k_s &= ceil\left(\sqrt{\frac{\min(length(P_m), length(P_n))}{\max(length(P_m), length(P_n))}}k\right) \\ k_l &= k - k_s \end{aligned} \tag{3.12}$$

The k–$\mathbb{AGOP}$ aggregation operator consists of the calculation of the quantity vector and the aggregation operation. First, $\mathbb{R}^m \times \mathbb{R}^n \rightarrow \mathbb{R}^k$ selects two sets of intervals from respective quantity vectors and uses them to calculate a k-dimensional quantity vector using the proposed quantity vector arithmetic. Then $\mathbb{R}^k \rightarrow \mathbb{R}$ aggregates the k-dimensional quantity vector into a unique numerical value using the $\mathbb{OWA}$ aggregation operators. In order to estimate the quantity vector boundary in the first-step mapping, the first and last elements of the k_s- or k_l-dimension quantity vectors are selected as the minimum and maximum elements of the m- or n-dimension quantity vectors. The remaining $k_s - 2$ and $k_l - 2$ elements can be chosen using a probability distribution or by other criteria. In this chapter, it is assumed that the data in a dataset is uniformly distributed across its dimensions, which results in:

$$P_m \to P_{k_m}\left(p^1_{\min}, \dots, p^i, \dots, p^{k_m}_{\max}\right)$$
$$P_n \to P_{k_n}\left(p^1_{\min}, \dots, p^s, \dots, p^{k_n}_{\max}\right)$$

where $i = j \times round\left(\frac{m}{k_m-1}\right)$, $s = t \times round\left(\frac{n}{k_m-1}\right)$ $for\, k_m \in \{k_s, k_l\}$ and $j = [2, \cdots, k_m - 1]$, $t = [2, \dots, k_m - 1]$. $P_{k_n}(p^1_{min})$ is the minimum of quantity vector P_{k_n} and $P_{k_n}(p^{k_m}_{min})$ is the maximum, it is similarly applied for P_{k_m}. Hence the first-stage mapping $\mathbb{R}^m \times \mathbb{R}^n \to \mathbb{R}^k$ can be decomposed into $\mathbb{R}^m \times \mathbb{R}^n \to \mathbb{R}^{k_m} \times \mathbb{R}^{k_m} \to \mathbb{R}^k$. It is obvious that the proposed k–$\mathbb{AGOP}$ aggregation operator satisfies the general requirements of the $\mathbb{OWA}$ operator such as monotonicity, commutativity and bounded conditions which are described in [16].

In order to implement the proposed aggregation operator to robotic motion behaviour description for intelligent robotics, control parameters λ_i have been presented for $P_i \in \mathbb{R}^k$, where $i \in (x,\ y,\ z)$.

$$\begin{bmatrix} x \\ y \\ z \end{bmatrix} = \begin{bmatrix} \lambda_x Max\left(P_x\right) + \left(1 - \lambda_x\right) Min\left(P_x\right) \\ \lambda_y Max\left(P_y\right) + \left(1 - \lambda_y\right) Min\left(P_y\right) \\ \lambda_z Max\left(P_z\right) + \left(1 - \lambda_z\right) Min\left(P_z\right) \end{bmatrix} \tag{3.13}$$

The control parameters represent the degree of the aggregated values related to the two boundary values, i.e., $Max(P_i)$ and $Min(P_i)$, $i \in (x,\ y,\ z)$. Knowledge-based methods and learning algorithms can be employed here to produce the control parameters, e.g., generate motion behaviour subject to observed environmental information.

3.4 Case Study

In order to demonstrate the ability of FQ robotic kinematics handling the conversion among numeric, symbolic and hybrid data in the context of intelligent robotics, we have implemented the proposed theory into a PUMA 560 robot whose parameters shown in Table 3.2. It is common in behaviour-related research to assume that robots have rigid bodies and to consider the joint angles as variables only. Hence we denote the joint angles as FQ variables and the other parameters for the PUMA 560 as numeric, i.e. $a_2 = 432$ mm, $a_3 = 20$ mm, $d_3 = 124$ mm and $d_4 = 432$ mm. The orientation parameter p is set as 60, the translation parameter q as 40 and the initial configuration is given as,

$$\begin{bmatrix} QS_a(\theta_1) \\ QS_a(\theta_2) \\ QS_a(\theta_3) \end{bmatrix} = \begin{bmatrix} 0.007, 0.015 \\ 0.391, 0.017 \\ 0.024, 0.016 \end{bmatrix}$$

$$\begin{bmatrix} QS_a(\dot{\theta}_1) \\ QS_a(\dot{\theta}_2) \\ QS_a(\dot{\theta}_3) \end{bmatrix} = \begin{bmatrix} -0.82, 0.12 \\ 0.65, 0.14 \\ 7.52, 0.12 \end{bmatrix} \tag{3.14}$$

A joint trajectory of a PUMA 560 robot is provided in Table 3.1, note that the joint trajectory is described in terms of qualitative states, which can be connected to symbols through domain knowledge.

After implementing Eq. 3.8 using the parameters in Table 3.2, the centre of the PUMA tool frame is obtained as

$$QS_d\left(\mathbf{p}_{tcp}\right) = \begin{bmatrix} C\theta_1\left[a_2C\theta_2 + a_3C\theta_{2+3} - d_4S\theta_{2+3}\right] - d_3S\theta_1 \\ S\theta_1\left[a_2C\theta_2 + a_3C\theta_{2+3} - d_4S\theta_{2+3}\right] + d_3C\theta_1 \\ -a_3S\theta_{2+3} - a_2S\theta_2 - d_4C\theta_{2+3} \end{bmatrix}, \tag{3.15}$$

where θ_{2+3} stands for $QS_a(\theta_2) + QS_a(\theta_3)$.

The simulation results of FQ positions are given in Figs. 3.3, 3.4 and 3.5. The positions of the initial robotic configuration in Eq. 3.14 are demonstrated in Fig. 3.3 in interval terms. It shows that Joint position $J(QS(1), QS(24), QS(2))$ leads to a qualitative state for the end-effector's position, which is then represented by a numeric value aggregated by the proposed aggregation operator in Eq. 3.13. Further, the corresponding end-effector's positions of the trajectory of qualitative states in

Table 3.1 A trajectory of joint qualitative states for a PUMA 560 robot

$QS\ No$	1	2	3	4	5	6
$QS(\theta_1)$	1	4	2	8	15	18
$QS(\theta_2)$	24	30	38	42	36	50
$QS(\theta_3)$	2	7	12	12	18	25

Table 3.2 Hybrid DH-parameters for a PUMA 560 robot

i	α_{i-1}	a_{i-1}	d_i	$QS_a(\theta_i)$
1	0	0	0	$QS_a(\theta_1)$
2	$-\frac{\pi}{2}$	0	0	$QS_a(\theta_2)$
3	0	a_2	d_3	$QS_a(\theta_3)$
4	$-\frac{\pi}{2}$	a_3	d_4	$QS_a(\theta_4)$
5	$\frac{\pi}{2}$	0	0	$QS_a(\theta_5)$
6	$-\frac{\pi}{2}$	0	0	$QS_a(\theta_6)$

Table 3.1 are given in Fig. 3.4 using the MR representation. It demonstrates that each qualitative position is composed of the centres of overlapped FQ intervals which are labelled using circles. It also shows that the qualitative positions do not have the same interval regions, which indicates that each qualitative position has different position behaviours. In order to extract a suitable behaviour from the qualitative trajectory, we employ max, min and mean aggregation operators and Eq. 3.13. Corresponding to Fig. 3.4, there are six positions of the end-effector illustrated in Fig. 3.5, each of which is represented in both terms of conventional robot kinematics and the proposed method. For instance, qualitative state QS_6, its qualitative state's profile is qualitatively described by symbols '$\star$'s, '$\triangle$' and '$\bigtriangledown$', which are positions based on combinatorial x,y,z coordinates in Eq. 3.13 with $\lambda_i \in \{0,\ 1\}$. Positions lablled by '$\bigtriangledown$' (where $\lambda_i = 1$) and '$\triangle$' (where $\lambda_i = 0$) are used to label the maximum and minimum positions of the qualitative state, and the qualitative state's mean position labelled by '$\square$'. Positions with symbols '$\bigcirc$' are those generated by conventional robot kinematics with defaulted α-centre values of input joint trajectory in Table 3.1. It shows that positions '$\bigcirc$' are within the profile of the qualitative states but not coincident with the mean values; a set of aggregated values labelled as '$\times$' calculated by Eq. 3.13 is provided with $\lambda_x = 0.2$, $\lambda_y = 0.5$ and $\lambda_z = 0.8$. The control parameters can be either assigned or learned by contextual information, e.g., recognised obstacle objects. Aggregated values can also be used for representing their corresponding qualitative states. This demonstrates the connection of numeric values and qualitative states through contextual information, the former can be employed to a low-level control system; the latter can be used to construct a symbol-based reasoning system. Additionally, each group of positions are lined up to show the trajectory sequence, trajectory I stands for the aggregated trajectory produced by the proposed method, trajectory II denotes the trajectory by conventional robot kinematics and trajectory

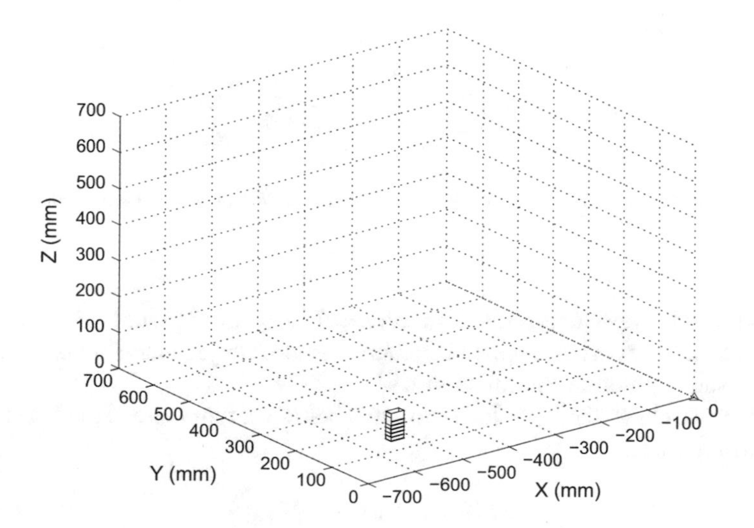

Fig. 3.3 An example of a fuzzy qualitative position for a PUMA robot

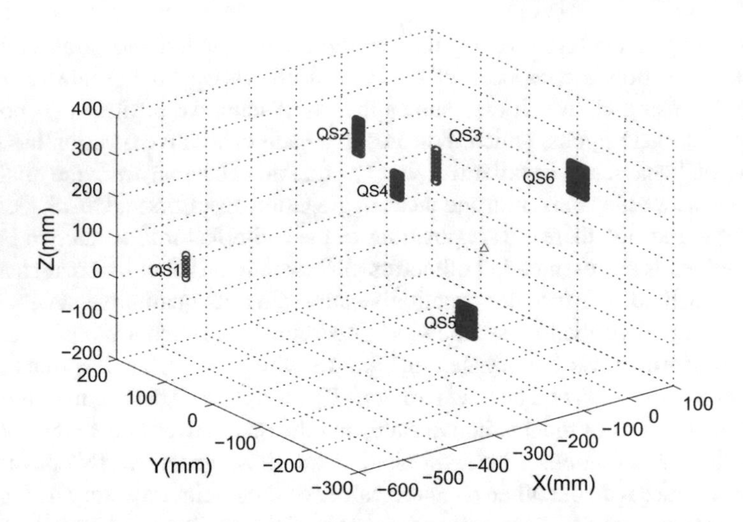

Fig. 3.4 An example of fuzzy qualitative positions for PUMA robot

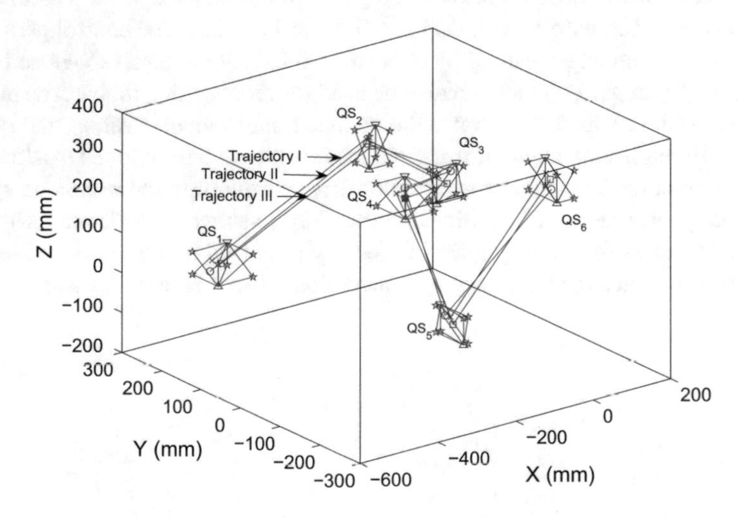

Fig. 3.5 Comparison study of the proposed method and conventional robot kinematics regarding to the position data in Fig. 3.4

III stands for the trajectory by the mean values of the qualitative states. Suitable interpolation techniques can be applied to generate smooth trajectories for non-symbolic modules such as motion control modules.

Further we obtain the FQ velocity formula below for the centre of the PUMA tool frame using Eq. 3.9:

$$QS_d\left(\dot{\mathbf{p}}_{tcp}\right) = \mathbf{J}_{3\times3} \cdot QS_d\left(\dot{\boldsymbol{\theta}}\right) \tag{3.16}$$

where $QS_a(\dot{\theta})$ is FQ vector which describes the angular velocity of the PUMA and is given by

$$QS_a(\dot{\theta}) = \left[QS_a(\dot{\theta}_1) \; QS_a(\dot{\theta}_2) \; QS_a(\dot{\theta}_3) \right]^T.$$

where $\mathbf{J}_{3\times 3}$ is its FQ Jacobian matrix which is defined as

$$\mathbf{J}_{3\times 3} = \left[\mathbf{J}_1 \; \mathbf{J}_2 \; \mathbf{J}_3 \right] \tag{3.17}$$

where

$$\mathbf{J}_1 = \begin{bmatrix} -S\theta_1 \left[a_2 C\theta_2 + a_3 C\theta_{2+3} - d_4 S\theta_{2+3} - d_3 C\theta_1 \right] \\ C\theta_1 \left[a_2 C\theta_2 + a_3 C\theta_{2+3} - d_4 S\theta_{2+3} - d_3 S\theta_1 \right] \\ 0 \end{bmatrix}$$

$$\mathbf{J}_2 = \begin{bmatrix} -C\theta_1 \left[a_2 S\theta_2 + a_3 S\theta_{2+3} + d_4 C\theta_{2+3} \right] \\ -S\theta_1 \left[a_2 S\theta_2 + a_3 S\theta_{2+3} + d_4 C\theta_{2+3} \right] \\ \left[-a_3 C\theta_{2+3} - a_2 C\theta_2 + d_4 S\theta_{2+3} \right] \end{bmatrix}$$

$$\mathbf{J}_3 = \begin{bmatrix} -C\theta_1 \left[a_3 S\theta_{2+3} + d_4 C\theta_{2+3} \right] \\ -S\theta_1 \left[a_3 S\theta_{2+3} + d_4 C\theta_{2+3} \right] \\ \left[d_4 S\theta_{2+3} - a_3 C\theta_{2+3} \right] \end{bmatrix}$$

The simulation result of the FQ velocity of the PUMA robot's tool frame is given in Fig. 3.6. It uses the same notation as in Fig. 3.5 but for the end-effector's velocity rather than its positions. Trajectory I denotes the aggregated-value based trajectory with $\lambda_x = 0.2$, $\lambda_y = 0.9$ and $\lambda_z = 0.5$, Trajectory II is based on the mean values of

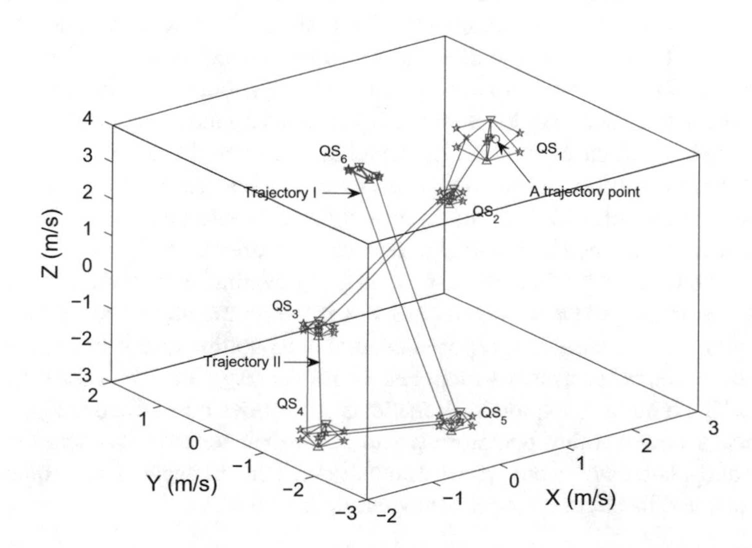

Fig. 3.6 Comparison study on the end-effector's velocity using the proposed method and conventional robot kinematics

the qualitative states, which provides a reference for the comparison. A trajectory point labelled by '○' illustrates that the velocity calculated by conventional robot kinematics is within its qualitative profile.

The proposed method can be easier adapted to robotic systems for a better solution to intelligent connection problem. For instance, Chan et al. [19] has adapted this method to model human motion in image sequences as stick model in order to analyse their motion behaviour. In addition, the proposed method potentially could be used to construct behaviour vocabularies for the development of skill-level interfaces for autonomous humanoid robots in [6].

3.5 Conclusion

This chapter proposed an extension named FQ trigonometry derivative to FQ trigonometry and implemented it into the XTRIG MATLAB toolbox. Based on the extension, a FQ version of robot kinematics in FQ terms is proposed. Ordered weighted averaging operators have been adapted to extract the behaviours of serial robots, an example of a PUMA robot has been presented to support the proposed methods.

The advantage of the proposed FQ robot kinematics is twofold. It can be used as a tool to measure robotic calibration issues, e.g., accuracy, repeatability and calibration [20]. More importantly, it provides an opportunity to connect numerical data and interval symbols together, these elements can work as atomic behaviour functions that could be used to build up behaviour vocabularies of a robot and its motion control. It paves the way for providing a feasible approach to intelligent connection problem. The connection problem is crucial for research in the field of AI and robotics because intelligent robots commonly comprises of non-symbolic subsystems and symbolic subsystems. The state of art in intelligent robotics including technical needs by a wider range of robots requires their symbolic or behaviour description. In practice, almost all existing robotic systems consisting of at least a motion control module and a reasoning/symbol-based module, hardcode methods are the dominant methods to connect these two types of modules; it is one of key barriers to intelligent robotics because symbols could have different meanings in various contexts. The proposed method provides a fundamental way to replace the hardcode methods because of its relation to conventional robot kinematics and fuzzy qualitative reasoning. In addition, our future work targets two research lines: one is the computational cost issue, which deals with how efficiently the proposed interval computation meets the requirements of robotic systems, or systems modelled as robots (e.g., stick models for human motion). The other is to explore new methods to improve how effective aggregation operators or new learning operators would be in their behaviour description. The priority of the latter would be given to optimally tune the propagated fuzzy qualitative parameters and its effect on robot behaviour description.

References

1. K. Kiguchi and T. Fukuda. Position/force control of robot manipulators for geometrically unknown objects using fuzzy neural networks. *IEEE Transactions on Industrial Electronics*, 47(3):641–649, 2000.
2. S.S. Ge, T.H. Lee, and C.J. Harris. Adaptive neural network control of robotic manipulators. *World Scientific Series in Robotics and Intelligent Systems*, 19, 1999.
3. R.C. Arkin. Behaviour-based robotics. *The MIT Press*, 1998.
4. R.R. Murphy. An introduction to ai robotics. *The MIT Press*, 2000.
5. S. Thrun, W. Burgard, and D. Fox. Probabilistic robotics. *The MIT Press*, 2005.
6. O.C. Jenkins and M.J. Mataric. Performance-derived behavior vocabularies: data-driven acquisition of skills from motion. *International Journal of Humanoid Robotics*, pages 237–288, 2004.
7. O. Khatib, L. Sentis, J. Park, and J. Warren. Whole-body dynamic behavior and control of human-like robots. *International Journal of Humanoid Robotics*, 1(01):29–43, 2004.
8. M. Brady. Artificial intelligence and robotics. *Artificial Intelligence*, 26:79–121, 1985.
9. P. Dpherty, G. Lakemeyer, and A.P. Del Pobil. *Proceedings of the Fourth International Cognitive Robotics Workshop*. CogRob, 2004.
10. H. Liu and G.M. Coghill. Fuzzy qualitative trigonometry. *Proc. IEEE International Conference on Systems, Man and Cybernetics, Hawaii, USA.*, 2005.
11. J. Craig. *Introduction to Robotics: Mechanics and Control*. Addison-Wesley Publishing Company, E.U.A., segunda edition, 1989.
12. G. Beliakov. Identification of general aggregation operators by lipschitz approximation. *Artificial Intelligence and Applications eds by Hamza*, 2005.
13. R.R. Yager. Hierarchical aggregation functions generated from belief structures. *IEEE Transactions on Fuzzy Systems*, 8(5):481–492, 2000.
14. D. Daney, N. Andreff, G. Chabert, and Y. Papegay. Interval methods for calibration of parallel robots: visioin-based experiments. *Meism and Machine Theory*, 41:929–944, 2006.
15. J. Fodor, J.L. Marichal, and M. Roubens. Characterization of the ordered weighted averaging operators. *IEEE Transactions on Fuzzy Systems*, 3(2):236–241, 1995.
16. R.R. Yager and D.P. Filev. Induced ordered weighted averaging operators. *IEEE Transactions on Systems, Man and Cybernetics*, 29(2):141–151, 1999.
17. R.R. Yager. On ordered weighted averaging aggregation operators in multi-criteria decision making. *IEEE Transactions on Systems, Man and Cybernetics*, 18:757–769, 1988.
18. E. Gregoire and S. Konieczny. Logic-based approaches to information fusion. *Information Fusion*, 7:4–18, 2006.
19. C.S. Chan, H. Liu, and D.J. Brown. Human motion classification using qualitative normalised templates. *Journal of Intelligent and Robotic Systems*, 48(1):in press, 2007.
20. T. Calvo, R. Mesiar, and R.R. Yager. Quantitative weights and aggregation. *IEEE Transactions on Fuzzy Systems*, 12(1):62–70, 2004.

Chapter 4
Fuzzy Qualitative Human Motion Analysis

This chapter proposes a fuzzy qualitative (FQ) approach to vision-based human motion analysis with an emphasis on human motion recognition. It achieves feasible computational cost for human motion recognition by combining FQ robot kinematics with human motion tracking and recognition algorithms. First, a data quantisation process is proposed to relax the computational complexity suffered from visual tracking algorithms. Secondly, a novel human motion representation, Qualitative Normalised Template (QNT), is developed in terms of the FQ robot kinematics framework to effectively represent human motion. The human skeleton is modelled as a complex kinematic chain, its motion is represented by a series of such models in terms of time. Finally, experiment results are provided to demonstrate the effectiveness of the proposed method. An empirical comparison with conventional Hidden Markov Model (HMM) and Fuzzy Hidden Markov Model (FHMM) shows that the proposed approach consistently outperforms both hidden Markov models in human motion recognition.

4.1 Introduction

Motion understanding is the ability to analyse human motion patterns, further to produce high-level interpretation of these patterns. Human motion in-depth understanding plays a crucial role in a diverse spectrum of applications from surveillance-based suspicious behaviour recognition to monitoring of daily health care for elderly people. Fundamentally, human motion analysis systems consist of description and recognition of human motion: first, *extracting* relevant information through visual tracking which involves the detection of regions of interest in image sequences that are changing with respect to time and/or finding frame to frame correspondence of each region so that features of each region can reliably be extracted. Secondly, *modelling* this information as an abstraction of sensory data that should reflect a

H. Liu et al., *Human Motion Sensing and Recognition*,
Studies in Computational Intelligence 675, DOI 10.1007/978-3-662-53692-6_4

real world situation. Finally, a *recognition* step aiming at determining the maximum similarity between an unobserved test sequence and pre-learned motion models [1]. Nevertheless, developing these algorithms is an immerse challenge as it is a problem that combines the uncertainty associated with computational vision and the added whimsy of human behaviour. For instance, the methodology of implementing such a system has been the focus of research in the past two decades, and many works have been conducted and published with a promising classification rate. Surprisingly, such a promising system has yet been employed in practice. The underlying issue is that these existing methodologies are not computationally feasible. That is, the complexity of such systems in terms of the processing time is too expensive to implement into real systems mainly due to the tracking precision.

In visual tracking, particle filters [2] and its variants [3, 4] have been well developed in human motion tracking research in the past decades; but its computational cost keeps them away from practical applications. It can be seen that a standard particle filter has a computational complexity of $\mathcal{O}(2N)$ and time complexity of $N \sum_{k=1}^{m} \tau_k$ where N is the number of particles and τ_k is the cost of calculation $p(z_k|x)$. This is a major drawback as practical human motion analysis systems require to be running in real-time, or near real-time frame rates. Whereas methods that do not use such approaches usually rely on the accuracy of motion sensors, but seldom provide a measure of confidence of the results which are crucial to discriminate similar events in a noisy environment. On the other hand, representation of the human motion is a very important and sometimes a difficult aspect of an intelligent system. The representation is an abstraction of sensory data that should reflect a real world situation and be compact and reliable. Probabilistic graph models have been the dominant methods in the field of human motion analysis systems [5]. Despite the fact that all these approaches have demonstrated success in modelling and recognising complex activities, there is, however, a tendency to use the parametrisation as a 'black box'. That is, these approaches highly depend on probabilities and intensive training to recognise all the activities. Thus, one needs to have a large number of training sequences with intensive training in order for each activity to be recognised correctly. This requires increasingly expensive computational power in that the complexity of these solutions in terms of processing time are proportional to the size of the training data.

Template matching is one of the earliest human motion analysis methods. Bobick and Davis [6] proposed a view-based approach to the representation and recognition of action using temporal templates. They made use of the binary Motion Energy Image (MEI) and Motion History Image (MHI) to interpret human movement in an image sequence. First, motion images in a sequence were extracted by differencing, and these motion images were accumulated in time to form MEI. Then, the MEI was enhanced into MHI which was a scalar-valued image. Taken together, the MEI and MHI were considered as a two-component version of a temporal template, a vector-valued image, in which each component of each pixel was some function of the motion at that pixel position. Finally, by representing the templates by its seven Hu-moments, a Mahalanobis distance was employed to classify the action of the subject by comparing it to the Hu-moments of pre-recorded actions. Bradski and Davis [7] further contributed to the idea of MHI by proposing timed MHI (tMHI) for motion segmentation. tMHI allows for determination of the normal optical flow. Motion is segmented relative to object boundaries and the motion orientation. Hu-moments are applied to the binary silhouette to recognise the pose. However, one of the main problems of template matching approaches are that the recognition rate of objects based on 2D image features is low, due to the non-linear distortion during perspective projection and the image variations with the viewpoint's movement. These algorithms are generally unable to recover the 3D pose of objects. Moreover, the stability of dealing effectively with occlusion, overlapping and interference of unrelated structures is generally poor. Alternatively, model based approaches have been adopted [8–11], in which a human body model is constructed with prior knowledge. Wang et al. [12] presented an approach where contours are extracted and a mean contour is computed to represent the static contour information. Dynamic information is extracted by using a detailed model composed of 14 rigid body parts, each one represented by a truncated cone. Particle filtering [2] is used to compute the likelihood of a pose given an input image. However there is a trade-off between tracking precision against computational cost where a number of systems are based on incremental updates or searching around a predicted value [2, 13, 14].

Hidden Markov Models (HMM), a family of popular parametric methods, have been formulated in human motion analysis systems. Common approaches consist of extracting low-level features by local spatio-temporal filtering on images and using the HMM on the collection of sequences of points to achieve activity recognition and classification tasks. The HMM considers this correlation between adjacent time instances by formulating a Markov process and assumes that the observation sequence is statistically determined by a hidden process which is composed of a fixed number of hidden states. For instance, Lv and Nevatia [15] decomposed the large joint space into a set feature space where each feature corresponds to a single joint or combination of related joints. An AdaBoost scheme was employed to detect and recognise each feature in the feature space, and follow by a HMM system to recognise each action class based on the detected features. Leo et al. [16] attempted to classify actions at an archaeological site. A system that uses binary patches and an unsupervised clustering algorithm to detect human body postures was proposed. A discrete HMM is used to classify the sequences of poses into a set of four different actions. In [17], a mixed state statistical model for the representation of motion had been proposed. That is, the work decomposes a human behaviour into multiple abstractions, and represents the high-level abstraction by HMMs built from phases of simple movements. Estimation and recognition of the human behaviour is performed with expectation maximisation approaches using particle filters [2] or structured variation inference techniques [18]. While all these solutions have demonstrated success in classifying complex activities, HMMs suffer from few drawbacks. First of all, a HMM is relying on stochastic learning, they require extensive training. Therefore, one needs to have a large number of training sequences for each activity to be recognised correctly. This is not feasible as practical human motion analysis systems must often work at real-time, or near real-time frame rates. Secondly, for each activity to be recognised, a separate HMM needs to be built. Hence, a solution which can flexibly handle the trade-off between human motion precision and its computation efficiency are the next important step in a human motion analysis system.

In this chapter, we propose a fuzzy qualitative (FQ) method to study human motion analysis in order to reduce the computational complexity encountered by existing solutions, enabling such systems to be used in practical situations. First of all, we propose a solution to handle the trade-off between computational efficiency and motion description precision in visual tracking algorithm by applying a data quantisation process. During this process, we consider the rigid motion of each human body joint in a FQ description. Secondly, a novel human motion representation, Qualitative Normalised Template (QNT), is developed based on the FQ robot kinematics in Chap. 3 to effectively represent human motion [19]. The QNT is a template-based method instead of a statistical learning method, hence large training datasets are not required. Instead strong discriminative features can be derived from just a couple of example activities. Finally, empirical results show that our proposed solution outperforms existing solutions in human motion recognition by flexibly handling the trade-off between human motion precision and its computational efficiency.

The remainder of the chapter is structured as follows. Section 4.2 derives the FQ human motion analysis, in particular how both the data quantisation and FQ robot kinematics framework are employed to study human motion in video sequences. Section 4.3 presents the experimental results and an empirical comparison with the HMM and FHMM in human motion classification. Section 4.4 concludes the chapter with discussions and future works.

4.2 FQ Human Motion Analysis

This section presents FQ description for human motion analysis in video sequences. The culmination of algorithms is represented by a distributed video collection and processing system in which the basic tasks of human body modelling, human motion tracking, representation and recognition are performed in support of a single, underlying task; that is human motion analysis.

4.2.1 Human Body Modelling

In principle, one must perceive a human motion before modelling and interpreting it. It means that an appearance model is needed. In this chapter, we simplify a human body into a collection of hierarchical structure skeleton that composed of segments and limbs linked together in a kinematic chain. The model parameters are given by $\phi_t = [(\phi_t^{pos})^T, (\phi_t^{vel})^T]^T = [t_t^T, \alpha_t^T, \theta_t^T, \hat{t}_t^T, \hat{\alpha}_t^T, \hat{\theta}_t^T]^T$ where t_t and α_t represent the translation and rotation that map the body into the world coordinates system and θ_t represents the relative angles between all pairs of connected limbs. Parameters $\hat{t}_t^T$, $\hat{\alpha}_t^T$, $\hat{\theta}_t^T$ represent the corresponding velocities. In addition, an event or action is defined as the temporal movement of a human body segment in a short time period and is represented by a state representation in a normalised FQ unit circle. Further an activity is defined as a combination ordered sequence of the entire participating body segment's movement over time, which is restricted by the motion constraints of the human body. For instance, the sequences of events for a walking activity include segment events for the foot, lower leg, and thigh, and joint events for the ankle, knee and hip. All these sequences occur in the leg that is moving forward, while the leg that supports the body will show no such events; and the walking activity is defined as a combination of these sequences of events.

4.2.2 *Human Motion Tracking*

The Condensation algorithm [2] has been employed to estimate the posterior probability distribution over human pose given a sequence of observation measurement. It is assumed that all elements $\phi_{i,t} \in \phi_t$ are independent and all parameters were initialised manually with a Gaussian distribution at time $t = 0$. In order to restrict the set of admissible motions and to reduce the ambiguities in the estimation, each joint is allowed only for one single degree of freedom (DOF), which is a rotation around its axis. This constraint is justified by the fact that typically the motion of the limbs can be approximated as planar around an axis perpendicular to the direction of the motion. Figure 4.1 shows examples of sets of trajectories from different limb motions in the case of real videos. The condensation algorithm is one of dominant tracking algorithms, for detailed human motion tracking algorithms please refer to [20–22].

The Condensation algorithm discussed herein has the advantage that the particle representation can represent distributions that are difficult to model analytically. The performances of such algorithms have not, however, been fully evaluated under circumstances specifically to real-time vision systems, where there exists a certain trade-off between motion description precision and computational efficiency. For instance the Condensation algorithm is an approximation technique by representing the posterior density as a set of samples of the state space with associated likelihood weights ω_t^i, $i \in \{1, \cdots, N\}$. The sample set approximation of the current posterior density $p(X_t|O_{1:t})$ can be obtained via:

$$p(X_t|O_{1:t}) \approx \sum_{i=1}^{N} \omega_t^i \delta(X_t - x_t^i) \tag{4.1}$$

Fig. 4.1 Sample human motions used from the database provided by [23, 24]. Trajectories from six landmarks (shoulder, elbow, wrist, hip, knee and ankle) on the human body are tracked over time using the Condensation algorithm.

where X_t denotes the multi-variant state at time t, $O_{1:t}$ denotes the sequence of observation measurements within the time range $[1, \ t]$, and $\delta(X_t - x_t^i)$ denotes the Dirac delta function and the prior is approximated as:

$$p(X_t|O_{1:t-1}) \approx \sum_{i=1}^{N} \omega_{t-1}^i \delta(X_t - x_{t-1}^i) \tag{4.2}$$

Weight ω_{t-1}^i is determined such that $\omega_{t-1}^i \propto p(O_{t-1}|X_{t-1}^i)$, $\sum_{i=1}^{N} \omega_{t-1}^i = 1$.

Looking into Eqs. 4.1 and 4.2, both equations are almost accurate if one employs a sufficient large number of particles. In other words, if the number of particles is infinite, i.e., $N \rightarrow \infty$, the right hand sides of Eqs. 4.1 and 4.2 become identical to the left hand sides. In reality, however, using an infinite number of particles is not practical, especially for real-time processing. Nevertheless, if relatively low number of particles are employed, the tracking system will fail and hence greatly affect human motion analysis systems during the representation and recognition stages.

4.2.3 *Data Quantisation*

Data quantisation is a process adopted from Liu and Coghill [25], in which the unit circle of the conventional trigonometry has been fuzzified by the introduction of FQ quantity spaces for its orientation and translation components; and the data quantisation is a process that mapped quantitative information qualitatively with respect to the configuration in the FQ unit circle. The FQ quantity space is a set of overlapped fuzzy numbers whose individual distance among them is defined by a predefined metric. Besides each fuzzy number is a finite and convex discretisation of the real number line by default. Four tuple fuzzy numbers (i.e., [a, b, α, β]) and its arithmetic are employed to describe the characteristic of each state in the FQ unit circle. Such a representation has been selected as it provides good compositionality and high resolution [26].

In this chapter, we employed the data quantisation process to represent the predicted motion parameters, ϕ_t, qualitatively. The advantage of the data quantisation process is that each of the FQ quantity space in the FQ unit circle which is finite and convex discretisation of the real number line will be able to model the tracking error when relatively low number of particles are selected to perform the visual tracking algorithm. That is, we consider the motion of each joint as a collection of time series describing the joint angles as they evolve over time. This is achieved through the visual tracking algorithm discussed in Sect. 4.2.2. However, as shown in Fig. 4.2 it is evident that about 1000 particles are needed in the Condensation algorithm, with approximation of 60 min of training time is required in order to achieve the required tracking precision. This is not feasible for a real-time vision-based human motion system.

The rigid motion of each joint is mapped into a FQ description through the data quantisation process, so each joint movement is represented by the FQ orientation states and FQ translation states in which the quantitative dynamic characteristic of motion data resides in the normalised FQ unit circle. A FQ unit circle is constructed using Eq. 4.3 where the orientation and translation components in the conventional unit circle are being replaced by FQ quantity spaces.

$$\begin{aligned} \lim_{s \to s_o = 12} C_t(s) &= QS(qp_l) \\ \lim_{r \to c_o = 16} C_o(r) &= QS(qp_\theta) \end{aligned} \tag{4.3}$$

where s is the number of states resides in the x-y translation while r is the number of states resides on the orientation in the FQ unit circle. That is, s and r represent the number of translation and orientation states employed in the quantity spaces to represent the FQ unit circle, respectively. As s $\rightarrow s_o$ and r $\rightarrow r_o$, the limits of $C_t(s)$ and $C_o(r)$ will approach to a set of s_o qualitative states for a translation component and a set of r_o qualitative states for an orientation component. The range of s and r are application dependent. Empirically, we selected the translation, s and orientation, r as $s = 12$ and $r = 16$ 4-tuple fuzzy numbers respectively as shown in Fig. 4.3.

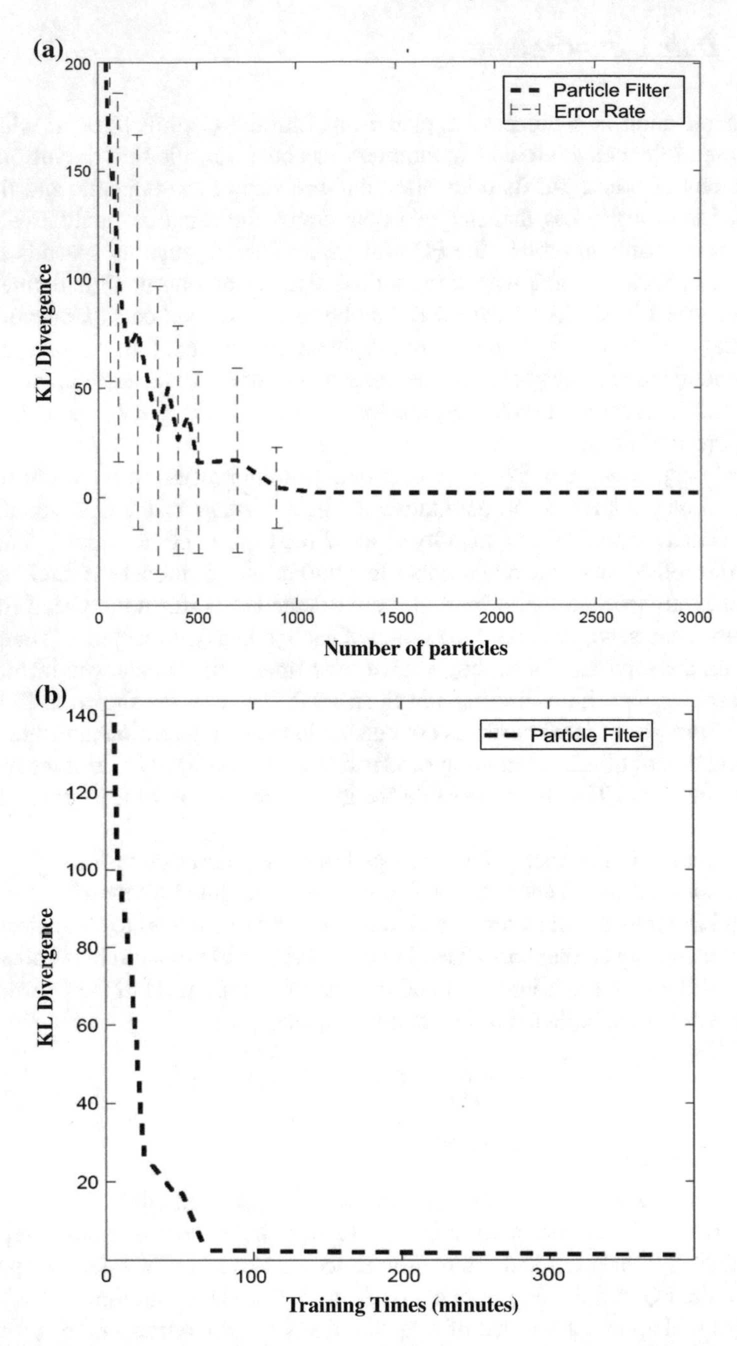

Fig. 4.2 Number of particles vs. Training time. **a** Referring to the error rate result, it is evident that about 1000 of particles are needed in order to achieve the required tracking precision **b** The training time of the Condensation algorithm employing 1000 particles requires about 60 min

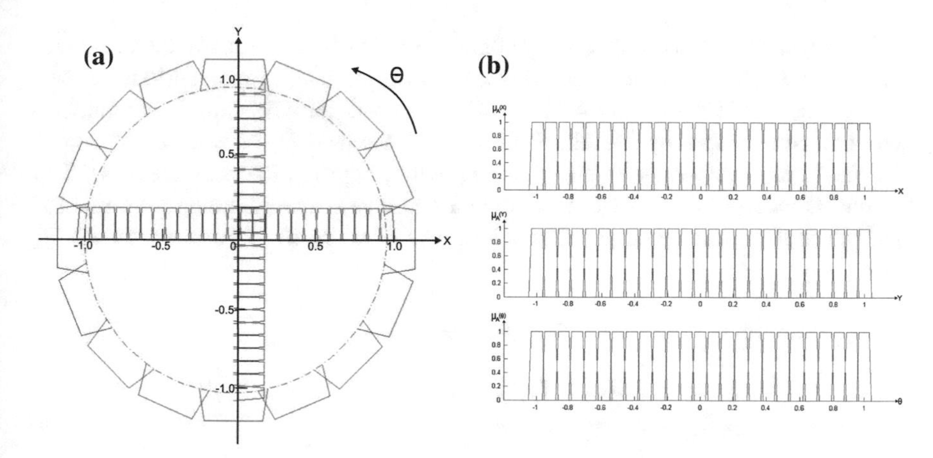

Fig. 4.3 FQ unit circle with resolution $s = 12$ and $r = 16$, respectively. **a** The description of the Cartesian translation and orientation in the conventional unit circle are replaced by a quantity space. **b** The element of the quantity space for every variable in the FQ unit circle is a finite and convex discretisation of the real number line

The FQ quantity space Q of the FQ unit circle consists of an orientation component Q^a and a translation component Q^d and it can be described as follows,

$$
\begin{aligned}
Q^a &= \{QS_a(\theta_i)\}, where\ i = 1, 2, \cdots, m \\
Q^d &= \{QS_d(l_j)\}, where\ j = 1, 2, \cdots, n
\end{aligned}
\tag{4.4}
$$

where $QS_a(\theta_i)$ denotes the state of an angle θ_i, $QS_d(l_j)$ denotes the state of a distance l_j, m and n are the number of the elements of the two components. The position measurement of $P(QS_a(\theta_i), QS_d(l_j))$ determined by both the characteristics of the fuzzy membership functions of $QS_a(\theta_i)$ and $QS_d(l_j)$. The geometric meaning of FQ trigonometry is demonstrated in a proposed FQ unit circle, in which the motion is described by an orientation component and a translation component. The level of resolution in the FQ unit circle can be adjusted by setting the fuzzy number.

Further, with respect to the constructed FQ unit circle, the predicted motion parameters, ϕ_t obtained from Sect. 4.2.2 are mapped into its corresponding FQ states as given in Eq. 4.5.

$$
\begin{cases}
qp_l^i | qp_l^i \in \left[0, l_{i1}, l_{i2}, ..., l_{i(r_i-1)}, l_{ir_i}\right] \\
qp_\theta^i | qp_\theta^i \in \left[0, q\theta_{i1}, q\theta_{i2}, ..., q\theta_{i(s_i-1)}, 2\pi\right]
\end{cases}
\tag{4.5}
$$

where

$$
\begin{aligned}
& qp_j^i = \tfrac{l_i j}{s_i}, q\theta_k^i = \tfrac{2\pi k}{r_i} \\
& 0 \le qp_1^i \le qp_2^i \le ... \le qp_{(s_i-1)}^i \le l_i \\
& 0 \le q\theta_1^i \le q\theta_2^i \le ... \le q\theta_{(r_i-1)}^i \le 2\pi
\end{aligned}
$$

Each of these states is the corresponding region in which the quantitative dynamic characteristic of motion data resides in the FQ unit circle. Due to the fact that a motion component comprises evenly distributed normalised numeric data, fuzzy numbers have the same shapes by default. For simplicity, we have $b - a = \kappa_0\alpha$ and $\alpha = \beta$, and the membership value of the crossing point of adjacent fuzzy numbers is 0.5 by default. Hence, a function for generating a quantity space including n normalised qualitative states can be obtained as below, where $[a_i, b_i, \alpha_i, \beta_i]$ denotes the ith FQ state $QS(i)$.

$$QS(i) = \begin{cases} [0, \kappa_0\alpha, 0, \beta], i = 1 \\ [(\kappa_0 + 1)(i - 1)\alpha, (\kappa_0 + 1)i\alpha - \alpha, 0, \beta], \\ i = 2, \ldots, n - 1 \\ [1 - \kappa_0\alpha, 1, \alpha, 0], i = n \end{cases} \tag{4.6}$$

where $\alpha = \frac{1}{n(\kappa_0+1)-1}\kappa_0$ is a threshold parameter to define the shape of the fuzzy numbers, κ_0 and n are chosen by applications or by a learning algorithm.

For each of the FQ state representation of the motion parameters, we normalised them within the FQ unit circle $[-1\ 1]$ using Eq. 4.7. Therefore, the qualitative and quantitative representations are linked together, it paves the way for connecting numerical image sequences with symbolic natural language description.

$$\begin{cases} QS(qp_l) = qp_l | qp_l \in [\frac{ql_1}{ql}, \frac{ql_2}{ql}, \cdots, \frac{ql_{s-1}}{ql}, 1] \\ QS(qp_\theta) = qp_\theta | qp_\theta \in [\frac{q\theta_1}{2\pi}, \frac{q\theta_2}{2\pi}, \cdots, \frac{q\theta_{r-1}}{2\pi}, 1] \end{cases} \tag{4.7}$$

where x-y translation states qp_l are normalised by the average length of the human body segment ql and the orientation states qp_θ are normalised to 2π. Hence, Eq. 4.5 can be rewritten as,

$$\begin{cases} qp_l^i | qp_l^i \in [\frac{qp_1^i}{\sum_{i=1}^{n} l_i}, \frac{qp_2^i}{\sum_{i=1}^{n} l_i}, \ldots, \frac{qp_{(s_i-1)}^i}{\sum_{i=1}^{n} l_i}, \frac{l_i}{\sum_{i=1}^{n} l_i}] \\ qp_\theta^i | qp_\theta^i \in [\frac{q\theta_1^i}{2\pi}, \frac{q\theta_2^i}{2\pi}, \ldots, \frac{q\theta_{(r_i-1)}^i}{2\pi}, 1] \end{cases} \tag{4.8}$$

The visual tracking parameters of each body joint are mapped into the normalised FQ unit circle in order to achieve the FQ description. It can be seen that each motion parameters ϕ_t is represented by the corresponding state region of a FQ state in the normalised unit circle. For example, the predicted joint motion parameters of J_1, J_2 and J_3 at time instant $t = 1$ are represented by the same FQ states after the data quantisation process as shown in Table 4.1. In other words, it means if we have sufficient enough number of particles to meet the resolution needed in the normalised FQ unit circle, the data quantisation process will be able to model the uncertainty

Table 4.1 Quantisation results with $s = 12$, $r = 16$

Joints	Before quantisation (°)	After quantisation (State)
$J_{1,t=1}$	40	$Q^a(2)$
$J_{2,t=1}$	48	$Q^a(2)$
$J_{3,t=1}$	46	$Q^a(2)$
$\vdots$	$\vdots$	$\vdots$
$J_{1,t=13}$	65	$Q^a(3)$
$J_{2,t=13}$	18	$Q^a(1)$
$J_{3,t=13}$	120	$Q^a(6)$
$\vdots$	$\vdots$	$\vdots$
$J_{1,t=24}$	46	$Q^a(2)$
$J_{2,t=24}$	46	$Q^a(2)$
$J_{3,t=24}$	50	$Q^a(2)$

under the constraints of the limitation in visual tracking algorithms, and to correctly preserve the underlying motion description for motion representation and recognition process.

In order to quantify the accuracy of the estimated motion parameter ϕ_t before and after the proposed data quantisation process, Table 4.2 shows a comparison of the estimated motion parameter ϕ_t obtained from the Condensation algorithm with $N = 800$ and $N = 2000$ to its ground truth position before and after the proposed data quantisation process. It can be seen that the accuracy of the estimated motion parameters ϕ_t after the proposed data quantisation process is typically much higher. Also, it can be noticed the error rates of both the Condensation algorithms ($N = 800$ and $N = 2000$) are similar. This demonstrates the effectiveness of the proposed data quantisation process in terms of flexibly handling the trade-off between human motion precision and its computational efficiency. This is an essential step in order to move the human motion analysis systems into practical use.

4.2.4 *Human Motion Representation*

The data quantisation process is addressed in previous sections with a focus on representing the motion of individual joints of the human body qualitatively in order to relax the computational complexity in visual tracking algorithms. It is evident that a human motion is a combination of ordered sequences of all the independent movements [27]. In order to build a reliable human motion template that reflects the real world situation, the FQ robot kinematics framework has been employed to construct a novel motion template called a qualitative normalised template (QNT).

Table 4.2 A comparison of the estimated motion parameters with the ground truth positions before and after data quantisation process

Particles *(N)*	Joint	Error rate		Computational complexity
		Before quantisation (%)	After quantisation (%)	
800	Thigh	6.5	**0**	$\mathcal{O}(2N)$
	Knee	15.6	**0**	
2000	Thigh	5.1	**0**	$\mathcal{O}(2N)$
	Knee	7.5	**0**	

We model a human skeleton structure as articulated rigid bodies with kinematic chains. It is assumed that the human motion is rigid so that the priority is given to solely on the joint movements and the work will not be distracted by affects of the muscle stretch and reflex during a movement.

Based on FQ robot kinematics [19], the motion of each body joint i is represented in terms of twist representation as shown in Eq. 4.9 sicne it provides a more simpler solution and leads to a compact 3D linear representation of a motion model [28].

$$\xi = \begin{bmatrix} \upsilon_1 \; \upsilon_2 \; \upsilon_3 \; \omega_1 \; \omega_2 \; \omega_3 \end{bmatrix} \tag{4.9}$$

where ξ is a 3D FQ unit vector that points in the direction ranges of the rotation axis. The amount of rotation is specific with a FQ angle state, θ, multiplied by the twist, $\xi\theta$. While the υ component determines the location range of the rotation axis and the amount of translation along this axis. For instance, in order to realise the motion performed by each human body at time t, first, we define a base body reference frame F_0 that is attached to the base body and a spatial reference frame F_a which is static and coincides with F_0 at time t. By considering a single kinematics chain of two human body segments connected to the base frame, we parametrise the orientation between these connected components in terms of the angle of rotation around the axis of the object coordinate frame θ. This rotation axis in the object frame can be represented by a 3D FQ unit vector ω_1 along the axis, and a FQ point q_1 on the axis. The twist representation can be described as below for revolute joint 1,

$$\xi_1 = \begin{bmatrix} -\omega_1 \times q_1 \\ \omega_1 \end{bmatrix} \tag{4.10}$$

The transformation of FQ point q_1 from F_a coordinates to the base frame F_0 can be obtained as,

$$g(\theta_1) = g_1.g(0) = e^{\xi_1 \theta_1}.g(0) \tag{4.11}$$

For a kinematics chain of K bodies, its motion of the kth body is represented by joint θ_k and each joint is described by a twist ξ_k, the forward kinematics $g_K(\theta_1, \theta_2, \cdots, \theta_k)$ therefore can be computed by the individual twist motion for each joint $e^{\xi_k \theta_k}$, and the transformation between the base frame g(0) and Frame F_k can be obtained as,

$$g(\theta_1, \theta_2, \cdots, \theta_k) = e^{\xi_1\theta_1 + \xi_2\theta_2 + \cdots + \xi_k\theta_k}.g(0) \tag{4.12}$$

As described, a typical human motion is a combination of ordered sequences of all the independent movements performed by each of the human body joint and can be considered as a function with respect to time. Thus, for a continuous representation in time period T where $T \in (1, \cdots, m)$, the compact version of the transformation can be written as:

$$g_K(\theta_1, \theta_2, \cdots \theta_k,) = \left[e^{\xi_1\theta_1 + \xi_2\theta_2 + \cdots + \xi_k\theta_k} \cdot g(0)\right]_{1\times m}$$

All the performed activities captured in the video data are front-to-parallel with the camera plane in this chapter; all the joints therefore have an axis orientation parallel to the Z-axis on the camera plane. Therefore, only half of the human model is employed to construct the QNT as shown in Fig. 4.4.

The base body reference frame F_0 is located at the hip, by employing the normalised qualitative representation and the concept of FQ robot kinematics [29], the product of exponential maps for the arm kinematics chains with respect to base frame $g(0)$ over time T can be obtained as below,

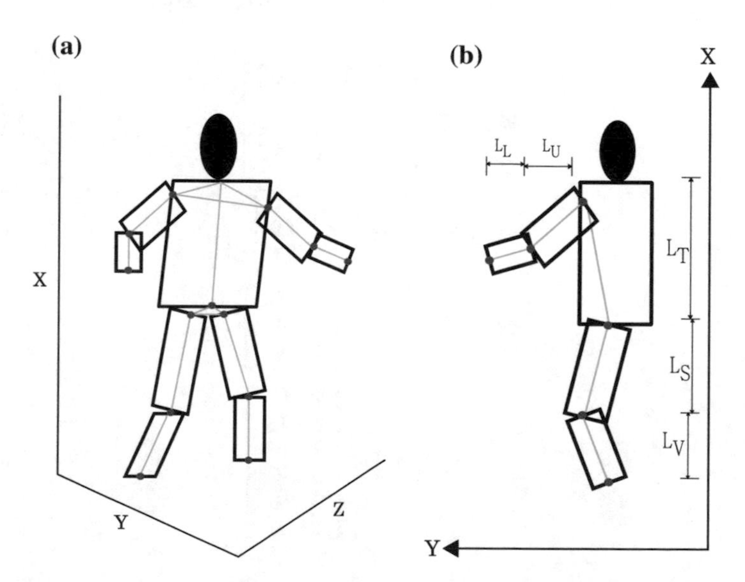

Fig. 4.4 The proposed human model where each segment (i.e., limb) of a human is represented by patches connected by joints. **a** 3D human body **b** 2D human body

$$g_{arm}\left(QS(\theta_1,\theta_2,\theta_3)\right)=\left[e^{\xi_1\theta_1+\xi_2\theta_2+\xi_3\theta_3}\cdot g_{arm}\left(0\right)\right]_{1\times m}$$

where

$$g_{arm}(0)=\begin{bmatrix}\mathbf{I} & \begin{bmatrix}\mathbf{L_T}\\ \mathbf{L_U}+\mathbf{L_L}\\ 0\end{bmatrix}\\ \mathbf{0} & \mathbf{1}\end{bmatrix} \tag{4.13a}$$

$$\omega_{arm}=\left[\,\omega_1\ \omega_2\ \omega_3\,\right]=\begin{bmatrix}\mathbf{0}\ \mathbf{0}\ \mathbf{0}\\ \mathbf{0}\ \mathbf{0}\ \mathbf{0}\\ \mathbf{1}\ \mathbf{1}\ \mathbf{1}\end{bmatrix} \tag{4.13b}$$

$$q_{arm}=\left[\,q_1\ q_2\ q_3\,\right]=\begin{bmatrix}\mathbf{0} & \mathbf{L_T} & \mathbf{L_T}\\ \mathbf{0} & \mathbf{0} & \mathbf{L_U}\\ \mathbf{0} & \mathbf{0} & \mathbf{0}\end{bmatrix} \tag{4.13c}$$

$$\xi_{arm}=\left[\,\xi_1\ \xi_2\ \xi_3\,\right]=\begin{bmatrix}\mathbf{0} & \mathbf{0} & -\mathbf{L_U}\\ \mathbf{0} & -\mathbf{L_T} & -\mathbf{L_T}\\ \mathbf{0} & \mathbf{0} & \mathbf{0}\\ \mathbf{0} & \mathbf{0} & \mathbf{0}\\ \mathbf{0} & \mathbf{0} & \mathbf{0}\\ \mathbf{1} & \mathbf{1} & \mathbf{1}\end{bmatrix} \tag{4.13d}$$

where

$$\mathbf{0}=\left[0\,0\,0\,0\right];\quad \mathbf{1}=\left[1\,1\,0\,0\right]$$

The product of exponential mapping for leg kinematics chains with respect to the same base frame over T is given as below,
where

$$g_{leg}\left(QS(\theta_4,\theta_5)\right)=\left[e^{\xi_4\theta_4+\xi_5\theta_5}\cdot g_{leg}\left(0\right)\right]_{1\times m}$$

$$g_{leg}(0)=\begin{bmatrix}\mathbf{I} & \begin{bmatrix}-\mathbf{L_S}-\mathbf{L_V}\\ \mathbf{0}\\ \mathbf{0}\end{bmatrix}\\ \mathbf{0} & \mathbf{1}\end{bmatrix} \tag{4.14a}$$

$$\omega_{leg}=\left[\,\omega_4\ \omega_5\,\right]=\begin{bmatrix}\mathbf{0}\ \mathbf{0}\\ \mathbf{0}\ \mathbf{0}\\ \mathbf{1}\ \mathbf{1}\end{bmatrix} \tag{4.14b}$$

$$q_{leg}=\left[\,q_4\ q_5\,\right]=\begin{bmatrix}\mathbf{0} & -\mathbf{L_S}\\ \mathbf{0} & \mathbf{0}\\ \mathbf{0} & \mathbf{0}\end{bmatrix} \tag{4.14c}$$

$$\xi_{leg}=\left[\,\xi_4\ \xi_5\,\right]=\begin{bmatrix}\mathbf{0} & \mathbf{0}\\ \mathbf{0} & -\mathbf{L_S}\\ \mathbf{0} & \mathbf{0}\\ \mathbf{0} & \mathbf{0}\\ \mathbf{0} & \mathbf{0}\\ \mathbf{1} & \mathbf{1}\end{bmatrix} \tag{4.14d}$$

An activity is defined as a combination of ordered sequences of all the independent movements in terms of the human body segments [27, 30], hence for any given activity, its corresponding QNT is derived as:

$$QNT = g_{arm} \oplus g_{leg} \tag{4.15}$$

It is evident that the QNT can be represented by strong discriminative features derived from small set of example activities, as is the advantage in terms of real-time performance over a statistical method which requires large training data. The proposed QNT method is a supervised learning method, its training phase algorithm is provided in Appendix B.

4.3 Experiments

In this section, we present the performance of the proposed approach under different conditions such as tracking errors, size of training data and the choice of training data, and comparisons with the Hidden Markov Model (i.e., HMM) and fuzzy hidden Markov model (i.e., FHMM).

4.3.1 Datasets and Preprocessing

We conducted experiments on two public databases: the KTH Database [24] and Weizmann Database [23]. Sample images from all the datasets are shown in Figs. 4.5 and 4.6, in which some activities are somewhat similar in the sense that limbs have similar motion paths and this high degree of similarity makes the discrimination more challenging. In addition, all the actors have different physical characteristics and performs activities differently in both motion styles and speeds.

Three datasets were created for the validation purpose. First, dataset S1: 225 video streams of 3 human motions in 3 planar view scenarios from each of the 25 subjects were selected from the KTH Database. The selected activities are walking, running and jogging. The aim here is to evaluate the efficacy of the QNT in that walking, running and jogging are motions that exhibit similar movements but are dramatically different in motion meaning. Secondly, dataset S2: 55 video streams from the six human motions are employed from the Weizmann Database. The selected activities are bending, walking, jacking, jumping, one hand waving (Wave1) and two hands waving (Wave2). The objective is to test the effectiveness of the proposed approach in distinguishing a wide variety of human motions that are performed by different subjects. Finally, dataset S3: All video streams of S1 and only the walking of S2 are selected. The purpose is to test the generality of the QNT in differentiating the same human motion from different environments.

Fig. 4.5 Example sequences from dataset with four human activities performed by eight different persons. Each activity in the dataset is repeated four times by each person, where for walking, jogging and running, each activity is repeated twice in each directions, i.e. form the left to the right and vice versa. The whole dataset contains $4 \times 8 \times 4 = 128$ sequences and is a subset of a larger database presented in [24]. All sequences were obtained with a stationary camera with frame rates of 25 fps and with sub-sampling of the spatial resolution to 160×120 pixels.

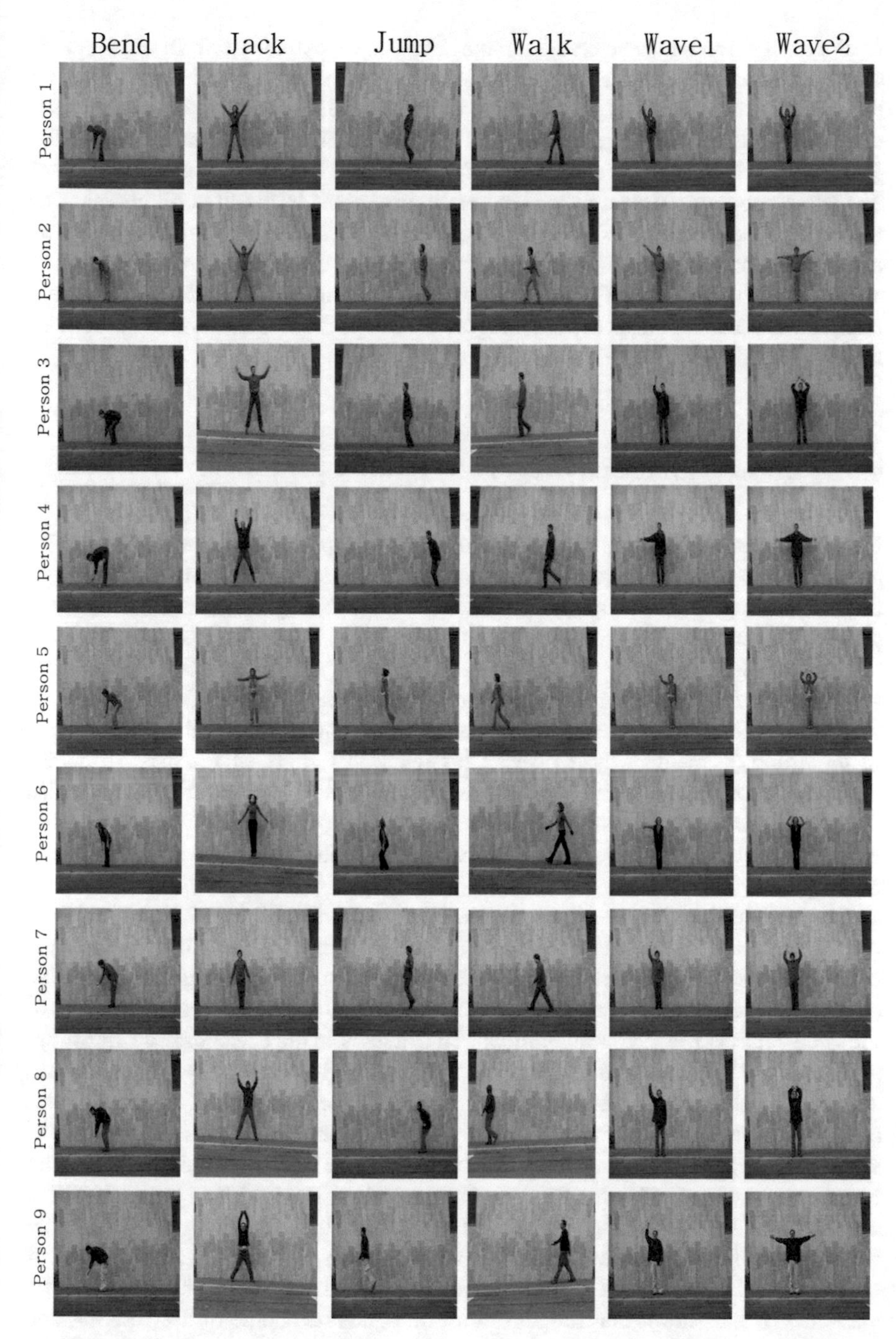

Fig. 4.6 Example sequences from dataset with six human activities performed by nine different persons. The whole dataset contains $6 \times 9 = 54$ sequences and is a subset of a larger database presented in [23].

For each video sequence created in the datasets, we defined a five DOF kinematic human skeleton structure. The motion parameter ϕ_t for the six landmarks points on the proposed human model as in Fig. 4.4 including the reference landmark point are obtained using a visual tracking approach [2]. The number of particles are selected as 400, 800 and 2000 respectively. Further, ϕ_t are mapped into their associated sequence discrete symbolic representation (i.e., qualitative states in the FQ unit circle). The level of resolution in the FQ unit circle is set to $s = 12$ and $r = 16$ for translation and orientation components, respectively. Finally, the QNTs for each human motion are constructed using the FQ robot kinematics algorithm [19] and human motion recognition is carried out afterwards. Figure 4.7 shows the visualisation of the activity model derived from eight different subjects.

4.3.2 Results and Analysis

Human motion recognition aims to recognise the type of activities performed by people in test sequences against results from training sequences. The datasets were divided into training sets and test sets with respect to subjects, such that the same person will not appear in the test and training sequences simultaneously. As the size of the datasets is relatively small, random permutations of the training and testing sets are considered and the recognition rates were averaged. For each of the datasets S1, S2, and S3, we repeated the process 50 times and evaluated the performance of the proposed method, the Correct Classification Rate (CCR) is used to justify the recognition rate. CCR is the ratio of correctly classified number of activities to the total number of the same activity. For activity classification, we adopted the nearest-neighbour classifier where the Euclidean metric was used as the distance measure.

The recognition results for each dataset are shown in Tables 4.3, 4.4 and 4.5, respectively. The conclusions are drawn as follows. First, the percentages of correct classification of the proposed approach are acceptable for all three datasets. The mean of classification accuracy for each dataset is higher than 80%. The recognition rates of the three datasets with different number of particles N in the Condensation algorithm are almost similar. This demonstrated that the proposed data quantisation process has sufficiently relaxed the trade-off between tracking precision and computational cost in the visual tracking algorithm. Secondly, the QNT are informative without necessary data lost, particularly in dataset S1 where the three activities exhibit very similar movement but have substantially different meanings. The QNT only mis-classified a small number of subjects given by the three tested human motions in S1. Actually the mis-classified data are even hardly distinguishable from a human perspective. A confusion matrix in Fig. 4.8 is shown to analyse which activity has been incorrectly classified. It has shown that the algorithm mis-classified some of the walking and running activities. Thirdly, we showed the True Positive Fraction (TPF) and False Positive Fraction (FPF) via the leave-one-out rule in the verification model. This is to test the capability of a pattern classifier to verify whether a new measurement

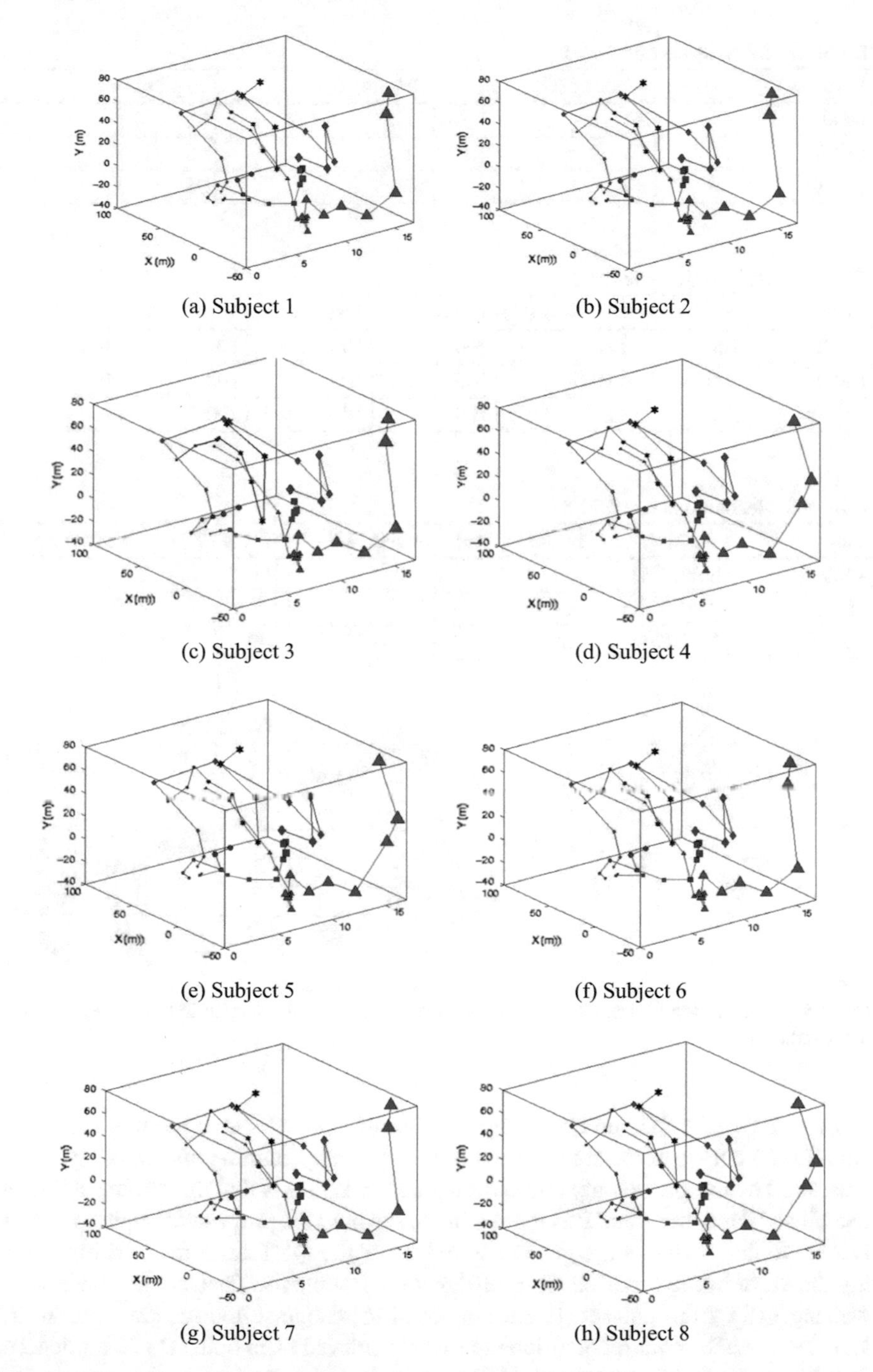

(a) Subject 1 (b) Subject 2

(c) Subject 3 (d) Subject 4

(e) Subject 5 (f) Subject 6

(g) Subject 7 (h) Subject 8

Fig. 4.7 Visualisation of activities manifold: Each of the five activities from eight subjects in quantity space.

Table 4.3 Recognition rate for S1

Particles	Walking (%)	Running (%)	Jogging (%)
(400)	80	81	92
(800)	82	81	92
(2000)	80	81	93

Table 4.4 Recognition rate for S2

Particles	Bending (%)	Walking (%)	Jacking (%)	Jumping (%)	Wave1 (%)	Wave2 (%)
(400)	100	100	100	100	100	100
(800)	100	100	100	100	100	100
(2000)	100	100	100	100	100	100

Table 4.5 Recognition rate for S3

Particles	Walking	Running (%)	Jogging (%)	Wave1 (%)	Wave2 (%)
(400)	86%(86%)	81	92	100	100
(800)	86%(86%)	81	92	100	100
(2000)	86%(86%)	81	92	100	100

KTH

	Jog	Walk	Run	Wave
Jog	88	10	02	0
Walk	18	76	06	0
Run	0	23	77	0
Wave	0	0	0	100

Weizmann

	Bend	Jack	Jump	Walk	Wave1	Wave2
Bend	100	0	0	0	0	0
Jack	0	100	0	0	0	0
Jump	0	0	100	0	0	0
Walk	0	0	0	100	0	0
Wave1	0	0	0	0	100	0
Wave2	0	0	0	0	0	100

Fig. 4.8 Confusion matrices: Comparison of activity classification results in the KTH and Weizmann database

belongs to certain claimed class. Figure 4.9 shows the receiver operating characteristic (ROC) curves of the three activities (i.e., jogging, running and walking) in the datasets. The reason we only chose these three activities is that these three activities exhibit similar actions and it is interesting to see how the proposed templates coped. Lastly, Table 4.5 shows the generality results of the QNT in terms of differentiating the same human motion from different environments. That is, the constructed walking QNT from dataset S1 are employed to recognise the walking data in S2 and vice versa. Similar recognition rates were achieved from both set of experiences. This illustrates that the proposed QNT are generic and insensitive to different motion styles, speeds across different human anatomy and environments.

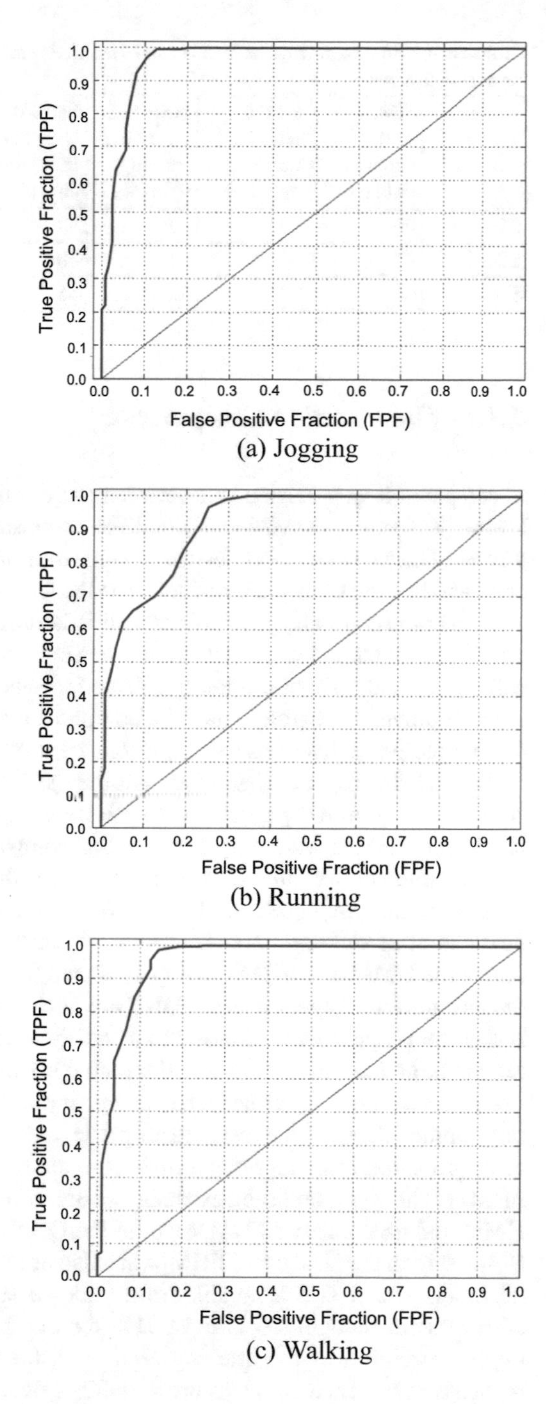

Fig. 4.9 ROC Curves for three activities which exhibit similar actions. The dotted diagonal line shows the random prediction. **a** ROC Curve for Kicking. The area under the ROC curve is 0.966026. **b** ROC Curve for Running. The area under the ROC curve is 0.923142. **c** ROC Curve for Walking. The area under the ROC curve is 0.957291

Table 4.6 Comparison with the HMM and FHMM. Average classification rate employing different training data sizes

	HMM with 1% training data (%)	HMM with 20% training data (%)	HMM with 50% training data (%)	FHMM with 1% training data (%)	FHMM with 20% training data (%)	FHMM with 50% training data (%)	QNT with 1% training data (%)
S1	54	75	77	58	77	78	85
S2	62	88	91	70	91	100	100
S3	68	72	72	75	78	82	88

4.3.3 Quantitative Comparison

A comparison was carried out between the dominant recognition methods (i.e., a HMM [31] and a FHMM [32]) and the proposed method. A four state left-right discrete HMM is selected for the comparison and the pre-processing steps were conducted as given in [33], and the FHMM was conducted as to [34]. The number of states is empirically determined and it is observed that an increase to a larger number of states did not result in any performance gains. Each model was trained using 1, 20 and 50% of randomly selected instances of human motions and the rest were employed as testing data. The comparison results are provided in Tables 4.6–4.8, respectively. It is observed that on the three data sets, the QNT outperforms both the HMM and FHMM. Note that the QNT employed in this experiment were constructed from 400 particles with 1% training data while the best models in the HMM and FHMM were employed for this comparison.

The simulation results have confirmed that the effectiveness of the HMM and FHMM models are significantly dependant on the accuracy of the training data and the amount of training data employed. For instance in Table 4.6, the classification rate of the HMM and FHMM using 1, 20 and 50% training data had a big impact of the recognition rate whereas the QNT are fairly consistent. One of the main reasons is that the proposed solution is not a statistical learning method hence it does not require large training data. Instead strong discriminative features are required from example activities. A further analysis by employing one subject sequentially as the only training data and the rest as testing data, results in Table 4.7 shows that the choice of the selection has a significant influence on the recognition rate in the HMM and FHMM. The worst and the best achieved are differed by approximately 50% for the HMM and 48% for the FHMM while the QNT only differ by approximately 2%. Besides, both the HMM and FHMM are also notoriously sensitive to the precision of the training data. This is notable from Table 4.8 as $N = 400$, the average percentage of successful recognition is only 51% for the three datasets. However at a much higher resolution of the tracking algorithm, the average percentage of successful recognition rate increases to more than 80%. Bear in mind that, the high recognition rate is achieved at the cost of significant computational power.

Table 4.7 Comparison with the HMM and FHMM. Average classification rate with using each subject as training data once, and tested against the remaining. The worst achieved are in brackets () and the best in square []

	HMM	FHMM	QNT
S1	54% (38%)[88%]	62% (45%)[88%]	85% (84%)[86%]
S2	75% (64%)[98%]	84% (72%)[100%]	100% (100%)[100%]
S3	67% (32%)[80%]	69% (36%)[84%]	88% (88%)[88%]

Table 4.8 Comparison with the HMM and FHMM. Average classification rate employing different tracking resolutions

	HMM with featured data 400 particles (%)	HMM with featured data 800 particles (%)	HMM with featured data 2000 particles (%)	FHMM featured data 400 particles (%)	FHMM featured data 800 particles (%)	FHMM featured data 2000 particles (%)	QNT with featured data 400 particles (%)
S1	38	78	82	39	81	83	85
S2	61	80	88	72	93	100	100
S3	54	83	88	51	84	88	88

4.3.4 Complexity Analysis

The Condensation algorithm is an approximation technique by representing the posterior density as a set of samples of the state space with associated likelihood weights $\omega_t^i, i \in \{1, \cdots, N\}$. The sample set approximation of the current posterior density $p(X_t|O_{1:t})$ can be obtained via:

$$p(X_t|O_{1:t}) \approx \sum_{i=1}^{N} \omega_t^i \delta(X_t - x_t^i) \tag{4.16}$$

where $\delta(X_t - x_t^i)$ denotes the Dirac delta function. Therefore, the complexity of a standard Condensation algorithm has a computational complexity of $\mathcal{O}(2N)$ where N is the number of particles.

In the meantime, the forward, backward and Viterbi algorithms are the central element of HMM training and testing. All have the same computational complexity [35], the equation for the standard forward algorithm is:

$$\alpha_j(t) = \left(\sum_i \alpha_i(t-1)a_{ij}\right) b_j(o_t), \forall j, t \tag{4.17}$$

Table 4.9 Comparison of algorithms computational complexities

Algorithm	Complexity
Particle filtering + Standard HMM	$\mathcal{O}(2N) + \mathcal{O}(M^2T)$
Particle filtering + FHMM	$\mathcal{O}(2N) + \mathcal{O}(M^2T)$
QNT	$\mathcal{O}(2N)$

Table 4.10 Comparison of algorithm computational complexities vs. Recognition rate. The best recognition rate achieved by each system is selected

	Algorithm	Complexity	Recognition rate (%)
S1	Particle filtering + Standard HMM	$\mathcal{O}(4048)$ $N{=}2000, M{=}4, T{=}3$	82
	Particle filtering + FHMM	$\mathcal{O}(4048)$ $N{=}2000, M{=}4, T{=}3$	83
	QNT	$\mathcal{O}(800)$ $N{=}400$	85
S2	Particle filtering + Standard HMM	$\mathcal{O}(4096)$ $N{=}2000, M{=}4, T{=}6$	88
	Particle filtering + FHMM	$\mathcal{O}(4096)$ $N{=}2000, M{=}4, T{=}6$	100
	QNT	$\mathcal{O}(800)$ $N{=}400$	100
S3	Particle filtering + Standard HMM	$\mathcal{O}(4048)$ $N{=}2000, M{=}4, T{=}3$	88
	Particle filtering + FHMM	$\mathcal{O}(4048)$ $N{=}2000, M{=}4, T{=}3$	88
	QNT	$\mathcal{O}(800)$ $N{=}400$	88

The complexity of the standard algorithm is normally given as $\mathcal{O}(M^2T)$ where M is total number of unique states in the HMM; T is the number of observations.

In Table 4.9, a comparison of computational complexity of the HMM, FHMM, and QNT method is shown. The QNT method outperformed both the HMM and FHMM in two folds. First of all, the proposed solution do not need to employ a large amount of particles, N to perform tracking as the proposed data quantisation process can account for tracking error as shown in Table 4.2. Secondly, the proposed method is not a statistical learning method, hence the algorithm does not need to loop until convergence. Furthermore, Table 4.10 shows a comparison of the recognition rate of the HMM, FHMM and our proposed method against the computational complexity. It is notable that in order to achieve the recognition rate with our proposed method on each data set S1, S2 and S3, the HMM and FHMM computational complexity are at least 5 times higher than our proposed method.

4.4 Conclusion

In this chapter, a FQ approach has been proposed to solve real-time vision-based human motion analysis. It has integrated human motion tracking and recognition algorithms with FQ robot kinematics. The simulation results have shown that the proposed method outperforms the dominant recognition methods, i.e., the hidden Markov models and fuzzy hidden Markov models. It is demonstrated that the proposed method can be applied to recognise a range of human motions for a real-time human motion recognition system from the computation perspective. The proposed algorithm has achieved average computational time at about 0.010591 seconds per individual human motion recognition, on the other hand, the frame rate of a CCTV image sequence is usually reduced to 10–15 frames per second when in use. That the proposed recognition method is at least 10 times faster than the frame rate of CCTV image sequences is effective enough to be implemented in a real-time vision-based human motion analysis. It is clearly evident that the proposed method can meet real-time requirement for human motion recognition. However, there are still some problems which keep the proposed method from real-world application. First, initial parameters of an image sequence are manually edited in human motion tracking. It has been confirmed that it is one of most challenging problems for the computer vision community. Secondly, issues related to object tracking algorithms such as partial occlusion, clutter, environmental lighting changes still remain as open problems for further applying the proposed method to real-world application. Additionally we are trying to overcome the above image-processing problems by understanding scenario context information without accurate human motion extraction. For instance, we are currently developing multiple qualitative normalised templates from different viewpoints to construct a pseudo 3D motion template, it allows reasoning the human motion analysis under environmental uncertainty such as lighting and clutter background, which relax the requirements on accurate initial image parameters and uncertainty caused by surroundings. We aim at developing a compositional model which uses predominantly knowledge-based techniques to translate among high-level human motion scenarios, Gaussian mixture models and filtered numerical data of human motions [36].

References

1. Chern Hong Lim, Ekta Vats, and Chee Seng Chan. Fuzzy human motion analysis: A review. *Pattern Recognition*, 48(5):1773–1796, 2015.
2. M. Isard and A. Blake. CONDENSATION conditional density propagation for visual tracking. *International Journal of Computer Vision*, 29(1):5–28, 1998.
3. Z. Jing and S. Sclaroff. Segmenting foreground objects from a dynamic textured background via a robust kalman filter. In *Proceedings of the Ninth IEEE International Conference on Computer Vision*, page 44, Washington, DC, USA, 2003.

4. Li Yuan, Ai Haizhou, T. Yamashita, Lao Shihong, and M. Kawade. Tracking in low frame rate video: A cascade particle filter with discriminative observers of different lifespans. In *Proceedings of the 2007 IEEE Computer Society Conference on Computer Vision and Pattern Recognition*, pages 1–8, Washington, DC, USA, 2007. IEEE Computer Society.
5. H. Buxton. Learning and understanding dyanmic scene activity. *Image and Vision Computing*, 21(1):125–136, 2003.
6. A.F. Bobick and J. Davis. Real-time recognition of activity using temporal templates. In *Proceedings of IEEE CS Workshop on Applications of Computer Vision*, pages 39–42, 1996.
7. G. Bradski and J. Davis. Motion segmentation and pose recognition with motion history gradients. *Machine Vision and Applications*, 13(3):174–184, July 2002.
8. Z. Chen and H.J. Lee. Knowledge-guided visual perception of 3-d human gait from a single image sequence. *IEEE Transactions on Systems, Man and Cybernetics*, 22:336–342, 1992.
9. S.X. Ju, M.J. Black, and Y. Yacoob. Cardboard people: A parameterized model of articulated image motion. In *Proceedings of the 2nd International Conference on Automatic Face and Gesture Recognition*, pages 38–44, Killington, Vermont, USA, 1996.
10. I.A. Karaulova, P.M. Hall, and A.D. Marshall. A hierachical model of dynamics for tracking people with a single video camera. In *Proceedings of the British Machine Vision Conference*, 2000.
11. C.S. Chan, H. Liu, and D.J. Brown. Recognition of human motion from qualitative normalised templates. *Journal of Intelligent and Robotic Systems*, 48(1):79–95, January 2007.
12. L. Wang, T. Tan, H. Ning, and W. Hu. Silhouette analysis-based gait recognition for human identification. *IEEE Transactions on Pattern Analysis and Machine Intelligence*, 25(12):1505–1518, 2003.
13. A. Sundaresan, R. Chellappa, and A. Roy Chowdhury. Multiple view tracking of humans modelled by kinematic chains. In *Proceedings of the International Conference on Image Processing*, volume 2, pages 1009–1012, Singapore, 2004.
14. H. Sidenbladh, M.J. Black, and D.J. Fleet. Stochastic tracking of 3D human figures using 2D image motion. *Proceedings of the European Conference on Computer Vision-Part II*, pages 702–718, 2000.
15. F. Lv and R. Nevatia. Recognition and segmentation of 3-D human action using HMM and multi-class AdaBoost. *Proceedings of European Conference on Computer Vision*, 4:359–372, 2006.
16. M. Leo, T. D'Orazio, I. Gnoni, P. Spagnolo, and A. Distante. Complex human activity recognition for monitoring wide outdoor environments. In *Proceedings of the International Conference on Pattern Recognition*, volume 4, pages 913–916. IEEE, 2004.
17. C. Bregler. Learning and recognizing human dynamics in video sequences. In *Proceedings of the IEEE Computer Society Conference on Computer Vision and Pattern Recognition*, page 568, Washington, DC, USA, 1997.
18. VI Pavlovic, R. Sharma, and TS Huang. Visual interpretation of hand gestures for human-computerinteraction: a review. *IEEE Transactions on Pattern Analysis and Machine Intelligence*, 19(7):677–695, 1997.
19. H. Liu, D.J. Brown, and G.M. Coghill. Fuzzy qualitative robot kinematics. *IEEE Transactions on Fuzzy Systems*, 16(3):802–822, 2008.
20. L. Wang, W. Hu, and T. Tan. Recent developments in human motion analysis. *IEEE Transactions on Systems, Man, and Cybernetics, Part C*, 34(3):334–352, 2004.
21. J.K. Aggarwal and Q. Cai. Human motion analysis: A review. *Computer Vision and Image Understanding*, 73(3):428–440, 1999.
22. T.B. Moeslund and E. Granum. A survey of computer vision-based human motion capture. *Computer Vision and Image Understanding*, 81(3):231–268, 2001.
23. M. Blank, L. Gorelick, E. Shechtman, M. Irani, and R. Basri. Action as space-time shapes. In *Proceedings of IEEE International Conference on Computer Vision*, volume 2, pages 1395–1402, Beijing, China, 2005.
24. C. Schuldt, I. Laptev, and B. Caputo. Recognizing human actions: A local svm approach. In *Proceedings of the International Conference on Pattern Recognition*, volume 3, pages 32–36, Hong Kong, 2004.

25. H. Liu and G.M. Coghill. Fuzzy qualitative trigonometry. *Proc. IEEE International Conference on Systems, Man and Cybernetics, Hawaii, USA.*, 2005.
26. Q. Shen and R. Leitch. Fuzzy qualitative simulation. *IEEE Transactions on Systems, Man, and Cybernetics*, 23(4):1038–1061, 1993.
27. A.F. Bobick and A. Wilson. A state-based technique for the summarization and recognition of gesture. In *Proceeding of the International Conference on Computer Vision*, pages 382–388, 1995.
28. Richard M. Murray, Shankar Shastry, Zexiang Li, and S. Shankar Sastry. *A Mathematical Introduction to Robotic Manipulation*. CRC Press, 1994.
29. H. Liu and D.J. Brown. An extension to fuzzy qualitative trigonometry and its application to robot kinematics. In *Proceedings of the IEEE International Conference on Fuzzy Systems*, pages 1111–1118, 2006.
30. G. Johansson. Visual motion perception. *Scientific American*, 232(6):76–88, 1975.
31. L.R. Rabiner. A tutorial on hidden markov models and selected applications in speech recognition. *Proceedings of the IEEE*, 77(2):257–284, February 1989.
32. M.A. Mohamed and P. Gader. Generalized hidden markov models. i. theoretical frameworks. *IEEE Transaction on Fuzzy Systems*, 8(1):67–81, 2000.
33. H. Qiang and C. Debrunner. Individual recognition from periodic activty using hidden markov models. In *Proceedings of the Workshop on Human Motion*, pages 47–52, 2000.
34. M.A. Mohamed and P. Gader. Generalized hidden markov models. ii. application to handwritten word recognition. *IEEE Transaction on Fuzzy Systems*, 8(1):82–94, 2000.
35. M.T. Johnson. Capacity and complexity of hmm duration modelling techniques. *IEEE Transactions on Signal Processing Letters*, 12(5):407–410, 2005.
36. H. Liu. A fuzzy qualitative framework for connecting robot qualitative and quantitative representations. *IEEE Transactions on Fuzzy Systems*, 16(6):1522–1530, 2008.

Chapter 5
Fuzzy Gaussian Mixture Models

5.1 Introduction

As one of the most statistically mature methods for clustering [1–7], Gaussian Mixture Models (GMMs) are used intensively in object tracking [8–10], background subtraction [11–13], feature selection [14, 15], signal analysis [16, 17] and learning and modelling [18–23]. Various kinds of GMMs based methods are developed for specific applications, such as Adapted GMMs [24], which are used for dealing with undesired effects of variations in speech characteristics, Mahalanobis Distance based GMMs [25], which are capable of splitting one component into two new components, and Wrapped Gaussian Mixture Models [26] using an expectation-maximisation algorithm suitable for circular vector data to model dispersion phases. However, more components are required when fitting the datasets with non-linear manifolds because of the intrinsic linearity of Gaussian model which leads to relative large fitting error. To solve this problem and approximate datasets with curve manifolds better, Zhang et al. [27] proposed active curve axis Gaussian Mixture Models (AcaGMMs) which are non-linear probability models. PCA and least-squares fitting methods are used to 'bend' AcaGMMs in the principal plane and points are considered by handling the projection points on the principal axis.

Fuzzy C-means (FCMs), also known as Fuzzy ISODATA, was developed by Dunn in 1973 [28] and improved by Bezdek in 1981 [29]. It is a popular and effective clustering method which employs fuzzy partitioning such that a data point can belong to all groups with different membership grades between 0 and 1. It employs a weighting exponent m on each fuzzy membership and distances between points and centres. The effects of the weighting exponent are discussed that optimal m may result in better performance or fast convergence in [30, 31], and approaches of determining the weighting exponent have also been presented in [30, 32, 33]. The algorithmic frameworks of FCM and GMMs are closely related [34, 35]. Based on FCM, Gustafson et al. [36] defined fuzzy covariance matrices of clusters which means that different clusters in the same dataset may have different geometric shapes. To define these different geometric shapes of clusters, Tran et al. [37] further made a modification of GMMs for speaker recognition, which refined the distances in the FCM functions as the negative of logarithms of density functions. Therefore, the relationship between the membership and distance is transferred from exponential relationship

H. Liu et al., *Human Motion Sensing and Recognition*,
Studies in Computational Intelligence 675, DOI 10.1007/978-3-662-53692-6_5

to linear relationship, which however misuses exponential distance parameter to formulate Gaussian density function. Hathaway [38] gave a general interpretation that the EM algorithm of GMMs is a penalised version of the hard means clustering algorithm. Ichihashi et al. [39] proposed a modified version of FCM with regularization by Kullaback-Leibler (KL) information in the fuzzy objective function. The relation and comparison between KLFCM and GMMs are discussed in [39, 40]. As a KLFCM algorithm variant, a generic methodology for finite mixture model fitting under a fuzzy clustering principle is proposed and applied to three types of finite mixture models in [41] to improve their performances.

In order to integrate the advantages of GMMs and FCM in the mathematical modelling, conventional GMMs are firstly generalised in such a way that generalised Gaussian model is equipped with non-linearity and a better performance, and then Fuzzy Gaussian Mixture Models (FGMMs) are proposed based on the generalised GMMs for a much faster convergence process which makes it more practical. The dissimilarity function in FCM maintaining the exponential relationship between membership and distance is refined for FGMMs with a degree of fuzziness in terms of the membership grades. Therefore, FGMMs not only possess non-linearity but also have a computationally inexpensive convergence process, which is testified by experiments comparing FGMMs with conventional GMMs and generalised GMMs in terms of fitting degree and convergence speed. The FGMMs are different from the type-2 FGMMs, which uses interval likelihoods to describe the observation uncertainty [42]. The type-2 FGMM is focused on the role of footprint of uncertainty in pattern classification to handle GMMs uncertain mean vector or uncertain covariance matrix, while the FGMM in this chapter focuses on the precise parameter estimation of GMMs based on the modified FCM algorithm. The aim of this chapter is to improve the conventional GMMs in terms of both performance and efficiency. This chapter is organised as follows: Sect. 5.2 introduces generalised GMMs; Sect. 5.3 proposes two types of FGMMs based on the generalised GMMs; Sect. 5.4 provides comparison results of the proposed FGMMs with conventional GMMs and generalised GMMs respectively on various kinds of datasets; Finally the chapter is concluded with remarks and future work.

5.2 Generalised GMMs

In this section, we extend the conventional GMMs into a generalised version which enables the GMMs to have capability of modelling curve datasets. Following a brief review of conventional GMMs, an EM algorithm is proposed for the generalised GMMs.

5.2.1 *Conventional GMMs*

The probability density function for a Gaussian distribution is given by the formula:

$$p(x|\theta) = \frac{1}{(2\pi)^{\frac{d}{2}}\sqrt{|\Sigma|}} \exp\left(-\frac{(x-\mu)^T \Sigma^{-1}(x-\mu)}{2}\right) \tag{5.1}$$

where the set of parameters has $\theta = (\mu, \Sigma)$, μ is the mean, Σ is the covariance matrix of the Gaussian, d is the dimension of vector x, and exp denotes the exponential function.

Let $\mathscr{X} = \{x_1, ..., x_n\}$ be a d-dimensional observed dataset of n vectors. If the distribution of $\mathscr{X}$ can be modelled by a mixture of k Gaussians, the density of each vector is:

$$p(x_t|\Theta) = \sum_{i=1}^{k} \alpha_i p_i(x_t|\theta_i) \tag{5.2}$$

where the parameters are $\Theta = (\alpha_1, \ldots, \alpha_k, \theta_1, \ldots, \theta_k)$ and $(\alpha_1, \ldots, \alpha_k)$ are the k mixing coefficients of the k mixed components such that $\sum_{i=1}^{k} \alpha_i = 1$; each p_i is a density function parameterised by θ_i. The resulting density for the samples is

$$p(\mathscr{X}|\Theta) = \prod_{t=1}^{n} p(x_t|\Theta) = \mathscr{L}(\Theta|\mathscr{X}) \tag{5.3}$$

The function $\mathscr{L}(\Theta|\mathscr{X})$ is called the likelihood of the parameters given the data, or the likelihood function. The likelihood is considered as a function of the parameters Θ where the data $\mathscr{X}$ is fixed. In the maximum likelihood problem, the objective is to estimate the parameters set Θ that maximises $\mathscr{L}$. That is to find Θ^* where

$$\Theta^* = \arg\max_{\Theta} \mathscr{L}(\Theta|\mathscr{X}) \tag{5.4}$$

Usually, the $\log(\mathscr{L}(\Theta|\mathscr{X}))$ is maximised instead because it is analytically easier. The log-likelihood expression is given by:

$$\log(\mathscr{L}(\Theta|\mathscr{X})) = \log\left(\prod_{t=1}^{n} p(x_t|\Theta)\right) = \sum_{t=1}^{n} \log\left(\sum_{i=1}^{k} \alpha_i p_i(x_t|\theta_i)\right) \tag{5.5}$$

Directly maximising the log-likelihood is difficult, hence an auxiliary objective function Q is taken into account:

$$Q = \sum_{t=1}^{n}\sum_{i=1}^{k} w_{it} \log[\alpha_i p_i(x_t|\theta_i)] \tag{5.6}$$

where w_{it} is a posteriori probability for individual class i, $i = 1, ..., k$, and it satisfies

$$w_{it} = \frac{\alpha_i p_i(x_t|\theta_i)}{\sum_{s=1}^{k} \alpha_s p_s(x_t|\theta_s)} \tag{5.7}$$

and

$$\sum_{i=1}^{k} w_{it} = 1 \tag{5.8}$$

Maximising Eq. 5.6 guarantees that $p(\mathscr{X}|\Theta)$ is maximised if it is performed by an EM algorithm (e.g., [43, 44]). The iteration of an EM algorithm estimating the new parameters in terms of the old parameters is given as follows:

- E-step: compute "expected" classes of all data points for each class using Eq. 5.7.
- M-step: compute maximum likelihood given the data's class membership distributions according to Eqs. 5.9–5.11.

$$\alpha_i^{new} = \frac{1}{n} \sum_{t=1}^{n} w_{it} \tag{5.9}$$

$$\mu_i^{new} = \frac{\sum_{t=1}^{n} w_{it} x_t}{\sum_{t=1}^{n} w_{it}} \tag{5.10}$$

$$\Sigma_i^{new} = \frac{\sum_{t=1}^{n} w_{it}(x_t - \mu_i^{new})(x_t - \mu_i^{new})^T}{\sum_{t=1}^{n} w_{it}} \tag{5.11}$$

When training GMMs, k-means is employed for initialisation before EM starts. The iteration of EM algorithm stops when the change value of log-likelihood is below a preset threshold.

5.2.2 *Generalized Gaussian Models*

The conventional Gaussian model has intrinsic linearity as its axes are all beelines, so more components are needed when fitting datasets with non-linear manifolds. Active curve axis Gaussian model (AcaG) has bent principal axis, which makes it powerful in modelling curve datasets [27].

In this chapter, the generalised Gaussian model is defined as the model including two modalities: one is the conventional Gaussian model with linear axes and the other is bent Gaussian or AcaG model with curve principal axis. Let $\mathscr{X} = \{x_1, \ldots, x_n\}$ be a d-dimensional observed dataset of n vectors. The distribution of $\mathscr{X}$ is based on one Gaussian or bent Gaussian.

First, the samples are mapped to the coordinate system determined by PCA:

$$y_i = Q \times (x_i - T) \tag{5.12}$$

where T and Q denote the translation vector and rotation matrix of PCA individually. Let $\mathscr{Y} = \{y_1, \ldots, y_n\}$ be the transferred coordinates of $\mathscr{X}$ by PCA. The d dimensions of $\mathscr{Y}$ are then denoted as $v = [v_1, \ldots, v_d]^T$.

Then least-squares fitting method [45] is used to fit the coordinates $[v_1, v_2]^T$, which are in $\mathscr{Y}$'s principal plane, with the standard curve which is defined as

$$y = ax^2 + b \tag{5.13}$$

The quadratic term coefficient a indicates the bend degree of the curve. The bigger a is, the more the principal axis bends. b is the offset of the curve. After fitted by the least-squares fitting method, the parameters a and b will be determined. According to the value of parameter a, the principal axis for generalized Gaussian model is defined as

$$y = \begin{cases} b & (|a| < \varepsilon) \\ ax^2 + b & (|a| \geq \varepsilon) \end{cases} \tag{5.14}$$

If $|a| < \varepsilon$ where ε is a preset small positive real number, the principal axis can be considered that $y = b$ which means the principal axis is beeline, and the generalised Gaussian model retrogresses to be conventional Gaussian model. If $|a| \geq \varepsilon$, the principal axis is bent to be $y = ax^2 + b$ and generalised Gaussian model with the bent axis is AcaG model. An example of generalized Gaussian model is given on the left of Fig. 5.1.

When the principal axis is beeline (i.e., $|a| < \varepsilon$), the probability density function of point x_t would be $p(x_t|\theta)$ in Eq. 5.2.

When the principal axis is bent (i.e., $|a| \geq \varepsilon$), the projective points of one sample can be thought as the points on the principal axis who have the local minimal distances with the sample. The distances between sample and points along the principal axis have zero derivatives at the sample's projective points. Given any sample $[v_{1t}, v_{2t}]^T$

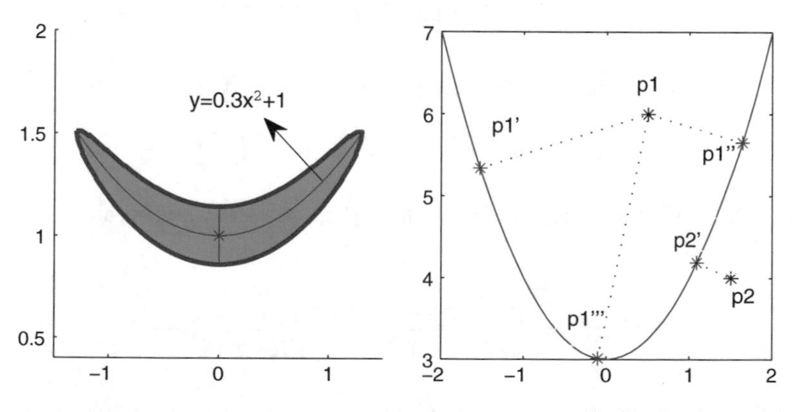

Fig. 5.1 Generalized Gaussian Model (*left*) and projection points on the principal curve axis (*right*)

which is the transferred coordinates from x_t, its distance to a point $z = [z_1, z_2]^T$ on the standard curve satisfies

$$d^2 = (v_{1t} - z_1)^2 + (v_{2t} - z_2)^2$$

Using the Lagrange method, the partial derivatives to z_1 and z_2 of $d^2 - \lambda(z_2 - az_1^2 - b)$ are equal to zero,

$$(z_1 - v_{1t}) - \lambda a z_1 = 0$$

$$2(z_2 - v_{2t}) - \lambda = 0$$

Eliminate λ,

$$z_1 - v_{1t} = 2az_1(z_2 - v_{2t}) \tag{5.15}$$

with the standard curve equation $z_2 = az_1^2 + b$, we can solve all the real roots of this equation, which are the projective points of sample $[v_{1t}, v_{2t}]$. In the principal plane, one point may have several projection points on the principal curve axis, because bending the principal axis may cause one point locates in several different normal directions of the curve axis as shown on the right of Fig. 5.1. Supposing $z = [z_{1j}, z_{2j}]^T$ is the jth projective point of the sample $[v_{1t}, v_{2t}]$, the arc length of z is formulated as

$$l_j(v_{1t}) = \int_{(0,b)}^{(z_{1j}, z_{2j})} \sqrt{(dz_{1j})^2 + (dz_{2j})^2}$$

z is on the standard curve, so it satisfies $z_2 = az_1^2 + b$. So we have

$$\begin{aligned} l_j(v_{1t}) &= \tfrac{1}{2} z_{1j} \sqrt{1 + 4a^2 z_{1j}^2} \\ &+ \tfrac{1}{4|a|} \ln\left(2|a|z_{1j} + \sqrt{1 + 4a^2 z_{1j}^2}\right) \end{aligned} \tag{5.16}$$

The distance between the sample $[v_{1t}, v_{2t}]^T$ and its projective point z is

$$l_j(v_{2t}) = \sqrt{(v_{1t} - z_{1j})^2 + (v_{2t} - z_{2j})^2} \tag{5.17}$$

Then the probability density function of point x_t is the sum probability of its projective points:

$$p(x_t|\theta) = \sum_{j=1}^{J_t} \prod_{s=1}^{2} \frac{\exp(-l_j^2(v_{st})/2\Sigma_s)}{\sqrt{2\pi|\Sigma_s|}} \prod_{s=3}^{d} \frac{\exp(-v_{st}^2/2\Sigma_s)}{\sqrt{2\pi|\Sigma_s|}} \tag{5.18}$$

where the parameter set $\theta = (\mu, \Sigma, C, Q, T)$; C denotes the control parameters of a standard curve $f : y = ax^2 + b$; J_t is the number of the projective points of sample x_t.

5.2.3 EM Algorithm for Generalized GMMs

Generalised GMMs combine conventional GMMs with AcaGMMs, both of whom employ EM algorithm for parameters estimation. With the renewed probability density function 5.18, the iteration of EM algorithm for generalised GMMs is presented:

- E-step: compute "expected" classes of all data points for each class. If $|a| < \varepsilon$, $p(x_t|\theta_i)$ is computed by Eq. 5.1; otherwise, Eq. 5.18 is employed. Then w_{it} is calculated by Eq. 5.7.
- M-step: compute maximum likelihood given the data's class membership distributions. α_i^{new} is estimated from Eq. 5.9, then the following are obtained,

$$\mathscr{X}_i' = [w_{i1}, \ldots, w_{in}] \cdot [x_1, \ldots, x_n] \tag{5.19}$$

$$(C_i^{new}, T_i^{new}, Q_i^{new}) = LSFM(PCA(\mathscr{X}_i')) \tag{5.20}$$

$PCA()$ is principal component analysis function for estimating the translation matrix T_i^{new} and rotation matrix U_i^{new}. $LSFM()$ is least-squares fitting method for estimating control parameters $C_i^{new} = (a, c)$ which shapes the curve axis with standard curve $y = ax^2 + b$; If $|a| < \varepsilon$; μ_i^{new} and Σ_i^{new} are assessed by Eqs. 5.10 and 5.11; otherwise, they can be computed by the following equations.

$$\mu_i^{new} = \frac{\sum_{t=1}^{n} w_{it} x_t}{\sum_{t=1}^{n} w_{it}} + (Q_i^{new})^{-1} \underbrace{[0, b, 0, \ldots, 0]^T}_{d} + T_i^{new} \tag{5.21}$$

$$\bar{L}_{te}^{(i)} = \frac{\sum_{j=1}^{J_t^{(i)}} p\left(l_j^{(i)}(v_{et})|N(0, \Sigma_{ie}^{old})\right) l_j^{(i)}(v_{et})}{\sum_{j=1}^{J_t^{(i)}} p\left(l_j^{(i)}(v_{et})|N(0, \Sigma_{ie}^{old})\right)} \quad (e = 1, 2) \tag{5.22}$$

$$\Sigma_{ie}^{new} = \frac{\sum_{t=1}^{n} w_{it} \bar{L}_{te}^{(i)}}{\sum_{t=1}^{n} w_{it}} \quad (e = 1, 2) \tag{5.23}$$

$$\Sigma_{i(3-d)}^{new} = \frac{\sum_{t=1}^{n} w_{it} (x_t - \mu_i^{new})_{(3-d)} (x_t - \mu_i^{new})_{(3-d)}^T}{\sum_{t=1}^{n} w_{it}} \tag{5.24}$$

If that sample point x_t has $J_t^{(i)}$ projective points on the ith component, the arc length of the jth projective point is $l_j^{(i)}(v_{1t})$ and the distance between jth projective point and x_t is $l_j^{(i)}(v_{2t})$. $\bar{L}_{t1}^{(i)}$ is the average arc length of x_t's projective points on the ith component, and $\bar{L}_{t2}^{(i)}$ is the average distance between x_t and its projective points on

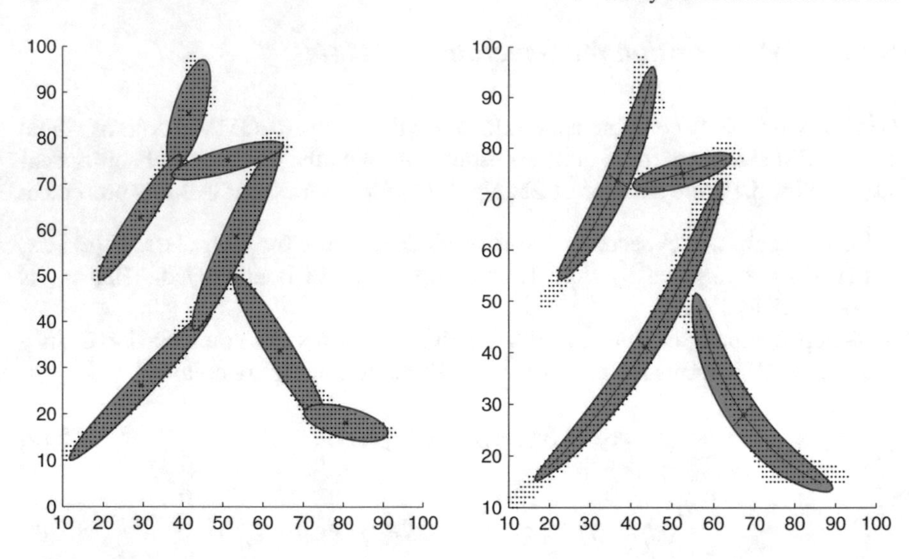

Fig. 5.2 Fitting Chinese character letter using conventional GMMs (*left*) with 7 components and generalized GMMs (*right*) with 4 components

the ith component. $p(x|N(\mu, \Sigma))$ is the probability density of x in the Gaussian with μ and Σ. Σ_{i1} and Σ_{i2} denote the two covariances in the ith principal axis, and $\Sigma_{i(3-d)}$ is the covariance in the $d - 3$ dimension space orthogonal to the principal plane.

When $|a|$ is very small in generalised GMMs, we turn AcaGMM to be GMM so that μ_i^{new} and Σ_i^{new} will be assessed only by Eqs. 5.10 and 5.11 instead of Eqs. 5.21–5.24, which would save the computational cost of AcaGMM. The switching Eq. 14 does not play a role in improving the fitting degree of AcaGMM, but it indeed saves unnecessary steps to estimate the new centres and covariances when $|a| < \varepsilon$. So that the proposed generalised GMMs are faster than the AcaGMMs especially when the principal axis of any component has a small degree of curvature.

Like conventional GMMs, the generalised GMMs are also trained by EM algorithm with k-means clustering initialisation; the number of generalised GMMs components in any dataset also needs to be selected properly. Thanks to its non-linear fitting, the components of the generalised GMMs is fewer than those of the conventional GMMs. An example is provided in Fig. 5.2. The iteration does not stop until the relative difference of the log-likelihood obtained by Eq. 5.5 reduces under the preset threshold.

5.3 Fuzzy Gaussian Mixture Models

Inspired from the mechanism of FCM, we introduce the weighting exponent on the fuzzy membership into generalised GMMs in order to improve the efficiency of convergence. It is evident that GMMs are mixtures of Gaussian distributions, the dissimilarities of individual cluster points are defined in form of certain exponential function of the distances. Two types of fuzzy generalised GMMs are proposed based on different dissimilarity functions denoted by the exponential distance, i.e., probability based FGMMs and distance based FGMMs,

5.3.1 *Fuzzy C-Means Clustering*

Let $\mathscr{X} = \{x_1, x_2, ..., x_n\}$ be the d dimensional observed dataset with n vectors, k be the number of clusters with $2 \leq k \leq n$, m be a weighting exponent on each fuzzy membership, and the degree of fuzziness, μ_i be the prototype of the centre of ith cluster, and $U = \{u_{it}\}$ where u_{it} is the degree of membership of x_t in the ith cluster and has

$$0 \leq u_{it} \leq 1, \sum_{i=1}^{k} u_{it} = 1 \tag{5.25}$$

$1 \leq i \leq k$ and $1 \leq t \leq n$. The dissimilarity function with a A norm distance measure between object x_t and cluster centre μ_i is

$$d_{it}^2 = ||x_t - \mu_i||_A^2 = (x_t - \mu_i)^T A(x_t - \mu_i) \tag{5.26}$$

The aim of FCM is to find the new cluster centres (centroids) that minimize a dissimilarity function, i.e. the weighted within group sum of squared error objective function $J_m(U, \mu, \mathscr{X})$:

$$J_m(U, \mu, \mathscr{X}) = \sum_{t=1}^{n} \sum_{i=1}^{k} (u_{it}^m d_{it}^2) \tag{5.27}$$

The processing of minimising object function J_m depends on how centres find their ways to the best positions, as the fuzzy memberships u_{it} and norm distance d_{it} would change along with the new centres' position. Giving the previous positions of the centres, Eq. 5.26 would produce the distance based dissimilarities. To minimise the object function 5.27, new positions of centres would be determined by the following equations:

$$u_{it} = \left[\sum_{j=1}^{k} \left(\frac{d_{it}}{d_{jt}}\right)^{\frac{2}{m-1}} \right]^{-1} \tag{5.28}$$

$$\mu_i^{new} = \frac{\sum_{t=1}^{n} u_{it}^m x_t}{\sum_{t=1}^{n} u_{it}^m} \tag{5.29}$$

where μ_i^{new} is the new position of ith centre estimated by the degree of membership with the weighting exponent m.

After a number of iterations using Eqs. 5.26–5.29, the centres of clusters are optimised to minimise the object function. The iteration stops when difference of the current value of object function and the previous value of object function is less than the preset threshold.

As FCM introduces a weighting exponent m, the degree of fuzziness, on each fuzzy membership in Eqs. 5.27 and 5.29, the points which are nearer to one cluster than other clusters are made more important for this cluster and at the same time more insignificant for other clusters.

5.3.2 *Probability Based FGMMs*

For a faster convergence speed of GMMs, a dissimilarity function for probability based FGMMs is defined as:

$$d_{it}^2 = \frac{1}{\alpha_i p_i(x_t|\theta_i)} \tag{5.30}$$

where α_i is the weight of ith component; $p_i(x_t|\theta_i)$ is from Eq. 5.1 for $|a| < \varepsilon$ or Eq. 5.18 for $|a| \geq \varepsilon$.

Compared to Eq. 5.26, the dissimilarity function defined by Eq. 5.30 consists of not only the distance but also the covariance and mixture weights, and it is in direct ratio with the exponent of the distance. The objective function is extended from Eq. 5.27 as:

$$J_m(U, \mathscr{X}, \mu, \Sigma) = \sum_{t=1}^{n} \sum_{i=1}^{k} (u_{it}^m d_{it}^2) \tag{5.31}$$

Minimizing J_m is performed by the same u_{it} operations of FCM using Eq. 5.28. The new mixture weights are estimated by Eq. 5.32:

$$\alpha_i^{new} = \frac{\sum_{t=1}^{n} u_{it}^m}{\sum_{i=1}^{k} \sum_{t=1}^{n} u_{it}^m} \tag{5.32}$$

The parameters of curve axis (a_i, b_i) in C_i can be achieved from Eq. 5.20. If $|a_i| < \varepsilon$ (i.e., the principal axis of component i is beeline) then the generalized Gaussian model regresses to be a conventional Gaussian model, the centre and the covariance can be obtained by Eqs. 5.29 and 5.33.

$$\Sigma_i^{new} = \frac{\sum_{t=1}^{n} u_{it}^m (x_t - \bar{\mu}_i)(x_t - \bar{\mu}_i)^T}{\sum_{t=1}^{n} u_{it}^m} \tag{5.33}$$

Otherwise when $|a_i| \geq \varepsilon$, the principal axis of component i would be shaped as standard curve $y = a_i x^2 + b_i$ and the generalised Gaussian model turns into an AcaG model. The parameters of standard curve (transaction and rotation matrices), centres and covariances can be computed through Eqs. 5.20, 5.34–5.36.

$$\mu_i^{new} = \frac{\sum_{t=1}^{n} u_{it}^m x_t}{\sum_{t=1}^{n} u_{it}^m} + (Q_i^{new})^{-1} \underbrace{[0, b, 0, \ldots, 0]^T}_{d} + T_i^{new} \tag{5.34}$$

$$\Sigma_{ie}^{new} = \frac{\sum_{t=1}^{n} u_{it}^m \bar{L}_{te}^{(i)}}{\sum_{t=1}^{n} u_{it}^m} \quad (e = 1, 2) \tag{5.35}$$

$$\Sigma_{i(3-d)}^{new} = \frac{\sum_{t=1}^{n} u_{it}^m (x_t - \mu_i^{new})_{(3-d)} (x_t - \mu_i^{new})_{(3-d)}^T}{\sum_{t=1}^{n} u_{it}^m} \tag{5.36}$$

The EM algorithm of probability based FGMMs is provided in algorithm 2 in C.1 where d_{it} is obtained from Eq. 5.30.

5.3.3 Distance Based FGMMs

Referring to Eq. 5.30, the degree of fuzziness in probability based FGMMs takes effect on the distance between any point and its centre point, and the mixture weights. On the other hand, we propose another dissimilarity function which focuses the effect of degree of fuzziness only on the distances between points and their component centres.

When the principal axis is bent, the distances in principal plane consists of arc length and normal length, Eq. 5.18 can be rewritten as,

$$p(x_t|\theta) = \left(\sum_{j=1}^{J_t} \prod_{s=1}^{2} \frac{\exp\left(\frac{-l_j^2(v_{st})m}{2\Sigma_s(m-1)}\right)}{\sqrt{2\pi|\Sigma_s|}} \prod_{s=3}^{d} \frac{\exp\left(\frac{-v_{st}^2 m}{2\Sigma_s(m-1)}\right)}{\sqrt{2\pi|\Sigma_s|}} \right)^{\frac{m-1}{m}} \tag{5.37}$$

Furthermore, based on above equation the dissimilarity function can be defined as follows,

$$\mathrm{d}_{\mathrm{it}}^2 = \begin{cases} \dfrac{\exp\left(\frac{(x_t-\mu_i)^T \Sigma_i^{-1} (x_t-\mu_i)}{2}\right)}{(\alpha_i (2\pi)^{-\frac{d}{2}} |\Sigma_i|^{-\frac{1}{2}})^{\frac{m-1}{m}}} & (|a_i| < \varepsilon) \\ \dfrac{1}{\alpha_i^{\frac{m-1}{m}} p_i(x_t|\theta_i)} & (|a_i| \geq \varepsilon) \end{cases} \tag{5.38}$$

where m is the degree of fuzziness; $p_i(x_t|\theta_i)$ is obtained from Eq. 5.37. This dissimilarity function is in form of direct ratio with the exponent of the distance. Its objective function, membership function and the functions estimating centre vectors, mixture weights and covariances are the same as probability based FGMMs, referring to Eqs. 5.34–5.36. The EM algorithm for distance based FGMMs is the same as probability based FGMMs shown in algorithm 2 in Appendix C.1, except that d_{it} is provided by Eq. 5.38. The proposed updating algorithm not only minimises the objective function 5.27 of FCM but also maximises the log-likelihood function 5.5 of GMMs. The theoretical justification is presented in Appendix C.2.

5.3.4 Comparison of Probability/Distance Based FGMMs

To clarify the difference between probability based FGMMS and distance based FGMMs, u_{it}^m can be obtained by their different dissimilarity functions. From Eqs. 5.28 and 5.30, u_{it}^m in probability based FGMMs can be obtained when $|a| < \varepsilon$:

$$u_{it}^m = \frac{[\alpha_i p_i(x_t|\theta_i)]^{\frac{m}{m-1}}}{\left[\sum_{j=1}^{k} (\alpha_j p_j(x_t|\theta_j))^{\frac{1}{m-1}}\right]^m} \tag{5.39}$$

so that

$$u_{it}^m \propto [\alpha_i p_i(x_t|\theta_i)]^{\frac{m}{m-1}} \tag{5.40}$$

which illustrates that the u_{it}^m is in direct ratio with the fuzzy probability.

In contrast, Eq. 5.38 is substituted to Eq. 5.28 and then u_{it}^m can be obtained for distance based FGMMs under condition of $|a| < \varepsilon$,

$$u_{it}^m = \frac{\alpha_i (2\pi)^{-\frac{d}{2}} |\Sigma_i|^{-\frac{1}{2}} \exp\left(\frac{-m}{m-1} \frac{(x_t-\mu_i)^T \Sigma_i^{-1} (x_t-\mu_i)}{2}\right)}{\left[\sum_{j=1}^{k} (\alpha_j (2\pi)^{-\frac{d}{2}} |\Sigma_j|^{-\frac{1}{2}})^{\frac{1}{m}} \exp\left(\frac{-1}{m-1} \frac{(x_t-\mu_j)^T \Sigma_j^{-1} (x_t-\mu_j)}{2}\right)\right]^m} \tag{5.41}$$

so that

$$u_{it}^m \propto \exp\left(\frac{-m}{m-1} \frac{(x_t-\mu_i)^T \Sigma_i^{-1} (x_t-\mu_i)}{2}\right) \tag{5.42}$$

Different from probability based FGMMs, the membership function of distance based FGMMs is only direct proportional to the exponent of the fuzzy distances. Similar comparison can be achieved for $|a| \geq \varepsilon$, so it is the reason that the former is named as probability based FGMMs and the later is called distance based FGMMs.

GMMs are closely related with FCM with K-L information term (KLFCM) [39] and KLFCM is considered to be a FCM-type counterpart of GMMs proposed by Ichihashi et al., who generalised the regularised objective function replacing the entropy term [34] with K-L information term [40]. As a KLFCM algorithm variant, a generic methodology for finite mixture model fitting under a fuzzy clustering principle is proposed in [41]. It is similar with the proposed FGMMs in the way that they both defined novel dissimilarity functions and have a parameter, the degree of fuzziness, in the $exp()$ of the fuzzy membership u_{it}. However, they are different in the following ways. Firstly, their dissimilarity functions are different. The algorithm in [41] defined the dissimilarity function as the posterior expectation of the negative log-likelihood of the mixture component with respect to the observable vector and the additional latent data; The probability based FGMMs adapt the dissimilarity function of multiplicative inverse of probability density function, while distance based FGMMs refine the dissimilarity function which limits the degree of fuzziness only on the exponential distance. Secondly, although 'λ' in KLFCM and 'm' in FGMMs all play a role in tuning the degree of fuzziness, the weighting exponent 'm' in FGMMs has the same meaning/effects as in standard FCM [29–31, 46] while 'λ' in KLFCM has not, so the methods of determining the weighting exponent in FCM [30, 32, 33] for better performances may also be used directly in distance based FGMMs. Thirdly, the objective and performance of introducing the dissimilarity functions are different. The algorithm in [41] was aimed to strengthen the fitting performance of the regular types of finite mixture models, and the dissimilarity function was employed to effectively incorporate the explicit information about the considered data objects' distributions into the fuzzy clustering procedure, while dissimilarity functions introduced for FGMMs are targeted at achieving a much faster convergence process.

5.4 Experiments

The proposed FGMMs are compared with both the conventional and the generalised GMMs, respectively, on Gaussian based datasets, analysis of characters and selected UCI datasets for demonstrating the individual performance. Since the initialisation of EM algorithm may cause different results of the same settings, in order to show the effectiveness of the fuzzy algorithms in fairness, the initialisations of all methods are implemented by FCM and each comparison has only one execution of FCM, which would trigger running of EM algorithms. For example, when comparing generalised GMMs with probability based FGMMs and distance based FGMMs, all of these three methods are initialised by FCM which is implemented once at the beginning of comparison and membership grades that it results would be used by all the three methods simultaneously. In addition, in order to have a fair comparison of conventional GMMs and the FGMMs, a is set as 0 for all components in probability based FGMMs and distance based FGMMs in the following experiments so that all the compared methods have elliptical components. For comparison of generalised GMMs and the FGMMs, the value of a is calculated from Eq. 5.20 where $|a| \geq 0$.

5.4.1 Gaussian Based Datasets

1000 2-dimensional points of 5 conventional Gaussian mixtures are generated for comparisons of conventional GMMs, probability based FGMMs and distance based FGMMs, as shown on the left of Fig. 5.3, and each mixture has 200 points. 1600 of 2-dimensional points are created according to the proposed theory of generalised Gaussian model in Sect. 5.2.2 for the comparisons, and the dataset has 5 mixtures as well on the right of Fig. 5.3.

First, we set the number of components as 5 to test the performance of the proposed methods with the suitable number of components. Structures of the most datasets are, however, usually unknown beforehand and obtaining general ideas of the structures would be useful for further data analysis. Hence, different numbers of components are considered in comparison with the 5-component constructed datasets in order to achieve a more comprehensive understanding of data structure estimation.

5.4.1.1 Five Components

Five components are set for conventional GMMs, probability based FGMMs and distance based FGMMs, the results can be seen in Fig. 5.4, where the locations of the five components of these three methods are similar.

Their convergence processes are shown in Fig. 5.5 in which the log-likelihoods tend to smooth-out with the iterations of EM algorithm. Numbers (i.e., 1–3) stand for different thresholds: '1' means the relative difference of log-likelihood, calculated by algorithm 2 between current iteration and the last iteration, is less than

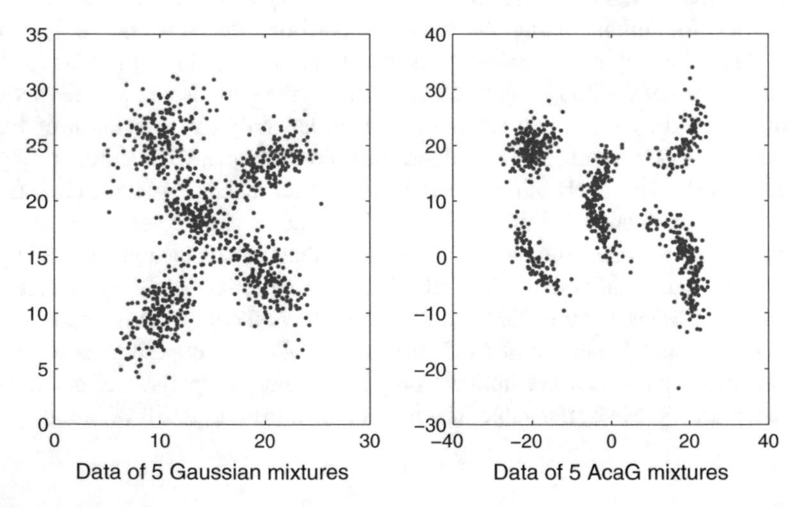

Fig. 5.3 Data of five conventional Gaussian mixtures (*left*) and data of five generalized Gaussian mixtures (*right*)

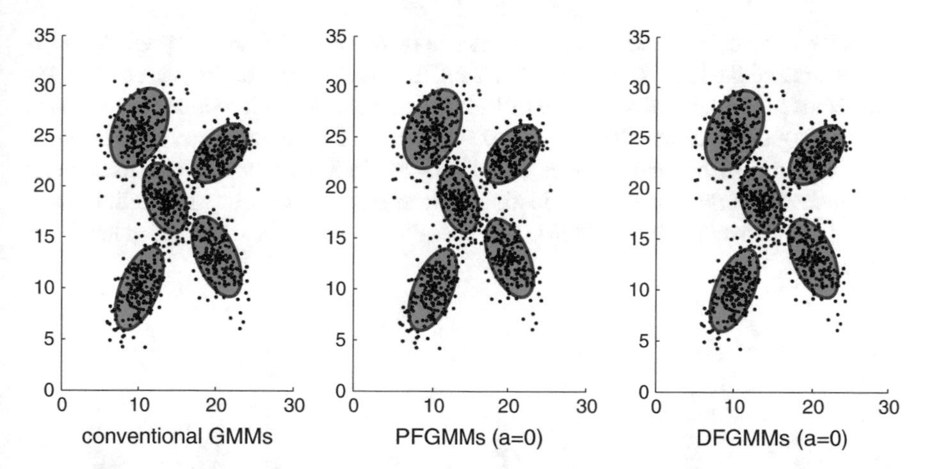

Fig. 5.4 Results of conventional GMMs using 56 iterations and 0.257 s (*left*), probability based FGMMs ($a = 0$) using 35 iterations and 0.232 s (*middle*) and distance based FGMMs ($a = 0$) using 48 iterations and 0.235 s (*right*) with 5 components on the dataset of five conventional Gaussian mixtures

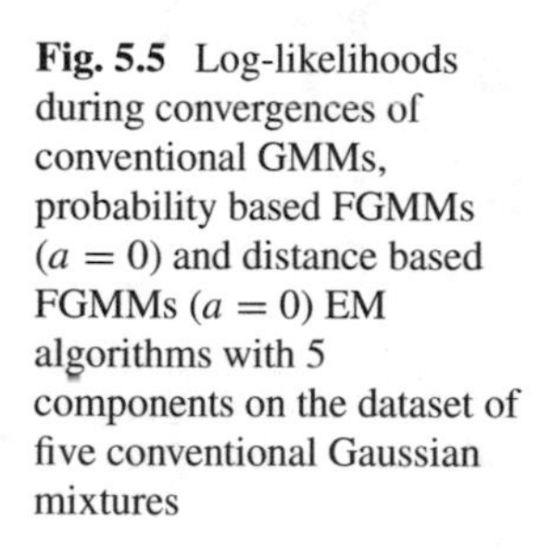

Fig. 5.5 Log-likelihoods during convergences of conventional GMMs, probability based FGMMs ($a = 0$) and distance based FGMMs ($a = 0$) EM algorithms with 5 components on the dataset of five conventional Gaussian mixtures

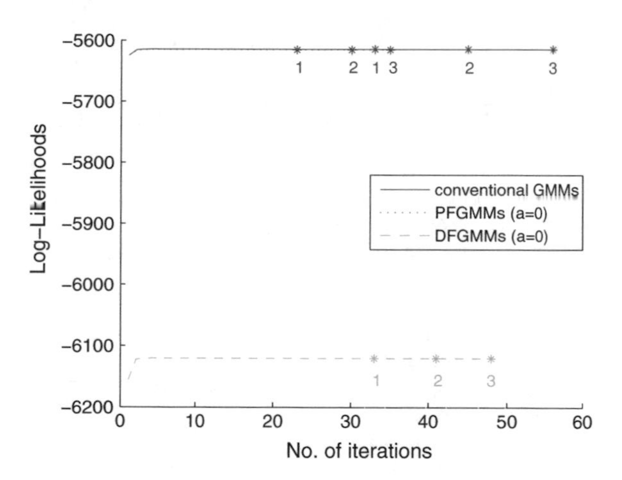

threshold of 10^{-8}; '2' is for the threshold of 10^{-9} and '3' is for the mostly used and the least threshold of 10^{-10}. In Fig. 5.5, these three methods have similar capabilities to fit the dataset since values of the three stabilised log-likelihoods are -5614.5, $-5615.3.1$ and -6120.5 for conventional GMMs, probability based FGMMs and distance based FGMMs, respectively. However, the numbers of iterations of probability based FGMMs and distance based FGMMs are 35 in 0.232 s and 48 in 0.235 s which are less than the iteration number used by GMMs (i.e., 56 in 0.257 s) in terms of threshold value 10^{-10}; the convergence speeds of probability based FGMMs and distance based FGMMs are faster than conventional GMMs from the distributions of these three thresholds.

For the data of five generalized Gaussian mixtures, Figs. 5.6 and 5.7 present the fitting results and the log-likelihoods of generalised GMMs, probability based FGMMs and distance based FGMMs, where little difference could be seen since the stabilised log-likelihoods locate at -10472, -10471 and -11272 for generalised GMMs, probability based FGMMs and distance based FGMMs individually. The distributions of the threshold markers of '1–3' illustrate convergence speeds of probability based FGMMs and distance based FGMMs evidently are faster than that of generalised GMMs.

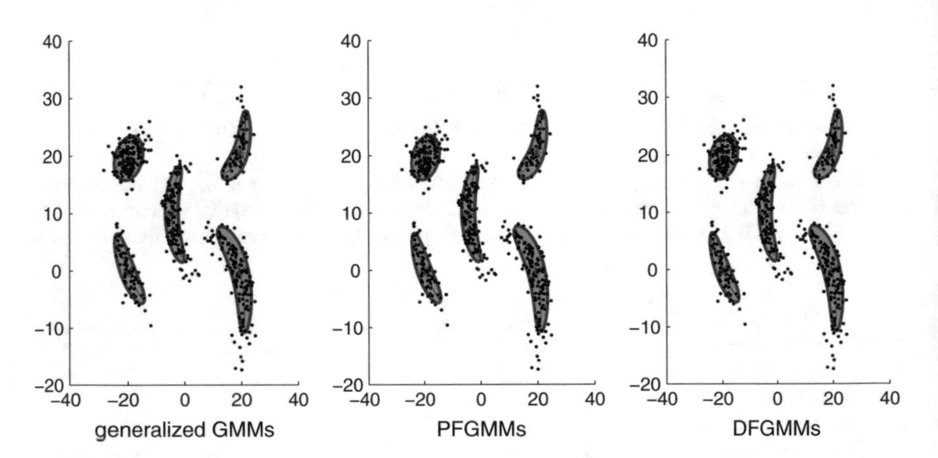

Fig. 5.6 Results of generalised GMMs using 11 iterations and 2.041 s (*left*), probability based FGMMs ($|a| \geq 0$) using 7 iterations and 1.562 s (*middle*) and distance based FGMMs ($|a| \geq 0$) using 8 iterations and 1.516 s (*right*) with 5 components on the dataset of five generalised Gaussian mixtures

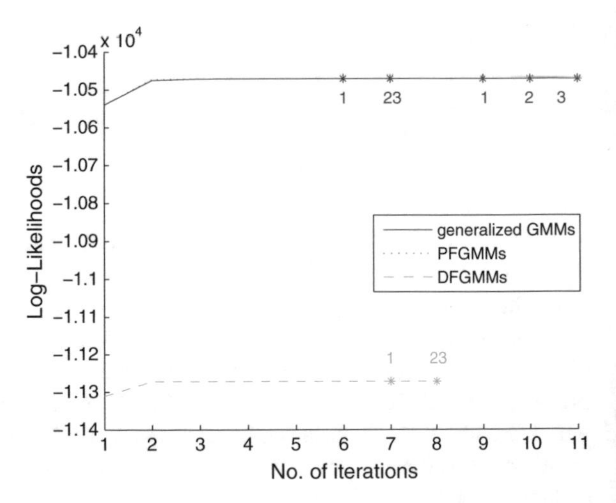

Fig. 5.7 Log-likelihoods during convergences of generalised GMMs, probability based FGMMs ($|a| \geq 0$) and distance based FGMMs ($|a| \geq 0$) EM algorithms with 5 components on the dataset of five generalised Gaussian mixtures

5.4.1.2 Various Components

We further investigate the comparison of these three methods' performances when setting the number k of components different value from 3 to 7; the performance criteria include log-likelihood and convergence speed. The convergence speed can be judged from the number of iterations and the computational time for the threshold of 10^{-10}. In the following tables, 'LL' stands for log-likelihoods, 'NI' for iteration number and 'Time' for the computational time.

Table 5.1 shows the results of NIs, LLs and time of conventional GMMs, probability based FGMMs and distance based FGMMs when setting k from 3 to 7 on the data of five conventional Gaussian mixtures. The results in the table demonstrate that, with same number of components, log-likelihoods of these three methods are similar to each other, but probability based FGMMs and distance based FGMMs have much smaller numbers of iterations and cost time in each row than conventional GMMs, which means the FGMMs not only have the GMMs' capability of fitting the data but also are capable of reducing the computational cost in comparison with GMMs. To have a general understanding of fuzzy GMMs' capabilities of computational efficiency, we calculate the NI average percentages ('AP' in the table) of probability based FGMMs and distance based FGMMs relative to conventional GMMs using the following formula:

Table 5.1 Results of conventional GMMs, probability based FGMMs ($a = 0$) and distance based FGMMs ($a = 0$) with 3–7 components on the dataset of five conventional Gaussian mixtures

		Conventional GMMs	Probability based FGMMs ($a = 0$)	Distance based FGMMs ($a = 0$)
k = 3	LL	−5857.1	−5920.6	−6425.3
	NI	172	75	68
	Time (s)	0.5	0.31	0.219
k = 4	LL	−5702.0	−5701.0	−6206.6
	NI	78	45	48
	Time (s)	0.312	0.202	0.171
k = 5	LL	−5614.5	−5615.4	−6120.5
	NI	56	35	48
	Time (s)	0.257	0.232	0.235
k = 6	LL	−5610.2	−5615.3	−6124.4
	NI	623	488	354
	Time (s)	3.282	3.175	2.203
k = 7	LL	−5600.2	−5609.9	−6105.3
	NI	811	611	480
	Time (s)	4.6	3.918	2.906
k = 3–7	AP	100%	63.49%	60.56%
	Time	100%	79.79%	64.07%

$$AP = \frac{\sum_{k=3}^{7} \frac{NI_k^F}{NI_k^G}}{5} \times 100\% \tag{5.43}$$

where NI_k^F denotes the number of iterations of fuzzy algorithms when setting number of component as k, NI_k^G stands for the number of iterations of Gaussian algorithms when setting number of component as k. The result of AP of conventional GMMs is set as 100% and the results of probability based FGMMs and distance based FGMMs are shown at the bottom of Table 5.1, which presents probability based FGMMs take only 63.49% of the iterations of conventional GMMs and distance based FGMMs take only 60.56%. The same operation is taken for the computational time, and the results are listed in the last row of the table. It shows that these two FGMMs have only 79.79 and 64.07% of the cost time of conventional GMMs.

Similarly, Table 5.2 shows the results of NIs, LLs and time of generalised GMMs, probability based FGMMs and distance based FGMMs when setting k from 3 to 7 on the data of five generalised Gaussian mixtures. Log-likelihoods of all the three methods are similar to each other, which proves the FGMMs have the similar capability of fitting datasets. On the other hand, the numbers of iterations and cost time of probability based FGMMs and distance based FGMMs are much less than those of generalised GMMs. The NI and time average percentages of probability based

Table 5.2 Results of generalized GMMs, probability based FGMMs ($|a| \geq 0$) and distance based FGMMs ($|a| \geq 0$) with 3–7 components on the dataset of five generalized Gaussian mixtures

		Generalized GMMs	Probability based FGMMs ($\|a\| \geq 0$)	Distance based FGMMs ($\|a\| \geq 0$)
k = 3	LL	−11435	−11430	−12217
	NI	85	48	44
	Time (s)	10.203	6.031	5.516
k = 4	LL	−10855	−10855	−12440
	NI	8	6	7
	Time (s)	1.313	1.047	1.234
k = 5	LL	−10472	−10471	−11272
	NI	11	7	8
	Time (s)	2.041	1.512	1.516
k = 6	LL	−10478	−10489	−12066
	NI	114	80	59
	Time (s)	18.047	19.528	13.953
k = 7	LL	−10479	−10495	−12080
	NI	191	95	84
	Time (s)	47.078	25.734	22.188
k = 3–7	AP	100%	63.11%	61.55%
	Time	100%	75.16%	69.35%

FGMMs and distance based FGMMs are computed according to Eq. 5.43 shown in the Table 5.2; it also demonstrates that the FGMMs are powerful in reducing the computational cost.

5.4.2 Structure Analysis of Characters

GMMs is one of the most dominate methods widely used in structure modelling and generalised GMMs show powerful functioning in modelling data with curve manifolds. Therefore, we choose the application for structure analysis of characters to test the effectiveness of our proposed fuzzy methods.

Figure 5.8 shows the results of fitting character 'R' by conventional GMMs and the FGMMs ($a = 0$), where k is set to be 9 for accepted recognition. Their log-likelihoods are -8995.3, -9047.6 and -9669.0, evidently whose difference is relatively small.

Their convergence processes are presented in Fig. 5.9, where the numbers of iterations are 1263, 520 and 473 for conventional GMMs, probability based FGMMs and distance based FGMMs, respectively, in terms of the threshold of 10^{-10}. The FGMMs take about half computational cost of conventional GMMs, since their computational time are 6.406 s and 4.047 s compared to 10.75 s by conventional GMMs.

Figure 5.10 shows that the results of fitting character 'R' by generalised GMMs and the FGMMs ($|a| \geq 0$). It demonstrates that generalised GMMs outperform conventional GMMs due to its capability of data fitting with curve manifolds.

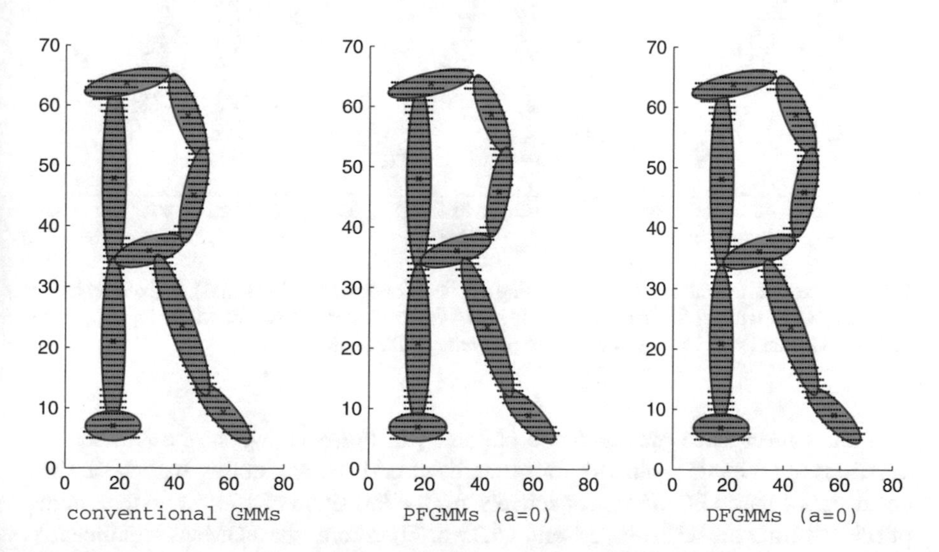

Fig. 5.8 Results of conventional GMMs using 1263 iterations and 10.75 s (*left*), probability based FGMMs ($a = 0$) using 520 iterations and 6.406 s (*middle*) and distance based FGMMs ($a = 0$) using 537 iterations and 4.047 s (*right*) with 9 components on the character 'R'

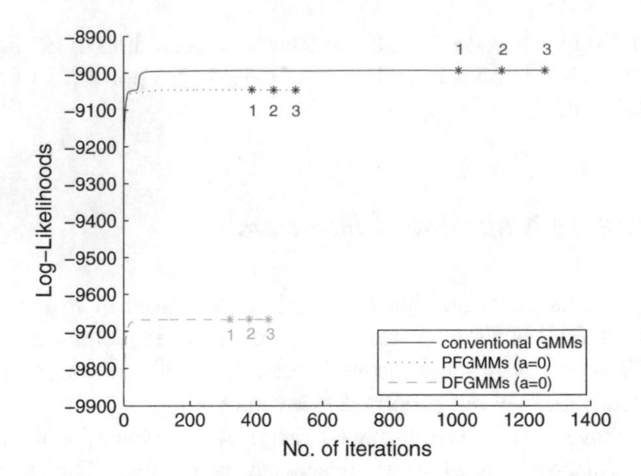

Fig. 5.9 Log-likelihoods during convergences of conventional GMMs, probability based FGMMs ($a = 0$) and distance based FGMMs ($a = 0$) EM algorithms with 9 components on the character 'R'

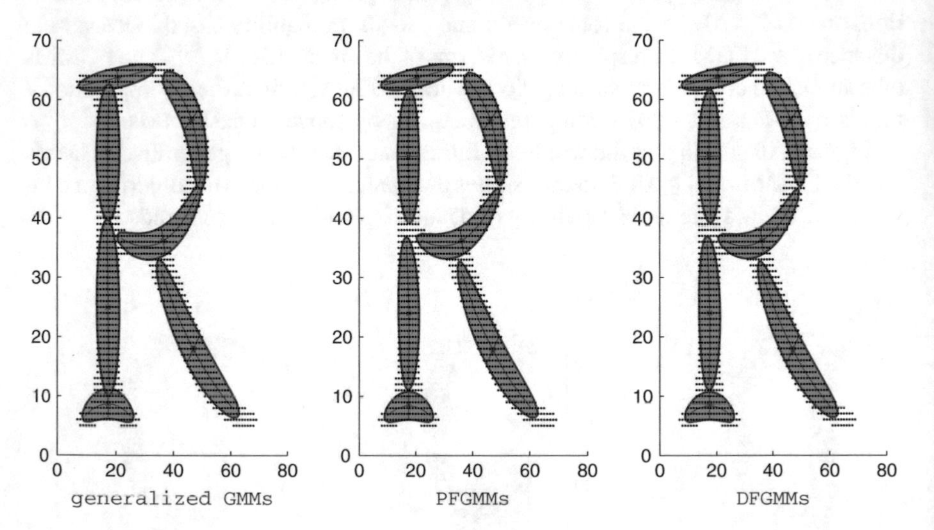

Fig. 5.10 Results of generalized GMMs using 649 iterations and 133.187 s (*left*), probability based FGMMs ($|a| \geq 0$) using 257 iterations and 55 s (*middle*) and distance based FGMMs ($|a| \geq 0$) using 197 iterations and 40.75 s (*right*) with 7 components on the character 'R'

Their convergence processes are presented in Fig. 5.11, where the numbers of iterations are 649, 257 and 197 for generalised GMMs, probability based FGMMs and distance based FGMMs, respectively, for the threshold of 10^{-10}, and their computational time are 133.186, 55 and 40.75 s. Therefore, the FGMMs significantly outperform generalised GMMs again.

Table 5.3 shows more examples including the datasets '6', '8', 'a', and 'B', whose datasets contain curve manifolds among the numbers and characters. The number

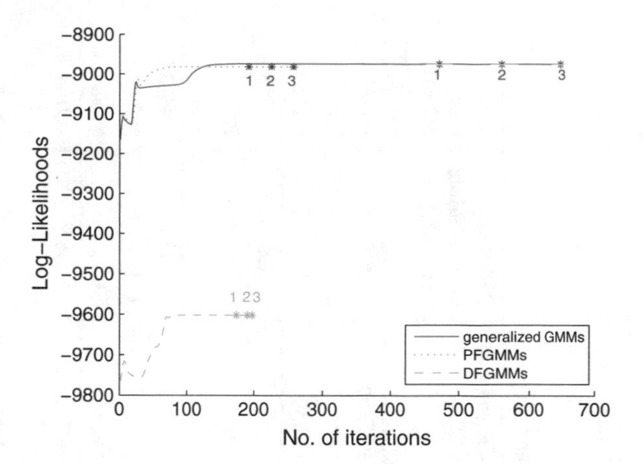

Fig. 5.11 Log-likelihoods during convergences of generalised GMMs, probability based FGMMs ($|a| \geq 0$) and distance based FGMMs ($|a| \geq 0$) EM algorithms with 7 components on the character 'R'

of components is chosen according to the dataset's distribution and facts such as ensuring datasets are fitted well and components are not too many. Degree of fuzziness is decided in accordance with [32] illustrating that optimal value of m locates around the range of [1.5–3.5]. The results of log-likelihoods do not change much when comparing the FGMMs, and the big changes of the number of iterations and also computational time indicate that the fuzzy algorithms are almost double efficient than conventional GMMs and generalised GMMs. The values of 'time' present that probability based FGMMs and distance based FGMMs use 69.29 and 47.63% computational cost of conventional GMMs, and 51.12 and 48.71% computational time of the generalized GMMs.

5.4.3 *Evaluation on UCI Datasets*

The proposed FGMMs are also evaluated on the datasets from the UCI Machine Learning Repository [47] including the 'Iris', 'Magic', 'Pima', 'Wine' and 'Satellite'. This experiment aims to investigate the fitting performance and efficiency of the proposed approach in the real-world scenarios in comparison of the conventional GMMs and generalised GMMs. The number of the components is chosen in accordance with the number of the clusters of the datasets, and degree of fuzziness is set in the same way as the experiments of the character structure analysis. From the results shown in Table 5.4, the log-likelihoods maintain almost the same, while the iterations of the FGMMs shown in 'AP' are half less than the conventional and generalised GMMs. The results also show that FGMMs cost nearly half time of the conventional and generalised GMMs, though each iteration of the FGMMs may cost more because of the introduced extra fuzzy operations. In addition, the FGMMs have higher log-likelihood values than those achieved from the method in [41] because of the nonlinear fitting capability.

Table 5.3 Results of all models with k components and m degree of fuzziness on datasets of '6', '8', 'a', 'B' and 'R'

Data	k	m	NI/LL	Conventional GMMs	Probability based FGMMs ($a = 0$)	Distance based FGMMs ($a = 0$)	Data	k	m	NI/LL	Generalised GMMs	Probability based FGMMs ($\|a\| \geq 0$)	Distance based FGMMs ($\|a\| \geq 0$)
'6'	6	3	LL	−14008	−14014	14932	'6'	5	3	LL	−13958	−13983	−14903
			NI	571	385	394				NI	578	278	288
			Time (s)	4.106	3.457	3.078				Time (s)	110.50	55.89	55.343
'8'	9	2	LL	−7397.3	−7404.4	−8450.9	'8'	6	4	LL	−7393.7	−7405.5	−7755.7
			NI	802	468	447				NI	666	211	224
			Time (s)	6.719	4.563	3.937				Time (s)	102.81	36	35.89
'a'	9	3	LL	−8816.5	−8830.7	−9450.6	'a'	7	3	LL	−8837.9	−8852.7	−9470.3
			NI	6530	2942	2092				NI	770	310	330
			Time (s)	53.61	32.02	19.31				Time (s)	150.3	63.98	64.98
'B'	10	4	LL	−8704.7	−8709.2	−9113.2	'B'	7	4	LL	−8753.4	−8761.8	−9164.2
			NI	2204	1204	533				NI	209	173	171
			Time (s)	22.14	17.42	6.844				Time (s)	41.84	36.05	35.45
'R'	9	3	LL	−8995.3	−9047.6	−9669.0	'R'	7	3	LL	−8974.2	−8982.1	−9601.9
			NI	1263	520	437				NI	649	257	197
			Time (s)	10.75	6.01	4.047				Time (s)	133.187	55	40.57
All	AP			100%	53.33%	43.11%	All	AP			100%	48.48%	47.70%
	Time			100%	69.29%	47.63%		Time			100%	51.12%	48.71%

Table 5.4 Results of all models with k components and m degree of fuzziness on UCI datasets of 'Iris'('Ir'), 'Magic'('Ma'), 'Pima'('Pi'), 'Wine'('Wi') and 'Satellite'('Sa')

Data	k	m	NI/LL	Conventional GMMs	Probability based FGMMs ($a = 0$)	Distance based FGMMs ($a = 0$)	Data	k	m	NI/LL	Generalised GMMs	Probability based FGMMs ($\|a\| \geq 0$)	Distance based FGMMs ($\|a\| \geq 0$)
'Ir'	3	2	LL	−180.19	−180.30	−201.05	'Ir'	3	3	LL	−130.29	−132.80	−149.93
			NI	42	29	18				NI	79	57	37
			Time (s)	0.094	0.078	0.062				Time (s)	4.0780	3.0620	2.094
'Ma'	2	4	LL	−540870	−540878 4	−571590	'Ma'	2	3	LL	−209380	−210740	−218020
			NI	69	59	39				NI	247	162	139
			Time (s)	4.5630	4.2567	2.9137				Time (s)	77.781	51.672	44.515
'Pi'	2	4	LL	−18577	−18577	−19539	'Pi'	2	3	LL	−8091.4	−8174.6	−8472.2
			NI	12	11	7				NI	129	50	46
			Time (s)	0.0310	0.0310	0.0274				Time (s)	6.7030	2.6880	2.4470
'Wi'	3	1.5	LL	−2901	−2908	−3096	'Wi'	3	2	LL	−1932	−1944	−2127
			NI	39	21	20				NI	206	59	50
			Time (s)	0.063	0.047	0.036				Time (s)	10.282	2.968	2.708
'Sa'	7	2	LL	−426820	−426970	−437560	'Sa'	7	2	LL	−46890	−47620	−48140
			NI	473	308	230				NI	568	346	291
			Time (s)	84.687	59.688	49.453				Time (s)	277.75	168.92	138.52
All	AP			100%	73.04%	51.52%	All	AP			100%	53.21%	42.85%
	Time			100%	84.72%	66.75%		Time			100%	54.26%	44.62%

The above experiments confirm that the proposed fuzzy algorithms can achieve similar level of likelihoods as conventional GMMs and generalised GMMs with double efficiency in terms of convergence. Comparing probability based FGMMs with distance based FGMM, though they both have abilities to fit the datasets with acceptable likelihoods, statistical result in Table 5.3 indicates that distance based FGMMs outperform the other in terms of computational cost.

In this chapter, one of our objectives is to save the computational time of GMMs, so the 'fast convergence' is said to be compared with the convergence speed of GMMs (including normal GMMs and generalised GMMs) rather than K-means whose objective function is linear function ('m = 1'). In GMMs (including normal GMMs and generalised GMMs in this chapter), the 'w_{it}' is the posteriori probability like the membership 'u_{it}' in FCM. After introducing a weighting component on the membership, in FGMMs, the relationship between 'w_{it}' and 'u_{it}' can be seen from above equations C.2–C.4. When 'm'($m > 1$) becomes larger, '$\frac{m}{m-1}$' becomes smaller; '$m \to \infty$', then '$\frac{m}{m-1} \to 1$'; at that time FGMM turns into GMM. Therefore, the smaller 'm ($m > 1$)' is, the larger 'u_{it}' changes, and the fuzzier FGMM becomes. However when the 'm' is too small, the fitness would become too low. Therefore, the balance between fitness (performance) and fuzziness (convergence speed) should be concerned. The experiments proved that proper 'm' can achieve a much faster convergence speed with a satisfied fitting degree.

5.5 Conclusion

This chapter proposed FGMMs in order to improve the conventional GMMs in terms of the performance and the efficiency. Firstly, generalised GMMs with an EM algorithm have been proposed by integrating conventional GMMs and active axis curve GMMs for fitting non-linear datasets. Then, based on the generalised GMMs, FGMMs have been presented including probability based FGMMs and distance based FGMMs by refining the dissimilarity functions. Probability based FGMMs adapt the dissimilarity function of multiplicative inverse of probability density function, while distance based FGMMs refine the dissimilarity function which limits the degree of fuzziness only on the exponential distance. Experiments have been conducted on conventional Gaussian based datasets, generalised Gaussian based datasets, characters datasets and real datasets from UCI Machine Learning Repository. Simulation results have shown that not only do the proposed fuzzy algorithms have the capabilities of fitting data as conventional GMMs or generalised GMMs but they also can reduce about half of their computational time. Comparing probability based FGMMs with distance based FGMM, the latter outperforms the former in terms of computational efficiency. Due to the nature of the fast computation and non-linear fitting of the proposed FGMMs, they would break new ground for extending GMMs' application area to the fields such as data mining with huge datasets and real-time systems. One of our future directions is to implement the proposed algorithms into human motion analysis and real-world surveillance applications [48–51].

References

1. I. Bilik, J. Tabrikian, and A. Cohen. Gmm-based target classification for ground surveillance doppler radar. *Aerospace and Electronic Systems, IEEE Transactions on*, 42(1):267–278, Jan. 2006.
2. S.C. Kim and T.J. Kang. Texture classification and segmentation using wavelet packet frame and Gaussian mixture model. *Pattern Recognition*, 40(4):1207–1221, 2007.
3. J. H. Song, K. H. Lee, J. H. Chang, J. K. Kim, and N. S. Kim. Analysis and improvement of speech/music classification for 3gpp2 smv based on gmm. *Signal Processing Letters, IEEE*, 15:103–106, 2008.
4. M. H. Siu, X. Yang, and H. Gish. Discriminatively trained gmms for language classification using boosting methods. *IEEE Transactions on Audio, Speech, and Language Processing*, 17(1):187–197, Jan. 2009.
5. Z. David, J. Samuelsson, and M. Nilsson. On entropy-constrained vector quantization using gaussian mixture models. *IEEE Transactions on Communications*, 56(12):2094–2104, 2008.
6. L. Lu, A. Ghoshal, and S. Renals. Regularized subspace gaussian mixture models for speech recognition. *IEEE Signal Processing Letters*, 18:419–422, 2011.
7. Abolfazl Mehranian and Habib Zaidi. Joint estimation of activity and attenuation in whole-body tof pet/mri using constrained gaussian mixture models. *IEEE transactions on medical imaging*, 34(9):1808–1821, 2015.
8. Y. Wang and B. Yuan. Human face tracking using gaussian mixture models and fuzzy shape analysis. *International Conference on Signal Processing*, 2:1322–1325, 2004.
9. X. Wang, Y. Wang, X. Feng, and M. Zhou. Adaptive gaussian mixture models based facial actions tracking. *International Conference on Computer Science and Software Engineering*, 2:923–926, Dec. 2008.
10. Yue Gao, Rongrong Ji, Longfei Zhang, and Alexander Hauptmann. Symbiotic tracker ensemble toward a unified tracking framework. *IEEE Transactions on Circuits and Systems for Video Technology*, 24(7):1122–1131, 2014.
11. S.P. Wilson, E.E. Kuruoglu, and E. Salerno. Fully bayesian source separation of astrophysical images modelled by mixture of gaussians. *IEEE Journal of Selected Topics in Signal Processing*, 2(5):685–696, Oct. 2008.
12. D. Persson, T. Eriksson, and P. Hedelin. Packet video error concealment with gaussian mixture models. *IEEE Transactions on Image Processing*, 17(2):145–154, Feb. 2008.
13. MA Balafar. Gaussian mixture model based segmentation methods for brain mri images. *Artificial Intelligence Review*, 41(3):429–439, 2014.
14. A. Wahab, Q. Chai, K. T. Chin, and K. Takeda. Driving profile modeling and recognition based on soft computing approach. *IEEE Transactions on Neural Networks*, 20(4):563–582, April 2009.
15. M. Vondra and R. Vich. Recognition of Emotions in German Speech Using Gaussian Mixture Models. *Multimodal Signals: Cognitive and Algorithmic Issues*, pages 256–261, 2009.
16. T. Routtenberg and J. Tabrikian. Mimo-ar system identification and blind source separation for gmm-distributed sources. *IEEE Transactions on Signal Processing*, 57(5):1717–1730, May 2009.
17. S. Takahashi and S. Tsukiyama. A New Statistical Timing Analysis Using Gaussian Mixture Models for Delay and Slew Propagated Together. *IEICE Transactions on Fundamentals of Electronics, Communications and Computer Sciences*, 92(3):900–911, 2009.

18. S. Allali, D. Blank, W. Müller, and A. Henrich. Image Data Source Selection Using Gaussian Mixture Models. *Lecture Notes In Computer Science*, pages 170–181, 2007.
19. Z. Ju, H. Liu, X. Zhu, and Y. Xiong. Dynamic Grasp Recognition Using Time Clustering, Gaussian Mixture Models and Hidden Markov Models. *Journal of Advanced Robotics*, 23:1359–1371, 2009.
20. S.O. Ba and J. M. Odobez. Recognizing visual focus of attention from head pose in natural meetings. *IEEE Transactions on Systems, Man, and Cybernetics, Part B: Cybernetics*, 39(1):16–33, Feb. 2009.
21. J. Hennebert, A. Humm, and R. Ingold. Modelling spoken signatures with gaussian mixture model adaptation. *IEEE International Conference on Acoustics, Speech and Signal Processing, Part II*, 2:229–232, April 2007.
22. M. Montagnuolo and A. Messina. Automatic genre classification of tv programmes using gaussian mixture models and neural networks. *International Conference on Database and Expert Systems Applications*, pages 99–103, Sept. 2007.
23. Ju-Chiang Wang, Yi-Hsuan Yang, Hsin-Min Wang, and Shyh-Kang Jeng. Modeling the affective content of music with a gaussian mixture model. *IEEE Transactions on Affective Computing*, 6(1):56–68, 2015.
24. A.S. Malegaonkar, A.M. Ariyaeeinia, and P. Sivakumaran. Efficient speaker change detection using adapted gaussian mixture models. *IEEE Transactions on Audio, Speech, and Language Processing*, 15(6):1859–1869, Aug. 2007.
25. D. Ververidis and C. Kotropoulos. Gaussian mixture modeling by exploiting the mahalanobis distance. *IEEE Transactions on Signal Processing*, 56(7):2797–2811, July 2008.
26. Y. Agiomyrgiannakis and Y. Stylianou. Wrapped gaussian mixture models for modeling and high-rate quantization of phase data of speech. *IEEE Transactions on Audio, Speech, and Language Processing*, 17(4):775–786, 2009.
27. B. Zhang, C. Zhang, and X. Yi. Active curve axis Gaussian mixture models. *Pattern Recognition*, 38(12):2351–2362, 2005.
28. J. C. Dunn. A fuzzy relative of the ISODATA process and its use in detecting compact well-separated clusters. *Cybernetics and Systems*, 3(3):32–57, 1973.
29. J. C. Bezdek. *Pattern recognition with fuzzy objective function algorithms*. Kluwer Academic Publishers Norwell, MA, USA, 1981.
30. N. R. Pal and J. C. Bezdek. On cluster validity for the fuzzy c-means model. *IEEE Transactions on Fuzzy Systems*, 3(3):370–379, 1995.
31. J. Yu, Q. Cheng, and H. Huang. Analysis of the weighting exponent in the fcm. *IEEE Transactions on Systems, Man, and Cybernetics, Part B: Cybernetics*, 34(1):634–639, Feb. 2004.
32. J. Pei, X. Yang, X. Gao, and W. Xie. On the weighting exponent m in fuzzy C-means(FCM) clustering algorithm. In *SPIE proceedings series*, pages 246–251. SPIE, 2001.
33. H. Choe and J. B. Jordan. On the optimal choice of parameters in a fuzzy c-means algorithm. In *IEEE International Conference on Fuzzy Systems*, pages 349–354, Mar 1992.
34. S. Miyamoto and M. Mukaidono. Fuzzy c-means as a regularization and maximum entropy approach. *In Proceedings of the International Fuzzy Systems Association World Congress*, pages 86–92, 1997.
35. I. Gath and A. B. Geva. Unsupervised optimal fuzzy clustering. *IEEE Transactions on Pattern Analysis and Machine Intelligence*, 11(7):773–780, 1989.
36. D. E. Gustafson and W. C. Kessel. Fuzzy clustering with a fuzzy covariance matrix. *IEEE Conference on Decision and Control*, 17:761–766, 1978.
37. D. Tran, T.V. Le, and M. Wagner. Fuzzy Gaussian Mixture Models For Speaker Recognition. In *Fifth International Conference on Spoken Language Processing*, pages 759–762. ISCA, 1998.
38. R.J. Hathaway. Another interpretation of the EM algorithm for mixture distributions. *Statistics & probability letters*, 4(2):53–56, 1986.
39. H. Ichihashi, K. Miyagishi, and K. Honda. Fuzzy c-means clustering with regularization by KL information. In *Proc. of 10th IEEE International Conference on Fuzzy Systems*, volume 2, pages 924–927, 2001.

40. K. Honda and H. Ichihashi. Regularized linear fuzzy clustering and probabilistic pca mixture models. *IEEE Transactions on Fuzzy Systems*, 13(4):508–516, 2005.
41. S. Chatzis. A method for training finite mixture models under a fuzzy clustering principle. *Fuzzy Sets and Systems*, 161:3000–3013, 2010.
42. J. Zeng, L. Xie, and Z.Q. Liu. Type-2 fuzzy Gaussian mixture models. *Pattern Recognition*, 41(12):3636–3643, 2008.
43. X. Huang, Y. Ariki, and M. Jack. *Hidden Markov Models for Speech Recognition*. Series Edinburgh University Press, 1990.
44. J. A. Bilmes. A gentle tutorial on the EM algorithm and its application to parameter estimation for gaussian mixture and hidden markov models. *CA: International Computer Science Institute, ICSI-TR-97-0*, pages 1–5, 1998.
45. A. Björck and Å. Björck. *Numerical methods for least squares problems*. Society for Industrial Mathematics, 1996.
46. R. L. Cannon, J. V. Dave, and J. C. Bezdek. Efficient implementation of the fuzzy c-means clustering algorithms. *IEEE Transactions on Pattern Analysis and Machine Intelligence*, 8(2):248–255, March 1986.
47. A. Frank and A. Asuncion. UCI machine learning repository, 2010.
48. C. S. Chan and H. Liu. Fuzzy Qualitative Human Motion Analysis. *IEEE Transactions on Fuzzy Systems*, 17(4):851–862, 2009.
49. X. Ji and H. Liu. Advances in View-Invariant Human Motion Analysis: A Review. *IEEE Transactions on Systems, Man and Cybernetics Part C*, 40(1):13–24, 2010.
50. H. Liu. Exploring human hand capabilities into embedded multifingered object manipulation. *IEEE Transactions on Industrial Informatics*, 7(3):389–398, 2011.
51. Z. Ju and H. Liu. A unified fuzzy framework for human hand motion recognition. *IEEE Transactions on Fuzzy Systems*, 19(5):901–913, 2011.

Chapter 6
Fuzzy Empirical Copula for Estimating Data Dependence Structure

6.1 Introduction

The information in the world is becoming more and more electronic. Due to the improvements in data collection and storage during the past decades, huge amounts of data can lead to the problem of information overload [1] to many researchers in domains such as engineering, economics and astronomy. The increase of the number of dimensions associated with each observation and growth of the sampling time points are the main reasons of information overload. In many cases, datasets contain not only useful messages but also considerable trivial and redundant information both in the dimensions (attributes) and samples. How to remove the redundant information and maintain the important information is crucial in many applications. Two important methods are normally employed to solve this problem: dimensionality reduction and clustering. The formal reduces trivial attributes, maintaining the number of samples, while the later eliminates the redundant samples without changing the number of attributes.

There are various traditional and current state of the art dimensionality reduction methods to solve the above problem. Principal Component Analysis (PCA) was invented in 1901 by Karl Pearson [2] and is mostly used for dimensionality reduction in a dataset by retaining the characteristics of the dataset that contribute most to its variance. It keeps lower-order principal components and ignores higher-order ones. Such lower-order components often contain the "most important" aspects. Like PCA, Factor analysis (FA) is another second-order method [3]. FA becomes essentially equivalent to PCA if the "errors" in the FA model are all assumed to have the same variance. These second-order methods require classical matrix manipulations and assumption that datasets are realisations from Gaussian distributions. For non-Gaussian datasets, higher-order dimension reduction methods such as Projection Pursuit (PP) [4] and Independent Component Analysis (ICA) [5] are introduced. Additionally, non-linear PCA can also deal with non-Gaussian datasets using non-linear objective functions to determine the optimal weights in principal [6]. Its resulting components are still linear combinations of the original variables, so it can be

H. Liu et al., *Human Motion Sensing and Recognition*,
Studies in Computational Intelligence 675, DOI 10.1007/978-3-662-53692-6_6

regarded as a special case of ICA. Other non-linear methods such as Principal Curves (PC) [7] and Self Organising Maps (SOM) [8] can be thought to be non-linear ICA [9] in that they replace the linear transformation of ICA with a real-valued non-linear vector function. Curvilinear Component Analysis (CCA) is a relatively new non-linear mapping method, being improved from Sammon's mapping by Jeanny Heault and Pierre Demartines [10]. It uses a new cost function able to unfold strongly non-linear or even closed structures, which significantly speeds up the calculation and interactively helps users control the minimised function. However, more parameters should be considered for most of these high-order and non-linear dimensionality reduction methods and their performances strongly depend on complex adjustments of these parameters, for instance there are three parameters in CCA: the projection space dimension and the two time decreasing parameters.

However, dimensionality reduction methods can not be used during estimating data dependence structure, because dependence structure includes all the interrelations of the attributes and high-order attributes are not supposed to be ignored. Clustering is the classification of objects into clusters so that objects from the same cluster are more similar to each other than objects from different clusters. It can effectively reduce the number of data samples, so it is suitable for reducing the redundant information when estimating data dependence structure. The most common algorithms include K-means [11], fuzzy C-means [12], and fuzzy C-means-derived clustering approaches such as fuzzy J-means [13] and fuzzy SOM [14], which construct clusters on the basis of pairwise distance between objects, so that they are incapable of capturing non-linear relationships and thereby fail to represent a dataset with non-linear structure. Hierarchical clustering is another important approach but suffers from lack of robustness, non-uniqueness, and inversion problems [15]. Gaussian Mixture Model (GMM) is based on the assumption that datasets are generated by a mixture of Gaussian distributions with certain probability. But this assumption is not always satisfied for general datasets even after various transformations aimed at improving the normality of the data distribution [16, 17].

Copula is a general way of formulating a multivariate distribution with uniform marginal distributions in such a way that various general types of dependence can be presented. The copula of a multivariate distribution can be considered as the part describing its dependence structure as opposed to the behaviour of each of its margins [18]. It is a good way of studying scale-free measures of dependences among variables and also a good starting point for constructing families of bivariate distributions [19]. Sklar's theorem [20] elucidates that a multivariate distribution function can be represented by a copula function which binds its univariate margins. Further, empirical copulas were introduced and first studied by Deheuvels in 1979 [21, 22], which can be used to study the interrelations of marginal variables with unknown underlying distributions. The copula approach has many advantages [23] and has been used widely in finance [24–27] and econometrics [28–30]. Kolesarova et al. [31] defined a new copula called discrete copulas on a grid of the unit square and showed that each discrete copula is naturally associated with a bistochastic matrix. Baets and Meyer [32] also presented a general framework for constructing copulas, which extended the diagonal construction to the orthogonal grid construction. Simultaneously, empirical copula has gained an increasing amount of attention recently. Dempster et al. [33] constructed an empirical copula for Collateralized debt obligation tranche pricing and achieved a better performance than the dominant base correlation approach in pricing non-standard tranches. Ma and Sun [34] proposed a Chow-Liu like method based on a dependence measure via empirical copulas to estimate maximum spanning product copula with only bivariate dependence relations, while Morettin et al. [35] proposed wavelet estimators based on empirical copulas which can be used for independent, identically distributed time series data.

It is evident, however, that the efficiency of empirical copula is outstandingly poor though it provides effective performance on data dependence structure estimation. It is common that natural datasets are represented by tremendous storage size, and it is impossible to process them using empirical copula in most cases. In order to overcome this problem, we propose an algorithm named fuzzy empirical copula which integrates fuzzy clustering with empirical copula. Fuzzy Clustering by Local Approximation of Memberships (FLAME) [17] is firstly extended into multi-dimensional space, then the FLAME$^+$ is utilised to reduce the number of sampling data and maintaining the interrelations at the same time before data dependence structure estimation takes over. The remainder of the chapter is organised as follows. Section 6.2 presents copula theory with a focus on dependence structure estimation using empirical copula. Section 6.3 proposes the Fuzzy empirical copula algorithm. Section 6.4 presents the experiments whose results demonstrate the effectiveness of the proposed fuzzy empirical copula. Concluding remarks and future work are found in Sect. 6.5.

6.2 Dependence Structure Estimation via Empirical Copula

As a general way of formulating a multivariate distribution, copula can be used to study various general types of dependence between variables. Other ways of formulating multivariate distributions include conceptually-based approaches in which the real-world meaning of the variables is used to imply what types of relationships might occur. In contrast, the approach via copulas might be considered as being more raw, but it does allow much more general types of dependencies to be included than would usually be invoked by a conceptual approach. Nelsen [19] has proven that these measures, such as Kendall's tau, Spearman's rho and Gini's gamma, can be re-expressed only in terms of copula. Though their direct calculation may have much less computational cost than when using copulas, copula summarises all the dependence relations and provides a natural way to study and measure dependence between variables in statistics. It is a very important approach since copula properties are invariant under strictly increasing transformations of the underlying random variables. Spearman's rho and Gini's gamma are considered in this chapter. In this section, we firstly revisit the theoretical foundation of copula and empirical copula, then introduce the theorem of calculating Spearman's rho and Gini's gamma using bivariate empirical copula, finally analyse the time complexity of the computation.

6.2.1 Copula

A n-dimensional copula is defined as a multivariate joint distribution on the n-dimensional unit cube $[0, 1]^n$ such that every marginal distribution is uniform on the interval [0, 1].

Definition 6.2.1.1 A n-dimensional copula is a function C from I^n to I with the following properties [19]:

1. C is grounded, i.e., for every $\mathbf{u}$ in I^n, $C(\mathbf{u}) = 0$ if at least one coordinate $u_j = 0$, $j = 1, \ldots, n$.
2. If all coordinates of $\mathbf{u}$ are 1 except for some u_j, $j = 1, \ldots, n$, then $C(\mathbf{u}) = C(1, \ldots, 1, u_j, 1, \ldots, 1) = u_j$.
3. C n-increasing, i.e., for each hyperrectangle $B = \times_{i=1}^{n}[x_i, y_i] \subseteq [0, 1]^n$

$$V_c(B) = \sum_{z \in \times_{i=1}^{n}\{x_i, y_i\}} (-1)^{N(z)} C(z) \geq 0 \tag{6.1}$$

where the $N(z) = card\{k \,|z_k = x_k\}$. $V_c(B)$ is the so called C-volume of B.

Sklar's Theorem [20] is central to the theory of copula and underlies most applications of the copula. It elucidates the role that copula plays in the relationship between multivariate distribution functions and their univariate margins.

Sklar's Theorem 6.2.1.1 *Let H be a joint distribution function with margins* $F_i(i = 1, 2, \ldots, n)$. *Then there exists a copula C such that for all* x_i *in* $\bar{R}$,

$$H(x_1, \ldots, x_n) = C(F_1(x_1), \ldots, F_n(x_n)) \tag{6.2}$$

where C is a n-dimensional copula, F_i *are marginal distribution function of* x_i.

If $F_i(i = 1, \ldots, n)$ are continuous, C is unique. If C is a n-dimensional copula and $F_i(i = 1, \ldots, n)$ are distribution functions, then the function H defined by Eq. 6.2 is a joint distribution function with margins $F_i(i = 1, \ldots, n)$. More details can be seen in [19, 23].

6.2.2 Empirical Copula and Dependence Estimation

The empirical copula is a characterisation of the dependence function between variables based on observational data using order statistics theory and it can reproduce any pattern found in the observed data. If the marginal distributions are normalised, the empirical copula is the empirical distribution function for the joint distribution. Priority has been given to bivariate empirical copula due to computational cost. The reason is twofold: one is that the interrelation between every two attributes is the basic relationship in most attributes, and it is practical to use bivariate empirical copula to construct the whole structure of every two attributes' dependence; the second is that the dependence structure of dataset X including r attributes would have $\binom{r}{2} = \frac{1}{2}r(r-1)$ bivariate interrelations. Bivariate empirical copula is given as follows.

Definition 6.2.2.1 Let $\{(x_k, y_k)\}_{k=1}^{n}$ denote a sample of size n from a continuous bivariate distribution. The empirical copula is the function C_n given by

$$C_n(\tfrac{i}{n}, \tfrac{j}{n}) = \frac{card\{(x,y): x \le x_{(i)}, y \le y_{(j)}\}}{n} \tag{6.3}$$

where $x_{(i)}$ and $y_{(j)}$, $1 \le i, j \le n$, denote order statistics from the sample [19].

The empirical copula frequency c_n is given by

$$c_n(\tfrac{i}{n}, \tfrac{j}{n}) = \begin{cases} \frac{1}{n}, & \text{if } (x_{(i)}, y_{(j)}) \text{ is an element of the sample} \\ 0, & \text{otherwise} \end{cases} \tag{6.4}$$

Note that C_n and c_n are related via

$$C_n(\frac{i}{n}, \frac{j}{n}) = \sum_{p=1}^{i}\sum_{q=1}^{j} c_n(\frac{p}{n}, \frac{q}{n}) \tag{6.5}$$

Theorem 6.2.2.1 *Let C_n and c_n denote, respectively, the empirical copula and the empirical copula frequency function for the sample $\{(x_k, y_k)\}_{k=1}^{n}$. If ρ and γ denote, respectively, the sample versions of Spearman's rho, and Gini's gamma [36, 37], then*

$$\rho = \frac{12}{n^2-1}\sum_{i=1}^{n}\sum_{j=1}^{n}\left[C_n\left(\tfrac{i}{n}\cdot\tfrac{j}{n}\right) - \tfrac{i}{n}\cdot\tfrac{j}{n}\right] \tag{6.6}$$

and

$$\gamma = \frac{2n}{\lfloor n^2/2\rfloor}\left\{\sum_{i=1}^{n-1} C_n\left(\tfrac{i}{n}, 1-\tfrac{i}{n}\right) - \sum_{i=1}^{n}\left[\tfrac{i}{n} - C_n\left(\tfrac{i}{n}, \tfrac{i}{n}\right)\right]\right\} \tag{6.7}$$

Spearman's rho and Gini's gamma are two ways of measuring two variables' association [19]. According to the definition and theorem, we can estimate correlations between variables using empirical copula and Spearman's rho & Gini's gamma. Suppose the number of objects is n and number of attributes is r. For $r << n$, according to the Eqs. 6.3, 6.6 and 6.7, the time complexity of Spearman's rho or Gini's gamma is $O(n^3)$.

6.3 Fuzzy Empirical Copula

In this section Fuzzy clustering by Local Approximation of Memberships (FLAME) is extended first in terms of dimension and distance functions, then is integrated into empirical copula to enhance its computational efficiency. FLAME was proposed to cluster DNA microarraydata [17]. It defines clusters in the relatively dense regions of a dataset and performs cluster assignment solely based on the neighbourhood relationships among objects. One of the FLAME algorithm features is that the memberships of neighbouring objects in the fuzzy membership space are set according to the neighbourhood relationships among neighbouring objects in the feature space. FLAME has been extended in terms of dimension and distance function (i.e., FLAME$^+$), which still consists of three main steps of FLAME algorithm: initialization, approximation and assignment.

6.3.1 *Initialisation*

The first step, initialization, is to classify three types of objects: Cluster Supporting Object (CSO), cluster outliers and the rest which are named Normal Points (NPs).

Let X be a r-dimensional dataset with n objects. The r-dimensional distance between two instances is

$$d_p(x, y) = (\sum_{i=1}^{r} |x_i - y_i|^p)^{(1/p)} \tag{6.8}$$

where $x, y \in X$; $1 \leq p \leq \infty$; d_1 is the Manhattan distance, d_2 is the familiar Euclidean distance, and d_∞ corresponds to the maximum distance in any dimension. Then the similarity of these two objects is calculated as:

$$s_{xy} = \frac{1}{d_p(x,y)} \tag{6.9}$$

Similarity is the degree of resemblance between two or more objects. There are different ways to calculate the similarity. Since "the density of each object is calculated as one over the average distance to the k-nearest neighbors" in the FLAME clustering algorithm [17], to make the relation between similarity and density more direct and simple, we choose Eq. 6.9 to calculate the similarity in the chapter.

The K-Nearest Neighbours (KNNs) for each object are defined as the k objects ($k \leq n$) with the k highest similarity. The density of object x with KNNs can be obtained

$$Den_p(x) = \frac{k}{\sum_{y \in knn(x)} d_p(x,y)} \tag{6.10}$$

where $knn(x)$ stands for the set of KNNs of the object x.

Subsequently, the set of CSOs is defined as the set of objects with local maximum density, i.e., with a density higher than that of every object in their KNNs. The higher k is, the less CSOs will be identified, then less clusters will be generated. A density threshold needs defining to find possible cluster outliers, so objects with densities below the threshold are defined as possible outliers.

Each object x is associated with a membership vector $p(x)$, in which each element $p_i(x)$ indicates the membership degree of x in cluster i

$$p(x) = (p_1(x), \ldots, p_m(x)), \tag{6.11}$$

where $0 \le p_i(x) \le 1$; $\sum_{i=1}^{m} p_i(x) = 1$; m is the total number of CSOs and the outlier cluster, i.e., $m = c + 1$ where c is the number of CSOs; Each element of membership vector takes value between 0 and 1, indicating how much percentage an object belonging to a cluster, or being an outlier.

Based on the density estimation, each CSO is assigned with fixed and full membership to itself to represent one cluster, for example $p(x) = (0, 1, \ldots 0)$ indicates that object x is the second CSO. Each outlier is assigned with fixed and full membership to the outlier group, $p(x) = (0, \ldots, 0, 1)$, and the NP is assigned with equal memberships to all clusters and the outlier group, $p(x) = (1/m, \ldots, 1/m)$.

6.3.2 Approximation

The second step is named local/neighbourhood approximation of fuzzy memberships, in which each NP's fuzzy membership is updated by a linear combination of the fuzzy memberships of its KNNs, while CSOs and outliers maintain the fixed and full memberships to themselves respectively.

The weights defining how much each neighbour will contribute to approximation of the fuzzy membership of that neighbour are estimated in Eq. 6.12, based on the fact that the neighbours that have higher similarities must have higher weights.

$$w_{xy} = \frac{s_{xy}}{\sum_{z \in knn(x)} s_{xz}} \tag{6.12}$$

where $y \in knn(x)$. The membership vector of each NP is approximated according to Eq. 6.13, minimising the overall difference between membership vectors and their approximations.

$$p^{t+1}(x) = \sum_{y \in knn(x)} w_{xy} p^t(y) \tag{6.13}$$

The overall local/neighbourhood approximation error is calculated by:

$$E(\{p\}) = \sum_{x \in X} \left\| p(x) - \sum_{y \in \text{knn}(x)} w_{xy} p(y) \right\|^2 \tag{6.14}$$

The iteration of Eq. 6.13 breaks under the condition that $E(\{p\})$ is less than a predetermined threshold.

6.3.3 Assignment

Finally, it is to assign each object to the cluster based on its fuzzy membership. Usually, one cluster contains the objects that have higher membership degrees in this cluster than other clusters.

An example of FLAME$^+$ is provided in Fig. 6.1, where a dataset with 600 objects is randomly generated from a 3 dimensional distribution. FLAME$^+$ is applied to this dataset and three groups of objects are clustered as outliers, CSOs and NPs.

Accurately calculating the entire time complexity of FLAME$^+$ is very challenging in that each iteration of local/neighbourhood approximation depends on the error threshold. However, it is necessary to analyse the complexity of the first step of the algorithm. Suppose the number of objects is n, number of attributes is r, its CSOs' number is c and number of nearest neighbours is k. For $r << n$ and $k << n$, the time complexity of the initialisation is $O(n^2)$. An empirical study of the time complexity

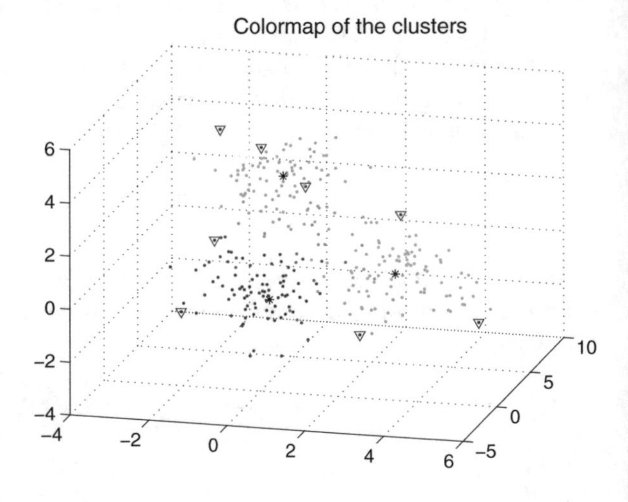

Fig. 6.1 Clustering random 3D Euclidean positions using FLAME$^+$. The star points in *black* are the centres of the clusters (CSO); points labeled with *triangles* are the outliers; the colour range of NPs represents their membership degrees

of FLAME$^+$ compared with other algorithms is performed to illustrate that FLAME$^+$ has significant computational advantage over hierarchical clustering, fuzzy C-means and fuzzy SOM, an exception is K-means [17]. In Sect. 6.4, FLAME+ and K-means are compared in the context of fuzzy empirical copula.

We aim at developing an algorithm which can efficiently reduce the computational cost of empirical copula by filtering out redundant information in the sample. In addition, this algorithm should also be capable of dealing with arbitrary-distributed datasets in order to inherit the main advantage of empirical copula for data structure estimation. FLAME algorithm is selected for this purpose in that it not only fulfils the above requirements but also possesses the merit of few parameters, i.e., the number of nearest neighbours and the value of the outlier's threshold.

It is evident that samples with higher densities are more reliable when used to represent whole samples in such a way that the main feature of the whole sample is maintained. The FLAME algorithms have the capability of identifying those "special" sampling points based on the objects' density analysis. The "special" points are represented by the CSOs with the highest densities in all clusters. Therefore, the fuzzy empirical copula algorithm is proposed for achieving the following: high dimension FLAME algorithm is employed to identify characteristic feature points, and dependence structure, on the other hand, it is estimated via empirical copula.

Let X be r-dimensional dataset with n objects:

$$X = \begin{pmatrix} x_{11} & \cdots & x_{1n} \\ \vdots & \ddots & \vdots \\ x_{r1} & \cdots & x_{rn} \end{pmatrix}$$

so the ith object is represented by the ith column in matrix X: $x_i = [x_{1i}, x_{2i}, \ldots, x_{ri}]^T$ and the jth attribute of X is defined as the jth row: $x(j) = [x_{j1}, x_{j2}, \ldots, x_{jn}]$. The dependence structure in this chapter is defined as the whole structure of every two attributes' dependence which can be calculated by bivariate empirical copula, because the interrelation between every two attributes is the most basic relationship in several attributes. Given interrelations between every two attributes, relations of three or more attributes would be derived from their dependence structure which would have $\binom{r}{2} = \frac{1}{2}r(r-1)$ interrelations (r is the number of attributes).

In fuzzy empirical copula, firstly FLAME$^+$ reduces the samples from n objects to c CSOs, and then empirical copula analyses the dependence of every two attributes in the derived CSO matrix. The first step can be considered as the operation on the column and the later on the row. For ideal performance of the proposed algorithm, one of outputs of FLAME$^+$, CSOs, is computed leading to efficient computation. That is to say we only have to implement the first step in the FLAME algorithm which has less time complexity than empirical copula and only one parameter, the number of neighbours, is required since the threshold works only for outliers. The Spearman's rho and Gini's gamma of the CSOs would be

$$\rho(u, v) = \frac{12}{c^2-1} \sum_{i=1}^{c} \sum_{j=1}^{c} \left[C_c^{(uv)} \left(\frac{i}{c}, \frac{j}{c} \right) - \frac{i}{c} \cdot \frac{j}{c} \right] \tag{6.15}$$

and

$$\gamma(u, v) = \frac{2n}{\lfloor n^2/2 \rfloor} \left\{ \sum_{i=1}^{c-1} C_c^{(uv)} \left(\frac{i}{c}, 1 - \frac{i}{c} \right) - \sum_{i=1}^{c} \left[\frac{i}{c} - C_c^{(uv)} \left(\frac{i}{c}, \frac{i}{c} \right) \right] \right\} \tag{6.16}$$

where $u \in [1, \ldots, r)$ and $v \in (u, \ldots, r]$; $C_c^{(uv)}$ is the bivariate empirical copula of the uth and vth attributes with c objects, and $C_c^{(uv)} = C_c^{(vu)}$.

The optimisation is designed to automatically identify the optimised number of neighbours with acceptable errors. The number of nearest neighbours is increased by one at every step during the optimisation until proper number of neighbours is identified. The optimisation stops when the overall error of Spearman's rho or Ginis gamma in equation is under the preset error threshold. The pseudo-code of fuzzy empirical copula is presented in Appendix D.1 and its 'EmpSG' function is in Appendix D.2.

6.4 Experiment and Discussions

Experiments are conducted in this section, and results and discussions are provided for evaluating the effectiveness and efficiency of fuzzy empirical copula. After a brief explanation of the datasets, empirical copula and fuzzy empirical copula are employed respectively to estimate the dependence structures of the datasets. The section is concluded with the roles that clustering algorithms play in fuzzy empirical copula.

6.4.1 Data

Abalone [38] and yeast [39] datasets from UCI machine learning repository [40] were selected to evaluate the proposed algorithm in this chapter. The abalone dataset was used to predict the age of abalone from the physical measurements such as weight and length, and it is not a trivial task to get their ages by counting the number of rings in their bodies through a microscope. 4177 abalone are sampled with 9 attributes in this dataset. Figure 6.2 shows interrelations of length, diameter, whole weight and shell weight. This dataset could be regarded as 9 dimensional data with 4177 samples in which some measurements are intrinsically interrelated. The yeast dataset was constructed for predicting the cellular localisation sites of proteins. It contains 1484 instances with 8 attributes for each instance. Both of these two datasets contain strong non-linear dependences between attributes. Priority herein is given to the sampling density and the interrelations among attributes. Supposing one dataset

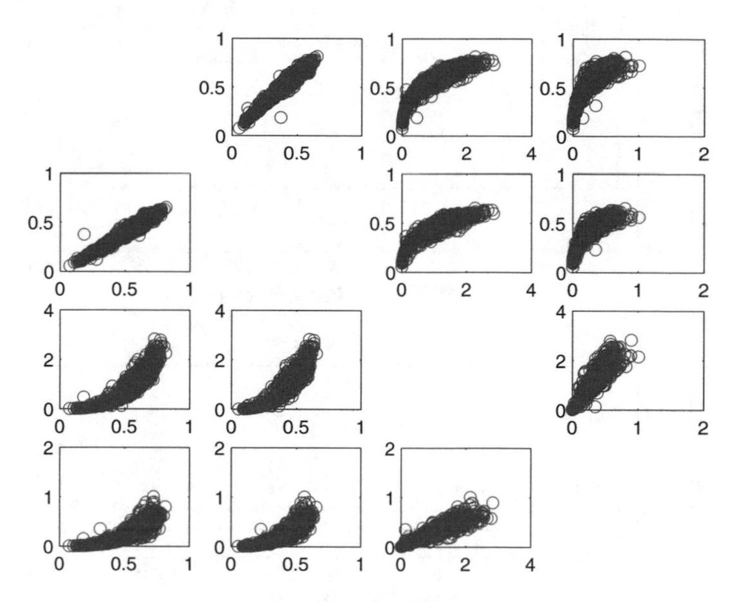

Fig. 6.2 The interrelations of length, diameter, whole weight and shell weight of abalone dataset

contains s attributes, its dependence structure includes $\binom{s}{2}$ interrelations of every two attributes among this dataset, which means the abalone and yeast datasets have two dependence structures of 36 and 28 interrelations to be analysed respectively.

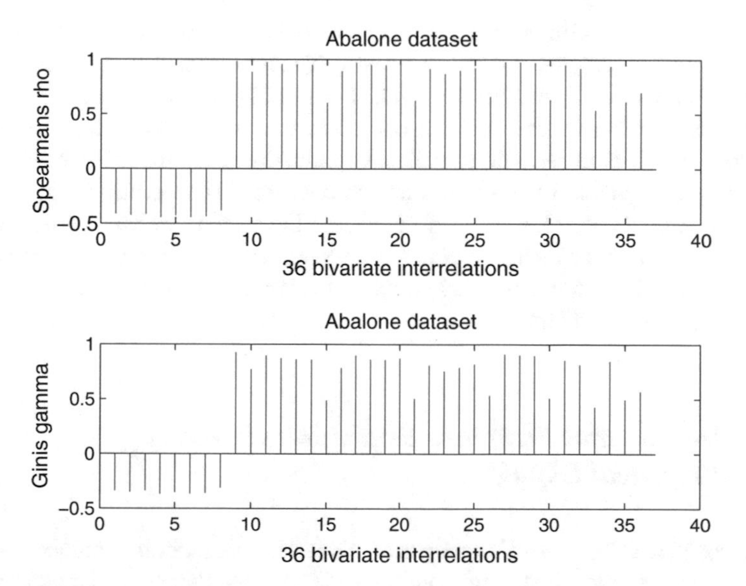

Fig. 6.3 The result of dependences of 36 correlations of the original abalone dataset

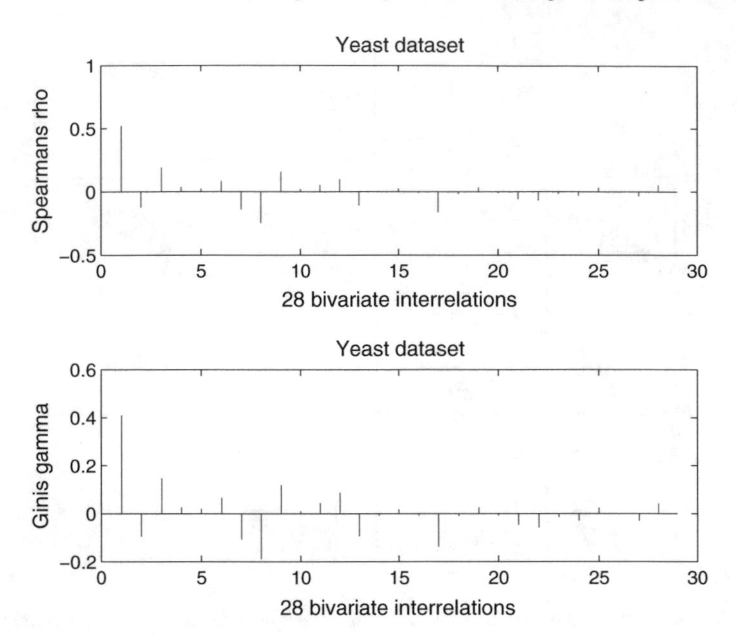

Fig. 6.4 The result of dependences of 28 correlations of the original yeast dataset

6.4.2 Dependence Structure Estimation via Empirical Copula

Spearman's rhos and Gini's gammas are calculated according to Eqs. 6.6 and 6.7 via empirical copula using the two above datasets. The results of abalone's dependences of 36 correlations for 9 attributes are listed in the Fig. 6.3, and yeast dataset of 28 correlations for 8 attributes in the Fig. 6.4. The whole computation time for abalone is 27226 s, which is unrealistic for related applications. On the other hand, though the yeast dataset has fewer instances its computational time is still high, at 2813 s. It should be noted that it has to carefully handle the tradeoff between efficiency and accuracy of fuzzy empirical copula. The more nearest neighbours are considered (e.g., Fig. 6.3), the fewer CSOs will appear, the more efficient the algorithm is. It also, however, leads to larger error.

6.4.3 Dependence Structure Estimation via Fuzzy Empirical Copula

The proposed fuzzy empirical copula is employed in this section to reduce computation time for dependence structure analysis of these two datasets. The more nearest neighbours are considered, the fewer CSOs will appear. It indicates that the calculation of fuzzy empirical copula will be more efficient at the cost of lower accuracy.

The proposed fuzzy empirical copula has the ability to identify the proper number of nearest neighbours which guarantees the fast computation with the overall error under the preset threshold. The threshold for Spearman's rho from Eq. 6.15 is predefined as,

$$threshold = p \times \binom{s}{2} = \frac{ps(s-1)}{2} \tag{6.17}$$

where $\binom{s}{2}$ is the number of interrelations, combining 2 attributes out of s attributes; p is the average error percentage for each interrelation. Different threshold results in different computational time, and p is defined to take a value in the range from 0.5 to 1% according to the different features of datasets. From the above results of the two datasets using empirical copula, p is predefined as 0.6 and 1% for abalone and yeast datasets respectively, which indicates that abalone and yeast datasets have the thresholds of 0.228 and 0.28.

6.4.3.1 Abalone Dataset

Under the overall error threshold of 0.228 for Spearman's rho, fuzzy empirical copula with 12 nearest neighbours has the lowest computation time. Thanks to the FLAME^{+}'s density sampling, when the number of nearest neighbours is 12, the number of data instances is reduced to 100 from 4177 depicted in Fig. 6.5. The

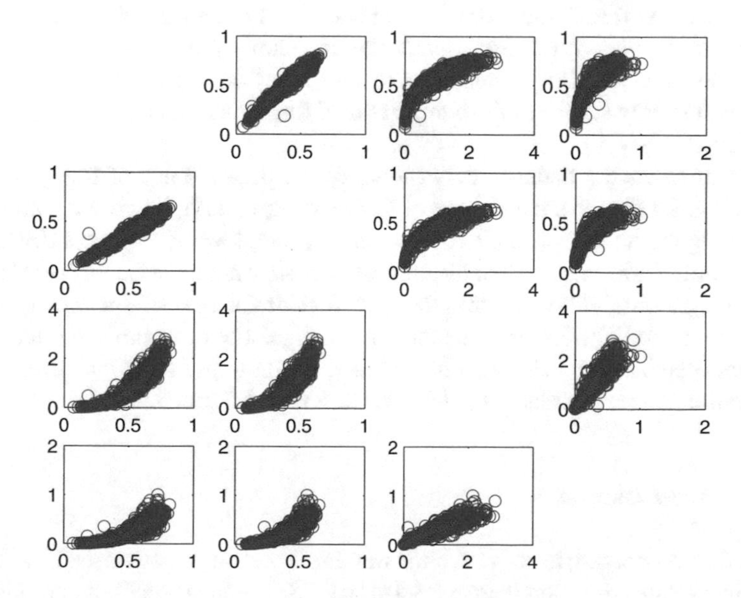

Fig. 6.5 Result of sampled abalone data using 12-neighbour-FLAME^{+}

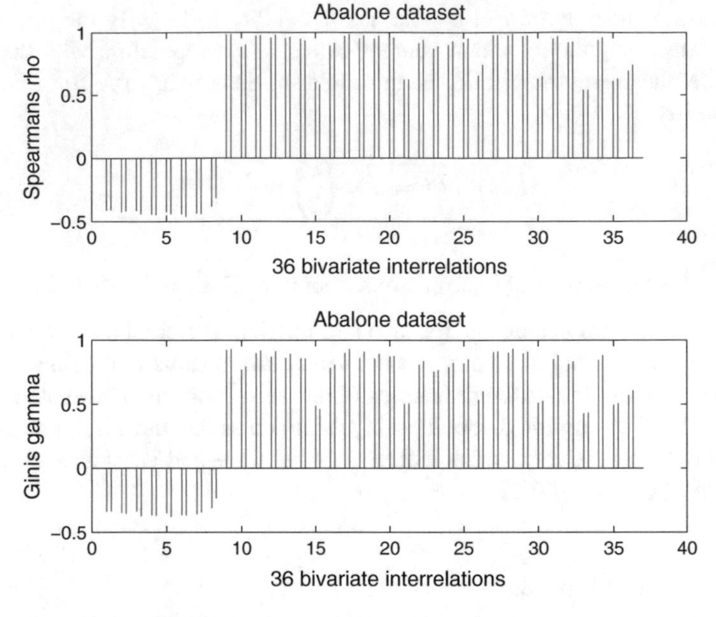

Fig. 6.6 Comparison of Spearman's rhos and Gini's gammas. *Blue lines* are the correlations of Empirical Copula algorithm while *red lines* are of Fuzzy Empirical Copula algorithm

36 interrelations of the 9 attributes of these 100 CSOs are estimated as shown in red line in Fig. 6.6, where the blue lines are results generated from empirical copula. It illustrates that fuzzy empirical copula with 12 nearest neighbours does not cause unacceptable error to Spearman's rho and Gini's gamma compared to empirical copula. However, the computation time of fuzzy empirical copula algorithm is 68 s, which is only 0.25% of the computational time conducted by empirical copula algorithm.

In order to have a better understanding of the performance of fuzzy empirical copula, Fig. 6.7 displays the change of CSO's number with the growing number of nearest neighbours from 1 to 20. It shows that the numbers of abalone's CSOs drops exponentially with the growth of the number of nearest neighbours. With the growing nearest neighbours, Fig. 6.8 shows the overall error changes of Spearmans rho and Ginis gamma and Fig. 6.9 presents the time change. The error threshold locates the place between 12 and 13 nearest neighbours. Given a threshold, the error and the computational time can easily be decided from Figs. 6.8 and 6.9.

6.4.3.2 Yeast Dataset

Similar data processing in above section was employed on yeast dataset. If the threshold of Spearman's rho overall error is set to be 0.28, 2 nearest neighbours would make the estimation perform well because of the relatively small error which is under the

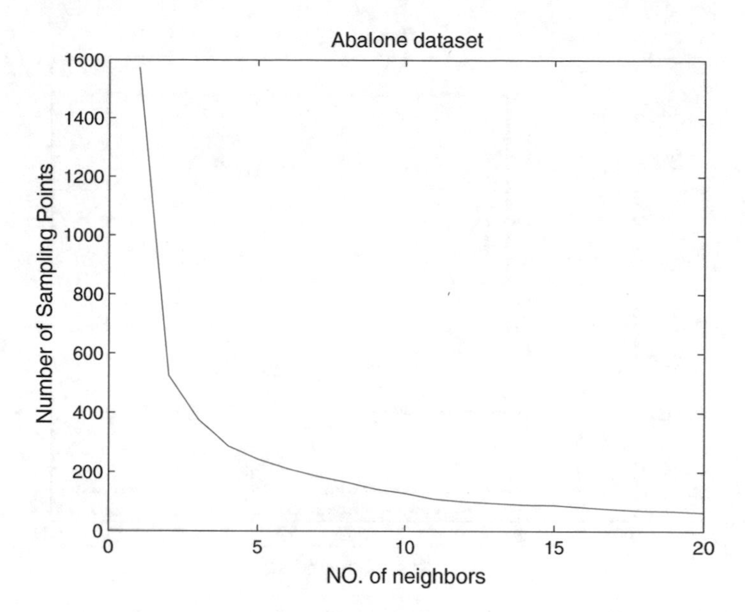

Fig. 6.7 The relationship between number of nearest neighbours and number of abalone's CSOs

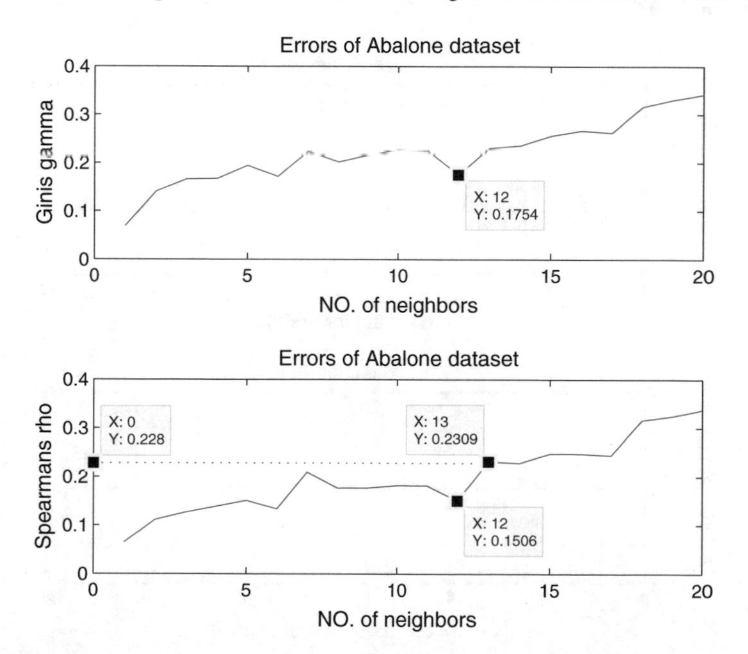

Fig. 6.8 Change of overall errors of Spearman's rhos and Gini's gammas with the growth of number of nearest neighbours in FLAME$^+$

threshold and fast computation. The result with 2 nearest neighbours is shown in red in Fig. 6.10, compared with the result in blue from Empirical Copula, which demonstrates that the dependence structure of yeast dataset is maintained. The errors are

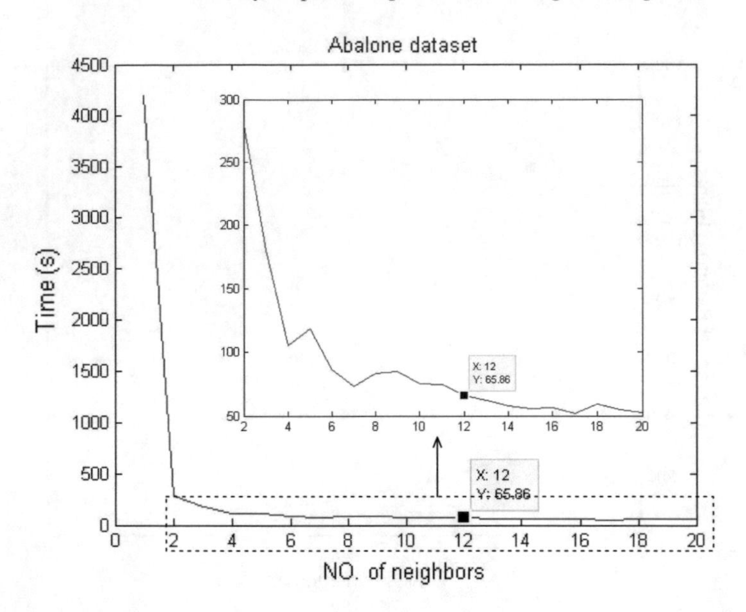

Fig. 6.9 Change of time costed by Fuzzy Empirical Copula with the growth of number of nearest neighbours in FLAME$^+$

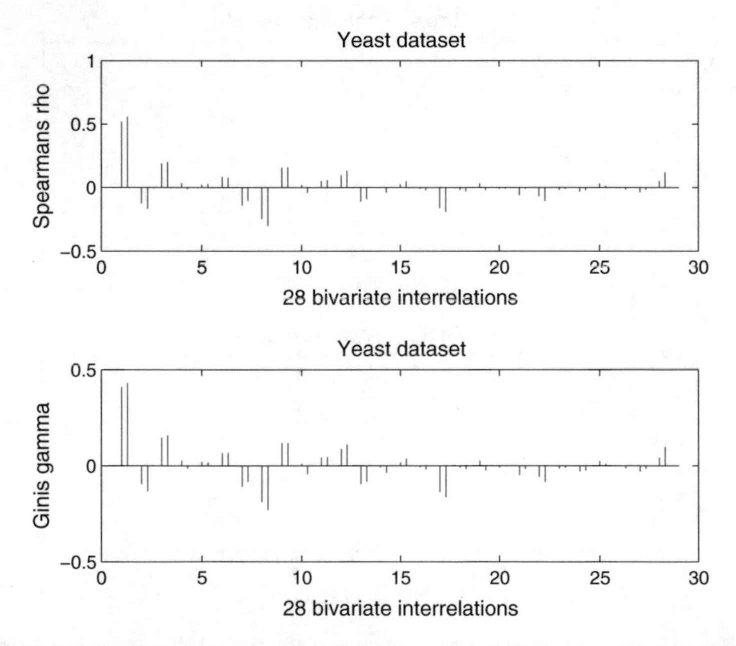

Fig. 6.10 Comparison of Spearman's rhos and Gini's gammas. *Blue lines* are the correlations of Empirical Copula algorithm while *red lines* are of Fuzzy Empirical Copula algorithm

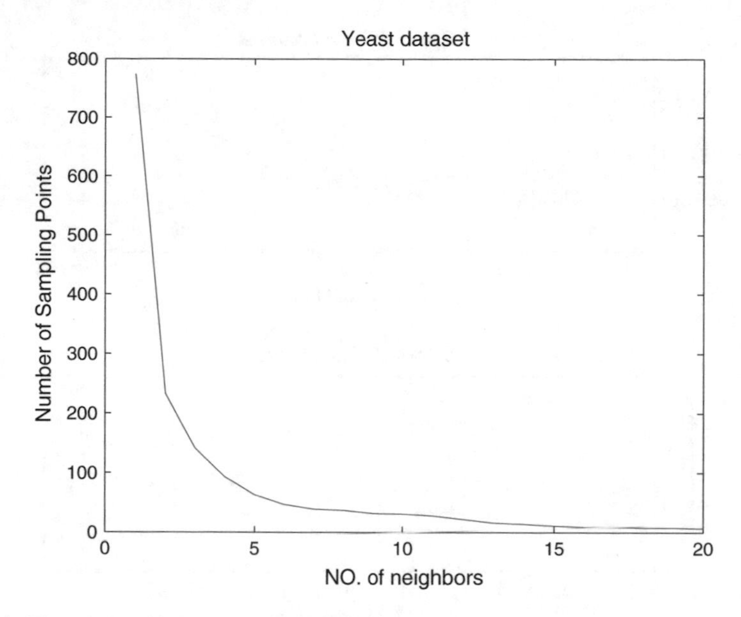

Fig. 6.11 The relationship between number of nearest neighbours and number of yeast's CSOs

only 0.28 for Spearman's rho and 0.24 for Gini's gamma in Fig. 6.10, and computation time is 26.3 s which is only 0.93% of the time cost by Empirical Copula.

The number of yeast's CSOs also drops exponentially with the growth of the number of nearest neighbours from 1 to 20 shown in Fig. 6.11. Figure 6.12 shows the changes of errors of Spearman's rho and Gini's gamma grows with the number of nearest neighbours, while Fig. 6.13 displays the changes of cost time.

However, for 5 or more nearest neighbours, the number of sampling data is reduced to less than 70 instances in Fig. 6.11, which are so few that the result becomes unacceptable with huge errors since the sampling data can not cover the main feature area of the whole dataset.

6.4.4 Comparison of FLAME$^+$ and K-Means in Fuzzy Empirical Copula

The reason to choose FLAME$^+$ as the fuzzy clustering algorithm instead of other algorithms (e.g., K-means) in Fuzzy Empirical Copula is on the basis of FLAME$^+$'s four main advantages. First it has the ability to capture non-linear relationships and non-globular clusters; secondly it can automatically define the number of clusters and identify cluster outliers; thirdly, compared with K-means and fuzzy C-means, the centres of FLAME$^+$ are real instances in the original dataset instead of the centroids of clusters with different traits which probably result in wrong dependence measures.

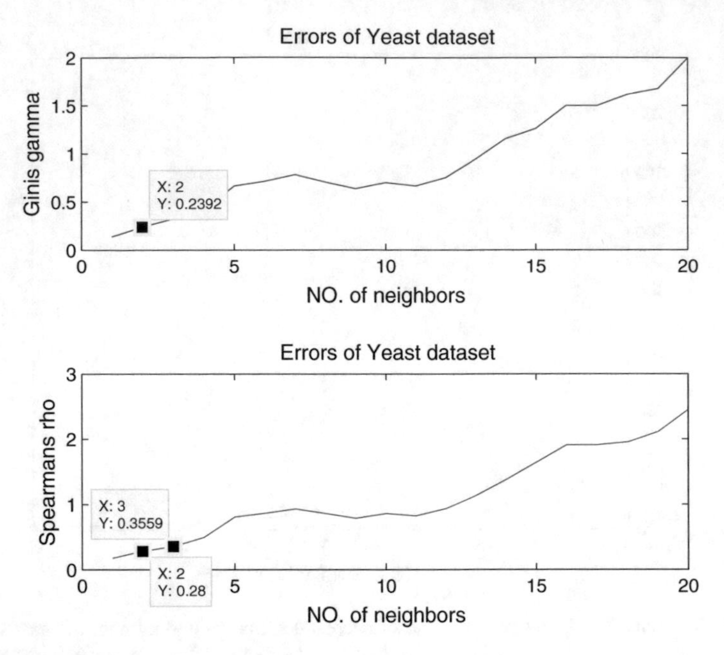

Fig. 6.12 Change of overall errors of Spearman's rhos and Gini's gammas with the growth of number of nearest neighbours in FLAME$^+$

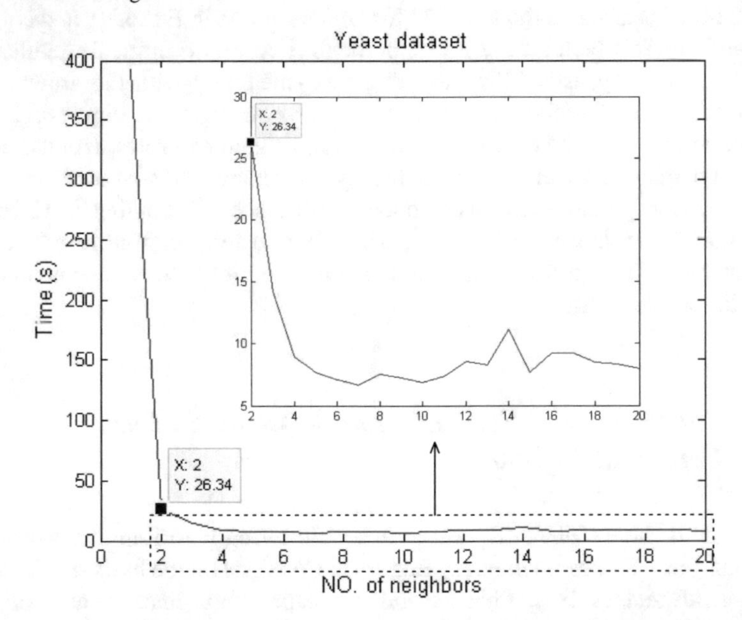

Fig. 6.13 Change of time cost by Fuzzy Empirical Copula with the growth of number of nearest neighbours in FLAME$^+$

Finally, FLAME$^+$ is also capable of dealing with a free-distributed dataset, which is not always true for algorithms like Gaussian Mixture Models.

In order to demonstrate the effectiveness of the proposed fuzzy empirical copula, we constructed K-means based Empirical Copula which employs the K-means algorithm to cluster the original dataset into numbers of subsets and uses centroids as the new dataset. Both of these two methods were applied to the abalone dataset. The comparison is based on the fact that cluster number of K-means is set to be the same as the abalone's CSOs number of FLAME$^+$ as listed in the Table 6.1.

Figure 6.14 demonstrates the comparison of computational cost by the proposed fuzzy empirical copula and K-means based empirical copula, where red curves are the changes of cost time by K-means based Empirical Copula while blue curves are by the proposed fuzzy empirical copula. It presents that both of the two algorithms achieve almost the same performance in saving the computational time.

In Fig. 6.15, with the decreasing number of clusters, the errors caused by K-means based Empirical Copula fluctuate violently and keep much higher than those by proposed fuzzy empirical copula. It illustrates that FLAME$^+$ outperforms K-means in maintaining the dependence structure though both of them have the almost same performance in reducing the cost time, and FLAME$^+$ is more suitable to be

Table 6.1 Number of clusters corresponding to number of nearest neighbors

Neighbors	1	2	3	4	5	6	7	8	9	10	11	12	13	14	15	16	17	18	19	20
Clusters	1574	525	377	287	243	211	185	165	142	128	109	100	94	89	88	81	74	69	67	63

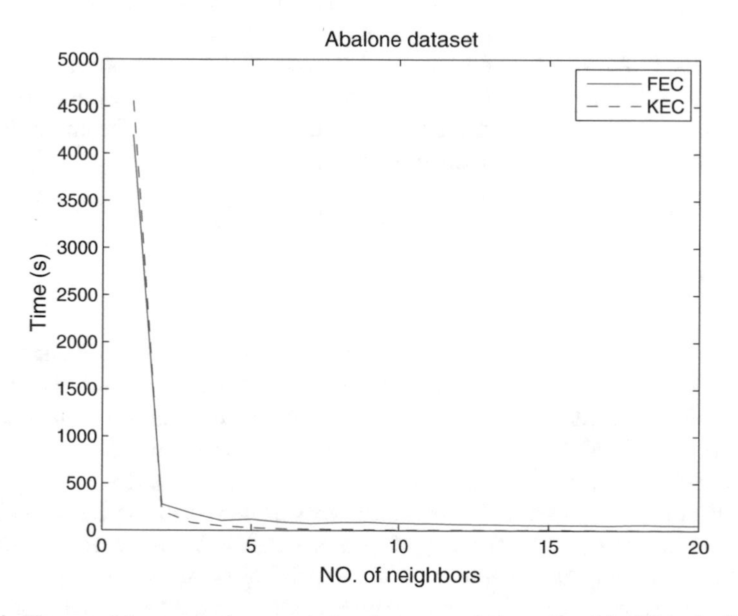

Fig. 6.14 Change of time cost when comparing the proposed Fuzzy Empirical Copula (FEC) and K-means based Empirical Copula (KEC)

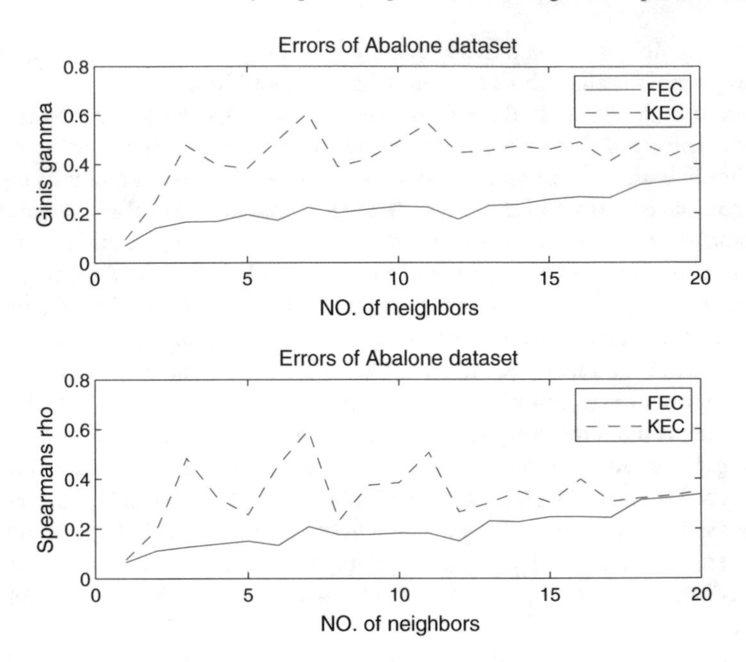

Fig. 6.15 Changes of overall errors of Spearman's rhos and Gini's gammas when comparing the proposed Fuzzy Empirical Copula (FEC) and K-means based Empirical Copula (KEC)

used in fuzzy empirical copula. One reason for the above results is that FLAME$^+$ is capable in dealing with non-linear relationships while K-means is not. Another reason is that FLAME$^+$ considers the real objects CSOs which are the samples in the datasets while K-means considers centroids which are virtual objects beyond the datasets. Centroids not belonging to the datasets may have different traits which probably result in wrong dependence measures.

6.5 Conclusion

Fuzzy empirical copula has been proposed to alleviate the computational burden of empirical copula. A high-dimensional FLAME$^+$ has been developed to identify the important objects containing the main features of the entire dataset, then empirical copula has been implemented to estimate the dependence structure of the objects. Abalone and yeast datasets from UCI machine learning repository are employed to evaluate the proposed method. The number of nearest neighbours is the tradeoff factor for handling accuracy and efficiency of data processing. With the preselected error threshold, fuzzy empirical copula has the capability of automatically identifying the optimised number of neighbours, which could be used to fast analyse similar datasets. Additionally nearest neighbours at the range of 0–20 have been used to demonstrate the overall error changes of Spearman's rho and Gini's gamma, and the change of computational time. The experimental results have shown that fuzzy empirical copula can substantially reduce the computation cost while features of the data are maintained with the preselected error threshold. In addition, we compare FLAME$^+$ with K-means to evaluate the clustering role in fuzzy empirical copula and the result has illustrated that FLAME$^+$ outperforms K-means in maintaining the dependence structure of the datasets.

Further work will be concerned with releasing the limitation of calculating the true value of original dataset and making the method more applicable. Though Copula has been widely applied to finance problems in the past decades, some areas such as intelligent robotics, artificial intelligence and automation require empirical copula and its variants being both effective approaches, and practical and efficient algorithms. Fuzzy Empirical Copula has succeeded in overcoming the problem of computation cost of dependence structure estimation via Empirical Copula. The algorithm will be evaluated in more real-time or near real-time applications, priority will be given to human hand gesture recognition and object manipulation using robotic hands [41–43].

References

1. A. Edmunds and A. Morris. The problem of information overload in business organisations: a review of the literature. *International Journal of Information Management*, 20(1):17–28, 2000.
2. K. Pearson. On lines and planes of closest fit to systems of points in space. *Philosophical Magazine*, 2(6):559–572, 1901.
3. K.V. Mardia, J.T. Kent, J.M. Bibby, et al. *Multivariate analysis*. Academic Press New York, 1979.
4. J.H. Friedman, S.L.A. Center, and J.W. Tukey. A project pursuit algorithm for exploratory data analysis. *The Collected Works of John W. Tukey: Graphics 1965-1985*, 1988.
5. P. Comon et al. Independent component analysis, a new concept. *Signal Processing*, 36(3):287–314, 1994.
6. J. Karhunen, P. Pajunen, and E. Oja. The nonlinear PCA criterion in blind source separation: Relations with other approaches. *Neurocomputing*, 22(1-3):5–20, 1998.
7. T. Hastie and W. Stuetzle. Principal Curves. *Journal of the American Statistical Association*, 84(406):502–516, 1989.
8. T. Kohonen. *Self-Organizing Maps*. Springer, 2001.
9. J. Karhunen. Nonlinear Independent Component Analysis. *ICA: Principles and Practice*, pages 113–134, 2001.
10. P. Demartines and J. Herault. Curvilinear component analysis: a self-organizing neural networkfor nonlinear mapping of data sets. *IEEE Transactions on Neural Networks*, 8(1):148–154, 1997.
11. JA Hartigan and MA Wong. A K-means clustering algorithm. *Journal of the Royal Statistical Society. Series C (Applied Statistics)*, 28:100–108, 1979.
12. J. C. Dunn. A fuzzy relative of the ISODATA process and its use in detecting compact well-separated clusters. *Cybernetics and Systems*, 3(3):32–57, 1973.
13. N. Belacel, P. Hansen, and N. Mladenovic. Fuzzy J-Means: a new heuristic for fuzzy clustering. *Pattern Recognition*, 35(10):2193–2200, 2002.
14. P. VUORIMAA. Fuzzy self-organizing map. *Fuzzy Sets and Systems*, 66(2):223–231, 1994.

15. P. Tamayo, D. Slonim, J. Mesirov, Q. Zhu, S. Kitareewan, E. Dmitrovsky, E.S. Lander, and T.R. Golub. Interpreting patterns of gene expression with self-organizing maps: Methods and application to hematopoietic differentiation. *Proceedings of the National Academy of Sciences*, 96:2907–2912, 1999.
16. KY Yeung, C. Fraley, A. Murua, AE Raftery, and WL Ruzzo. Model-based clustering and data transformations for gene expression data. *Bioinformatics*, 17(10):977–987, 2001.
17. L. Fu and E. Medico. FLAME, a novel fuzzy clustering method for the analysis of DNA microarray data. *BMC Bioinformatics*, 8:3, 2007.
18. O. Scaillet and J.D. Fermanian. Nonparametric estimation of copulas for time series. *FAME Research paper*, -(57), 2002.
19. R.B. Nelsen. *An introduction to copulas*. Springer Verlag, 2006.
20. A. Sklar. Fonctions de répartition a n dimensions et leurs marges. *Publ Inst Statist Univ Paris*, 8:229–231, 1959.
21. P. Deheuvels. La fonction de dépendance empirique et ses propriétés: Un test non paramétrique dindépendance. *Bulletin de lAcadémie royale de Belgique: Classe des sciences*, 65:274–292, 1979.
22. P. Deheuvels. A non parametric test for independence. *Publ. Inst. Statist. Univ. Paris*, 26(2):29–50, 1981.
23. N. Kolev, U. Anjos, and B.V.M. Mendes. Copulas: A Review and Recent Developments. *Stochastic Models*, 22(4):617–660, 2006.
24. A. Dias and P. Embrechts. Dynamic copula models for multivariate high-frequency data in finance. *Manuscript, ETH Zurich*, 2004.
25. P. Embrechts, F. Lindskog, and A. McNeil. Modelling dependence with copulas and applications to risk management. *Handbook of Heavy Tailed Distributions in Finance*, 8:329–384, 2003.
26. Harry Joe. *Dependence modeling with copulas*. CRC Press, 2014.
27. Dong Hwan Oh and Andrew J Patton. (in press) modelling dependence in high dimensions with factor copulas. *Journal of Business & Economic Statistics*, 2015.
28. P.K. Trivedi and D.M. Zimmer. Copula Modeling: An Introduction for Practitioners. *Foundations and Trends® in Econometrics*, 1(1):1–111, 2006.
29. L. Hu. Dependence patterns across financial markets: a mixed copula approach. *Applied Financial Economics*, 16(10):717–729, 2006.
30. Kunlapath Sukcharoen, Tatevik Zohrabyan, David Leatham, and Ximing Wu. Interdependence of oil prices and stock market indices: A copula approach. *Energy Economics*, 44:331–339, 2014.
31. A. Kolesarova, R. Mesiar, J. Mordelova, and C. Sempi. Discrete Copulas. *IEEE Transactions on Fuzzy Systems*, 14(5):698–705, 2006.
32. B. De Baets and H. De Meyer. Orthogonal Grid Constructions of Copulas. *IEEE Transactions on Fuzzy Systems*, 15(6):1053–1062, 2007.
33. M.A.H. Dempster, E.A. Medova, and S.W. Yang. Empirical Copulas for CDO Tranche Pricing Using Relative Entropy. *International Journal of Theoretical and Applied Finance*, 10(4):679, 2007.
34. J. Ma and Z. Sun. Dependence Structure Estimation via Copula. *Arxiv preprint* arXiv:0804.4451, 2008.
35. P.A. Morettin, C. Toloi, C. Chiann, and J. Miranda. Wavelet-smoothed empirical copula estimators. *Brazilian Review of Finance*, 8(3):263–281, 2010.
36. W.H. Kruskal. Ordinal measures of association. *Journal of the American Statistical Association*, 53(284):814–861, 1958.
37. E.L. Lehmann. Some concepts of dependence. *The Annals of Mathematical Statistics*, 37(1):137–1, 1966.
38. W.J. Nash, T.L. Sellers, S.R. Talbot, A.J. Cawthorn, and W.B. Ford. The population biology of abalone (Haliotis species) in Tasmania. I. Blacklip abalone (H. rubra) from the north coast and the Furneaux group of islands. *Sea Fisheries Division Technical Report*, 48:1–69, 1994.

39. P. Horton and K. Nakai. A probabilistic classification system for predicting the cellular localization sites of proteins. In *Ismb*, volume 4, pages 109–115, 1996.
40. A. Asuncion and D.J. Newman. UCI machine learning repository, 2007.
41. H. Liu. A fuzzy qualitative framework for connecting robot qualitative and quantitative representations. *IEEE Transactions on Fuzzy Systems*, 16(6):1522–1530, 2008.
42. H. Liu, D.J. Brown, and G.M. Coghill. Fuzzy qualitative robot kinematics. *IEEE Transactions on Fuzzy Systems*, 16(3):802–822, 2008.
43. Z. Ju, H. Liu, X. Zhu, and Y. Xiong. Dynamic Grasp Recognition Using Time Clustering, Gaussian Mixture Models and Hidden Markov Models. *Journal of Advanced Robotics*, 23:1359–1371, 2009.

Chapter 7
A Unified Fuzzy Framework for Human Hand Motion Recognition

7.1 Introduction

The human hand is capable of fulfilling various everyday-life tasks using the combination of biological mechanisms, sensors and controls. It is attracting more and more research interest to build human-like robotic hands with similar dexterous manipulation, which can be of great help in humanoid robots, industry and healthcare. Recent innovations in motor technology and robotics have achieved impressive results in the hardware of robotic hands such as the Southampton Hand, DLR hands, Robonout hand, Barret hand, Emolution hand and Shadow hand [1–4]. Especially, the ACT hand [5] has not only the same kinematics but also the similar anatomical structure with the human hand, providing a good start for the new generation of anatomical robotic hands. However, autonomously controlling multifingered robots is still a challenge, which holds the key to related multidisciplinary research and a wide spectrum of applications in intelligent robotics.

Human-robot skill transfer is of great importance in order to develop advanced multifingered manipulation planning and control systems [6–8]. Zollner et al. [9] and Chella et al. [10] confirmed that prior knowledge should be introduced in order to achieve fast and reusable learning in behavioural features, and integrate it in the overall knowledge base of a system, in order to fulfil requirements such as reusability, scalability, explainability and software architecture [11, 12]. Bae et al. [13] also proved that characteristics of human grasping could enhance dexterity in Robotic Grasping. Robertsson et al. [14] and Carrozza et al. [15] designed artificial hand systems for dexterous manipulation; they showed that human-like functionality could be achieved even if the structure of the system is not completely biologically inspired. Learning from human hand motions is preferred for human-robot skill transfer in that, unlike teleoperation-related methods, it provides non-contact skill transfer from human motions to robot motions by a paradigm that can endow artefacts with the ability for skill growth and life-long adaptation without detailed programming [16]. In principle, not only does it provide a natural, user-friendly means of implicitly programming the robot, but it also makes the learning problem become significantly more tractable by separating redundancies from important characteristics of a task [17]. The learning process for multifingered manipulation is segmented into four main

H. Liu et al., *Human Motion Sensing and Recognition*,
Studies in Computational Intelligence 675, DOI 10.1007/978-3-662-53692-6_7

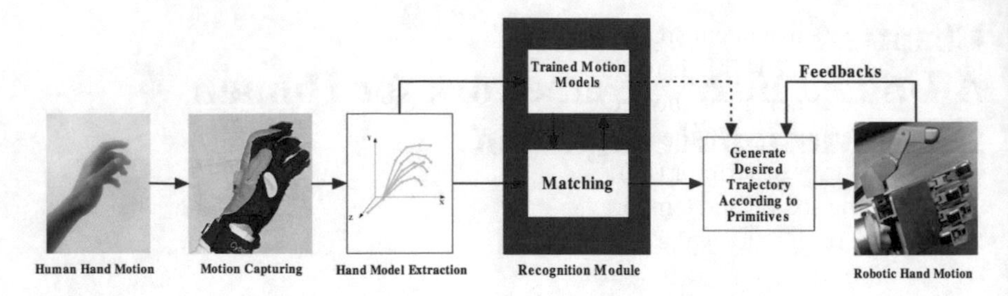

Fig. 7.1 Exploring human hand capabilities into multifingered robot manipulation

fundamental phases: data capturing, hand model extraction, motion recognition and robot hand control, as shown in Fig. 7.1.

Due to the complexity and dexterity of the human hand, hand motion recognition is still an open problem though it has been investigated over the past two decades. The attractiveness of applying neural network could be due to the nonlinearity of the network. Neural network approaches demonstrated its powerful grasp recognition in Glove-Talk project [18]. Sato et al. [19] identified predetermined gestures in a fast and robust manner by using a neural network which has been properly trained beforehand. Harding et al. [20] applied Fourier analysis to the hand gesture data and then input the complex harmonic data to a Probabilistic Neural Network for gesture classification. Support vector machine (SVM) was employed by Zoellner et al. [21] to classify dynamic grasps in the framework of Programming by Demonstration. Chuanjun et al. [22] used the properties of singular value decomposition to reduce the motion data and then applied SVM to classify these achieved feature vectors. In addition, David et al. [23] used a combination of discriminative (SVM, conditional random fields) and generative approaches, Hidden Markov Model (HMMs), for the modeling and recognition of actions through motor primitives. Similar combination of discriminative and generative methods could be seen in [24]. However, the choice of the SVM kernel function has a significant effect on its performance and the best choice is application dependent. Bedregal et al. [25] developed fuzzy rule-based methods for the recognition of hand gestures, however the methods are highly dependent on a detailed previous analysis of the recognised gesture features with manual transfer involved. Note that Heumer et al. [26] attempted to provide a performance criteria for hand grasp classification including 28 algorithms provided by Weka data mining software package. However these algorithms usually require more parameters such as finger angular speed instead of position trajectories only, they have also paid less attention to the context of intelligent robotics, for instance, real-time recognition requirement.

In addition, state of the art in human hand motion recognition can be represented by Gaussian Mixture Model (GMMs), HMM and their variants, since they are very rich in mathematical structure and hence their theoretical basis can be adopted for a wide range of applications. Calinon and Billard [27] proposed an approach to teach incrementally human gestures to a humanoid robot, which consists of projecting the movement data in a latent space and encoding the resulting signals in a GMM. Kaustubh et al. [28] used GMMs to classify one-handed gestures based on ultrasonic sensors, and the experiment showed simplified one-handed gestures can be recognised with a high accuracy. On the other hand, in [29], the moving hand is tracked and analyzed by Fourier descriptor (FD) from image sequences, and then the feature vector consisting the spatial and temporal features is processed by HMMs for recognition of the hand gesture. In [30], the recognition strategy used a combination of static shape recognition, Kalman filter based hand tracking and a HMM based temporal characterisation scheme. Bernardin et al. [31] presented a method using HMMs to recognise continuously executed sequences of grasping gestures. Both finger-joint angles and information about contact surfaces, captured by a data glove and tactile sensors, are used. By using the Kamakura taxonomy, most of all grasps used by humans in everyday life can be classified, increasing the system's application domain to general manipulation tasks. For 12 grasp classes, and with very little training data, a recognition accuracy of up to 89% for a single-user system and 95% for a multiple-user system could be reached. Ju et al. [32] compared Time Clustering (TC) with HMMs and GMMs on the recognition rate of 13 types of different grasps. Just et al. [33] provided an empirical comparison of two state-of-the-art techniques, HMMs and Input/Output HMMs, for temporal event modelling combined with specific features on two different datasets. Binh et al. [34] introduced pseudo three-dimensional Hidden Markov Model (P3DHMM) to recognise hand motion, which gained higher rate of recognition than P2DHMMs [35].

In this chapter, a novel fuzzy framework consisting of a set of recognition methods: TC, Fuzzy Gaussian Mixture Models (FGMM) and fuzzy Empirical Copula (FEC), has been proposed for optimally real-time recognising human hand motions. The system performances are compared with HMM and GMM based on a thorough investigation covering all aspects on human hand recognition under different datasets, different subjects and different training samples. This chapter is organised as follows: Sect. 7.2 presents TC, which can directly learn repeated motions from the same subject or similar gestures from various subjects with numerical value as output. Section 7.3 introduces FGMM, which is capable of fast modelling non-linear datasets as abstract Gaussian patterns. Section 7.4 explains a novel method to learn human hand motions through FEC which studies the data dependence structure among the finger joint angles. The experiments are evaluated in Sect. 7.5. Finally the chapter is concluded with remarks and future work.

7.2 Time Clustering Recognition

Time clustering was firstly proposed by Rainer Palm et al. in [36, 37] to recognise human grasps, and applied to develop a next-state-planner for programming-by-demonstration of reaching motions by Alexander Skoglund et al. in [38]. It has been demonstrated that the method is capable of identifying the segments of the occurrence of grasps and recognising the grasp types. This session proposes an expansion to the standard TC algorithm. The degree of membership of the model has been extended with both time instance weights and different learning motion weights. The extended degree of membership has enabled the TC method to be more comprehensive to learn repeated motions from the same subject or similar gestures from various subjects.

7.2.1 Motion Model Construction

The nonlinear functions of the finger angle trajectories are described as

$$\mathbf{y}_j(t) = \mathbf{f}_j(t) \tag{7.1}$$

where $t \in \Re^+$ is the time instant; $\mathbf{y}_j(t) \in \Re^k$ is expected angle trajectories from jth modelling motion; k is the number of finger angles of the hand; $j = 1, 2, \ldots z$; z is the total number of modelling motions; $\mathbf{f}_j : \Re \to \Re^k$.

The Eq. 7.1 is linearised at the selected time points t as

$$\mathbf{y}_j(t) = \mathbf{y}_j(t_i) + \frac{\Delta \mathbf{f}_j(t)}{\Delta t}|_{t_i} \cdot (t - t_i) \tag{7.2}$$

where $i = 1, 2, \ldots, c$ is the selected time instance; $t \in [t_i, t_{i+1}]$; $\mathbf{y}_j(t_1)$ is the initial known angle value from jth modelling motion at t_1 time instance. A local linear equation is derived in t,

$$\mathbf{y}_j(t) = \mathbf{A}_{ij} \cdot t + \mathbf{B}_{ij} \tag{7.3}$$

where $\mathbf{A}_{ij} = \frac{\Delta \mathbf{f}_j(t)}{\Delta t}|_{t_i} \in \Re^k$ and $\mathbf{B}_{ij} = \mathbf{y}_j(t_i) - \frac{\Delta \mathbf{f}_j(t)}{\Delta t}|_{t_i} \cdot t_i \in \Re^k$. According to the fuzzy modelling by Takagi and Sugeno [39], the output model from z training motions is constructed by summing selected local linear models,

$$\mathbf{y}(t) = \sum_{j=1}^{z} \sum_{i=1}^{c} \omega_{ij}(t)(\mathbf{A}_{ij} \cdot t + \mathbf{B}_{ij}) \tag{7.4}$$

$\omega_{ij}(t) \in [0, 1]$ is the degree of membership of the time point t to the selected time instance t_i from jth learning motion, and $\sum_{j=1}^{z} \sum_{i=1}^{c} \omega_{ij}(t) = 1$. The degree of membership is determined by

$$\omega_{ij}(t) = d_j \cdot c_i(t) \tag{7.5}$$

where $d_j \in [0, 1]$ is the weight of the jth training motion, and $\sum_{j=1}^{z} d_j = 1$; $c_i(t) \in [0, 1]$ is the degree of membership of the time point t to the selected time instance t_i, and $\sum_{j=1}^{z} c_i(t) = 1$. d_j is different for varied training motions according to validity of each training dataset. For example, if the dataset is more reliable more weight will be assigned. $c_i(t)$ is calculated by

$$c_i(t) = \frac{a_i(t)}{\sum_{i=1}^{c} a_i(t)} \tag{7.6}$$

where

$$a_i(t) = \frac{1}{\sum_{j=1}^{c} \left(\frac{(t-t_i)^T M_i (t-t_i)}{(t-t_j)^T Mj(t-t_j)}\right)^{\frac{1}{m-1}}} \tag{7.7}$$

$m > 1$ is the degree of fuzziness of the time instance t_i contributing to the time point t; T is the transpose function; M_i define the induced matrices of the input clusters C_i $(i = 1 \ldots c)$ and the general clustering and modelling steps can be seen in [40]. The Eq. 7.6 is for normalisation to ensure $\sum_{j=1}^{z} c_i(t) = 1$.

7.2.2 Motion Recognition

We propose a norm function for hand motion recognition. Basically, given a combination of grasp sequences, the dissimilarity between the models and grasps at the time instance t is

$$D(t) = ||\mathbf{V}_{model} - \mathbf{V}_{grasp}||_p = \left(\sum_{i=1}^{s} |v_{model}(i) - v_{grasp}(t+i-1)|^p\right)^{\frac{1}{p}} \tag{7.8}$$

where $D(t)$ is p-norm distance; usually we take $t \in \{1, 2, \infty\}$ that D_1 is the taxicab norm, D_2 is the Euclidean norm and D_∞ is the maximum norm [41]. V_{model} is the model trajectory with $s \geq 1$ time instances; V_{grasp} is the combination of grasp sequence with any order and may include any number of different groups.

The grasp sequences are separated by an 'intermediate state' which in our experiment means keeping the hand open and flat. The 'intermediate state' of the hand results in a stable dissimilarity when the model is shifted along the time series of the combined grasp sequence by Eq. 7.8, since the 'intermediate state' does not change much. The shifting and comparing are taking place in steps of time instances $\{t, t+1, \ldots, t+s-1\}$ of the combination grasp sequence. When the model begins to overlap a grasp in the combination sequence, the dissimilarity starts to change and reaches a local minimum with a highest overlap. By calculating the dissimilarities with local minimum between models and combined grasp sequences, the most similar grasp in the combination to the model is identified with the smallest local minimum. Figure 7.2b shows the dissimilarities when shifting five grasp models along the time series of a grasp sequence. The five local minimum have been identified with star point and the smallest minimum is distinct from other minima.

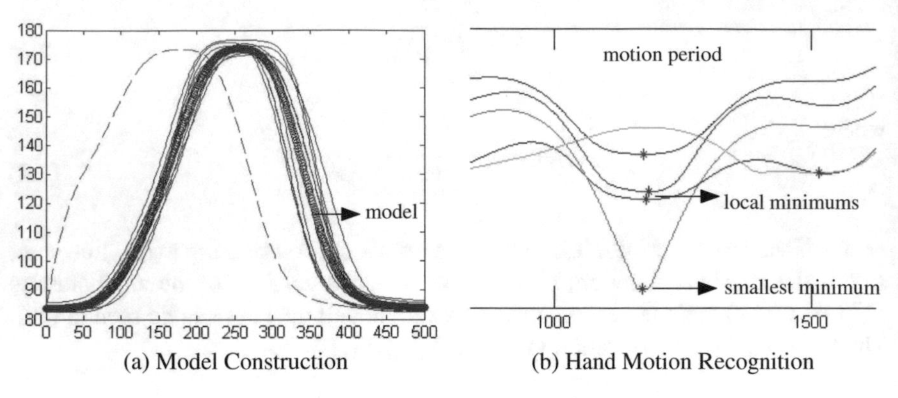

Fig. 7.2 Time clustering based recognition

7.3 Fuzzy Gaussian Mixture Models Recognition

Gaussian Model is confined to normal distributed clusters, so that more Gaussian components are needed for approximating the data with curve manifolds [42, 43]. In order to solve this problem, Active curve axis Gaussian Model (AcaGM), proposed by Baibo Zhang et al. [44], is a new non-linear probability model imagined as Gaussian Model being bent at the first principal axis. Inspired from the mechanism of Fuzzy C-means (FCM), a fuzzy EM algorithm was applied to GMMs with a fast convergence in our previous work [45]. FGMMs and their algorithms can be found in Chap. 5, this chapter will employ FGMMs to recognise the human hand motions.

The algorithm of Eqs. 7.9 and 7.10 are constructed to recognise the testing motions. Supposing there are a attributes in the human hand motion dataset. The dissimilarity of the testing motion and the trained model of FGMM is defined as

$$Dis = \frac{1}{a}\sum_{g=1}^{a} AvgDis^{(g)} \tag{7.9}$$

where g is the number of the attributes; $AvgDis^{(g)}$ is the average distance of the gth attribute and is calculated by

$$AvrgDis^{(g)} = \frac{1}{Cnum^{(g)}}\sum_{i=1}^{Cnum^{(g)}} |(Tmotion^{(g)}(t_i) - Center^{(g)}(i))| \tag{7.10}$$

where $Cnum$ is the number of the Gaussian components which is set before applying EM algorithm; $i = 1, \ldots, Cnum$ is the index of the Gaussian components; t_i is the time point of ith Gaussian component; $Tmotion(t_i)$ is the angle value or fingertip position of the testing hand motion at t_i; $Center(i)$ is the ith model centre. The corresponding motion points at the time t_i are identified according to the time points of the model centres, and then the average distances between the centres and these

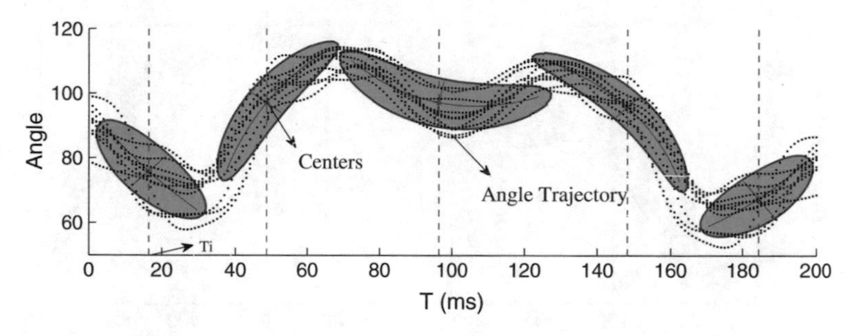

Fig. 7.3 An example of FGMMs based in-hand manipulation motion recognition

points which are in the same time as these FGMM's centres along the dash line are calculated, as shown in the Fig. 7.3. It calculates the average distance between testing motions and FGMM centres. The testing sample is recognised as the corresponding grasp class with the minimum dissimilarity.

7.4 Recognizing with Fuzzy Empirical Copula

Chapter 6 presented the theoretical foundation of Empirical Copula (EC) and dependence estimation; this chapter will propose the mechanism of the recognition algorithm using fuzzy Empirical Copula (FEC).

7.4.1 Re-sampling

Suppose the number of sampling points is n and number of attributes is m, for $m << n$ (m is far less than n), according to the Eqs. 6.3 and 6.6, the time complexity of Spearman's rho is $O(n^3)$. If number of samples is huge, computational time would be too long to get the result, which has been proven by our previous study [46]. To save the computational time, re-sampling processing is also employed in this study. We used the same strategy which is to take the samples at equal interval on the trajectories, e.g., in Fig. 7.4, sampling points of the trajectories is set to be 11 instead of 500, which alleviates the burden of the huge computational cost.

7.4.2 One-to-One Correlation and Motion Template

Suppose there are m variables which could be finger angles or positions in every motion, C_m^2 is the total number of the one-to-one correlations. Let ρ_{ij} be the Spearman's rho between ith and jth variables, and the motion template is defined as the matrix $\mathbf{P}$ of Spearman's rhos:

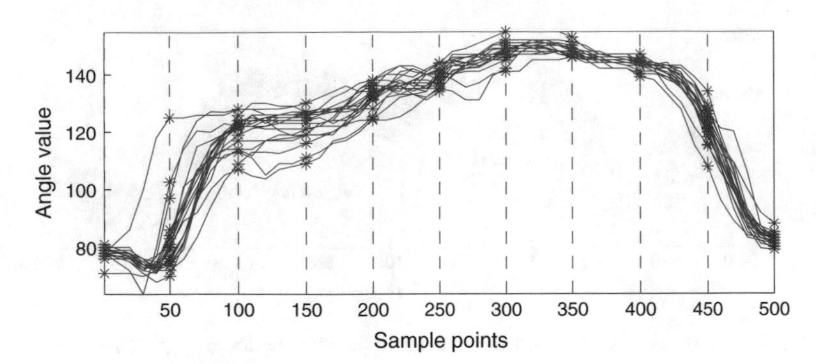

Fig. 7.4 Re-sampling the data at equal interval

$$\mathbf{P} = \begin{pmatrix} \rho_{11} & \cdots & \rho_{1m} \\ \vdots & \ddots & \vdots \\ \rho_{m1} & \cdots & \rho_{mm} \end{pmatrix}$$

where $\rho_{ij} = \rho_{ji}$ and $\rho_{ij} = 1$ if $i = j$. Given s observations for one motion, the template is trained by taking the average of all Spearman's rho matrices.

$$\widehat{P} = \frac{\sum_{i=1}^{s} w_i \mathbf{P}_i}{\sum_{i=1}^{s} w_i} \tag{7.11}$$

where $\mathbf{w} = [w_1, \ldots, w_s]$ is a weight vector used to store the relative difference of each observation in the estimated template, so that more valid observation may carry larger weight than those with more uncertainties, which may be caused by noise, capturing devices, software and the environment. The factors effecting the observation uncertainty include the calibration system, capturing device, software, conditions of the experiment and so on, which would be quantified by different ways. The way used in our experiment is based on the assumption that the observation obtained from better equipments is considered to be more valid. Motions with a higher validity are assigned to a higher weight in the weight vector. If the observations are obtained under the same experimental environment, equal weights will be assigned to them. Figure 7.5 shows an example of the motion template representing the one-to-one correlations among the finger angles when grasping a big ball.

The matrix $\mathbf{P}$ effectively aggregates the dependence relations of m variables into just one $m \times m$ matrix, a highly reduced dimensionality of feature space. The relation matrix is naturally uniformed that the matrix is not dependent on differently sampled trials associated with specific speeds. This makes direct comparisons of relation matrices with differently sampled data feasible and computationally efficient.

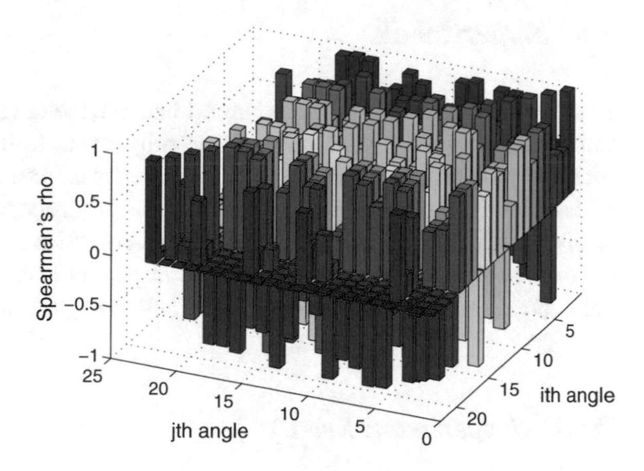

Fig. 7.5 An example of the motion template

7.4.3 Motion Recognition

Motion recognition is straightforward with the proposed template. It is achieved by finding the best match between an observed motion template and pre-trained motion templates. The proposed algorithm is applied on an observed motion to generate its motion template $\mathbf{U} = \{\rho_{ij}|i, j = 1, \ldots, m\}$. Its dissimilarity with the pre-trained template is achieved by

$$D_t = \|\mathbf{U} - \mathbf{P}\|_t = \left(\sum_{i=1}^{m}\sum_{j=1}^{m} |\rho_{ij} - \rho_{ij}|^t\right)^{\frac{1}{t}} \tag{7.12}$$

D_t is t-norm distance; $t \geq 1$ and is a real number; usually we take $t \in \{1, 2, \infty\}$ that D_1 is the taxicab norm, D_2 is the Euclidean norm and D_∞ is the maximum norm. The derived D_t norm infers the dissimilarity between the observed motion and the trained motions. The threshold of the template $\mathbf{P}$ is defined as

$$th_{\mathbf{P}} = \frac{\sum_{i=1}^{m}\sum_{j=1}^{m} |\rho_{ij}|}{\alpha}; \tag{7.13}$$

where $\alpha \geq 1$, which indicates the threshold is $100/\alpha\,\%$ of the whole absolute value of the template. Different datasets may have different α values. In this chapter, we set $\alpha = 10$, then the threshold of the template in Fig. 7.5 would be 8.33. The matching criterion of the motion recognition is that if $D_t \leq th_{\mathbf{P}}$, the observed motion is recognised as belonging to the trained motion.

7.5 Experiments

In order to evaluate the proposed algorithms a wide range of scenarios are used: (a) training with one sample from a single subject; (b) training with one sample from multiple subjects; (c) training with various samples from a single subject; (d) training with various samples from multiple subjects. The proposed algorithms are compared with classical HMMs and GMMs under the conditions, which have been employed to learn and recognise human gestures in papers [27, 30, 31]. In this study, two datasets, 13 grasps and 10 in-hand manipulations, are employed.

7.5.1 Experiment Set-Up

A fully instrumented data glove, CyberGlove from Immersion Corporation, has been employed to measure the finger joint angles during the hand movement. It can accurately transform hand and finger motions into real-time digital joint angle data by proprietary resistive bend-sensing technology. The sampling frequency of CyberGlove system in our experiments was set to 100 Hz without any filter. The joint angles are measured by 22 bend sensors, which include three flexion sensors per finger, four abduction sensors, a palm-arch sensor, and sensors to measure flexion and abduction, as shown in Fig. 7.6. Each sensor is extremely thin and flexible and is virtually undetectable in the lightweight elastic glove. CyberGlove is calibrated through the software from the company, and displayed by VirtualHand Studio software which converts the data into a graphical hand mirroring the subtle movements of the physical hand.

Six (2 female, 4 male) healthy right-handed subjects volunteered for the study, their ages range from 23 to 35 with the average of 29.5 years. All subjects were trained to manipulate different objects, each grasp or in-hand manipulations is repeated 20 times for every subject. Both grasps and in-hand manipulations are considered in the study of object manipulation.

7.5.1.1 Grasp

According to the human hand grasp taxonomy [47], hand grasps are classified into 6 atomic different types, for instance, power grasps, precision grasps, and circular/prismatic grasps, etc. We selected 13 different grasp gestures as shown in Fig. 7.7 to test the algorithm recognition ability. A cup has been selected three times respectively in motions 3, 4 and 5 with different strategies. In motion 3, only three fingers are used to precisely contact the cup top; in motion 4, five fingers are laterally attached to the cup and cup handle is grasped with three fingers in motion 5.

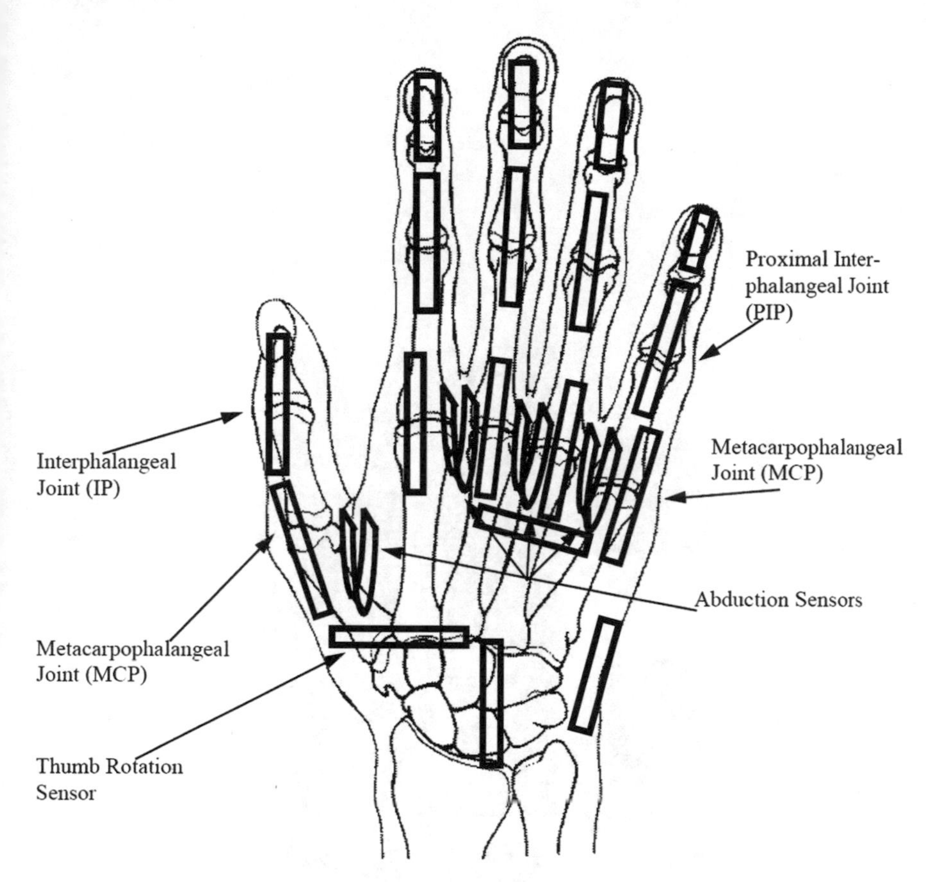

Fig. 7.6 Positions of the recorded hand joint angles

7.5.1.2 In-Hand Manipulations

The other dataset is for in-hand manipulations, which is much more complex than simple grasp motion and associated with the most complex human motor skills. It's the ability to change the position/orientation or adjust an object within one hand. Ten types of manipulations are recorded also by Cyberglove as listed in Table 7.1, and a few examples are shown in Fig. 7.8.

7.5.2 Experiment 1—Training with One Sample from a Single Subject

The first experiment is designed to investigate the recognition performances of the proposed algorithms with only one sample from a single subject for training models.

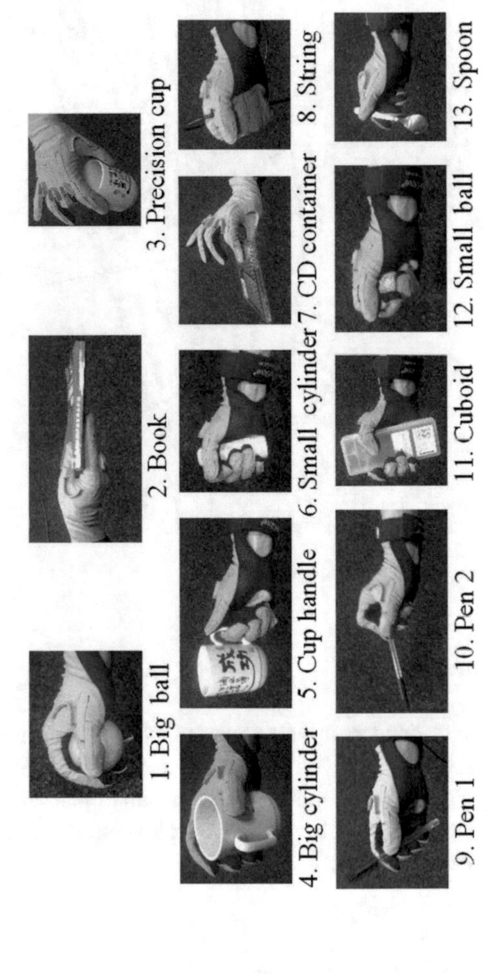

Fig. 7.7 Selected grasping tasks

Table 7.1 10 types of in-hand manipulations

1	Open a mobile phone and then close it
2	Screw to open a small bottle using only thumb, index finger and middle finger
3	Pick up a coin and move it from the fingertip to the palm
4	Remove the pencil from back to front for writing, as shown in Fig. 7.8a pencil walking
5	Pick up a pencil and simply rotate to write, as show in Fig. 7.8c simple rotation
6	Pick up a pencil and complexly rotate to write, as show in Fig. 7.8d complex rotation
7	Screw to open a big bottle using all five fingers
8	Roll a small cylinder
9	Pick up a scissor and cut paper
10	Pencil flips, as shown in Fig. 7.8b pencil flips

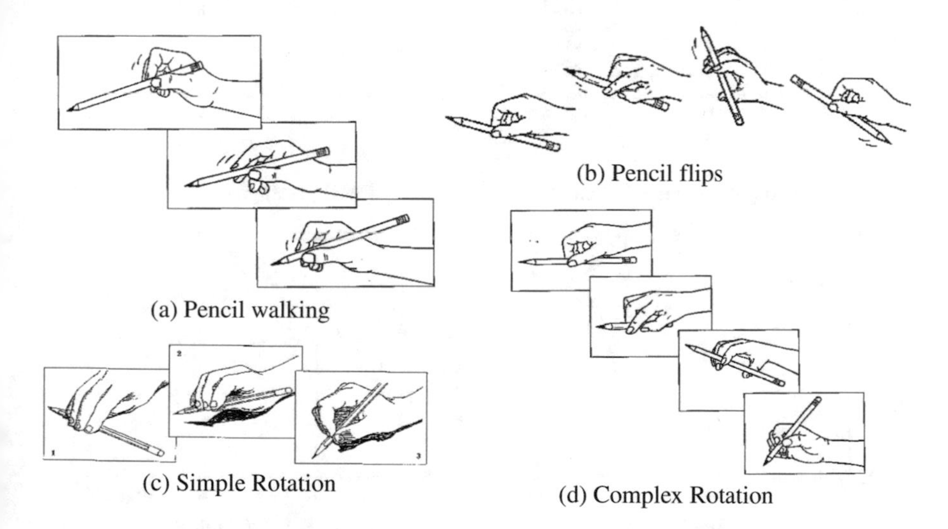

Fig. 7.8 Four examples of pencil in-hand manipulations

The trained model is used to identify the other samples from the corresponding subject. For example, one grasp model is trained from the first sample (totally 20 samples) of subject 1 when he is grasping a big ball, and then the model is used to test the other grasps of subject 1 only. In this experiment, 13 grasp models and 10 in-hand manipulation models have been trained for each subject. The models of every subject are only used for the subject itself. No cross testing is taken.

The recognition results are summarised in Table 7.2, showing that FEC approach performs better than the other four methods. FEC achieved 89.06 and 94.82% accuracy for identifying the motions in these two datasets. TC scores the second high

Table 7.2 Recognition results of FEC, TC, FGMM, GMM and HMM with only one sample from a single subject as training data

	Grasp dataset				
Method	FEC	TC	FGMM	GMM	HMM
Recognition Rate (%)	89.06	75.37	59.65	52.67	0
	In-hand manipulation dataset				
Method	FEC	TC	FGMM	GMM	HMM
Recognition Rate (%)	94.82	81.80	65.61	61.11	0

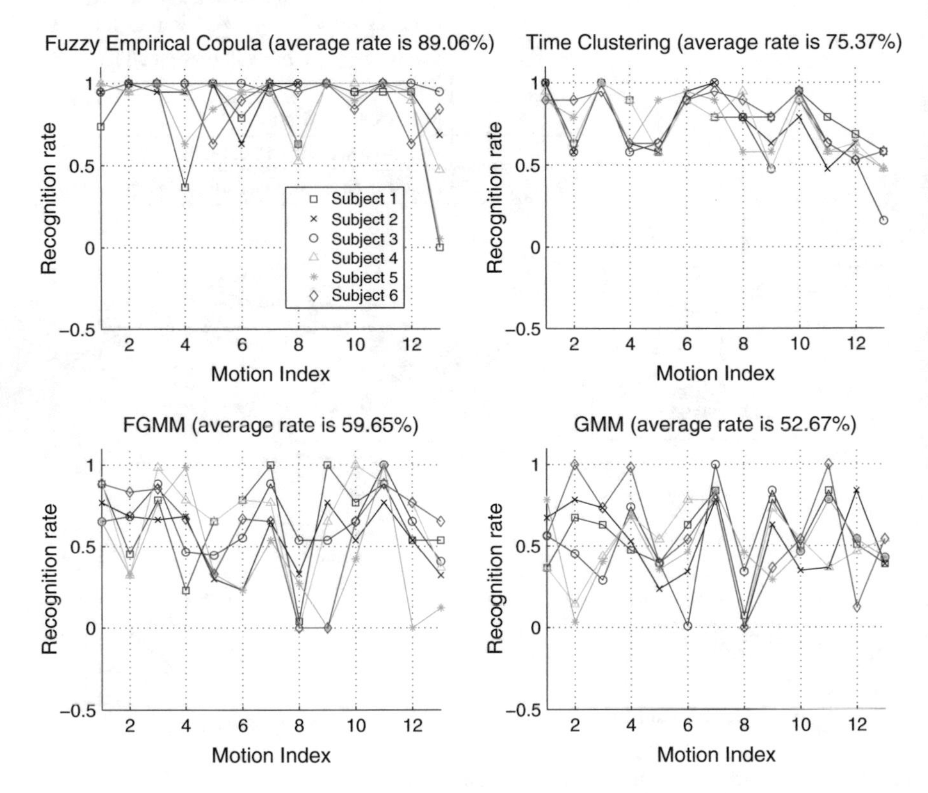

Fig. 7.9 Recognition results of different motions from different subjects on grasp dataset with one sample from a single subject as training data

recognition rate, followed by FGMM and GMM. Note that FGMM performs better than GMM in each dataset, HMM was not able to identify any motion correctly with only one set of training samples. Figures 7.9 and 7.10 present the detailed recognition results of different motions from different subjects for grasps and in-hand motion respectively. In Fig. 7.9, grasps 3 and 7 have relatively high accuracy for all the methods except HMM, while grasps 4, 8 and 13 are difficult to recognise.

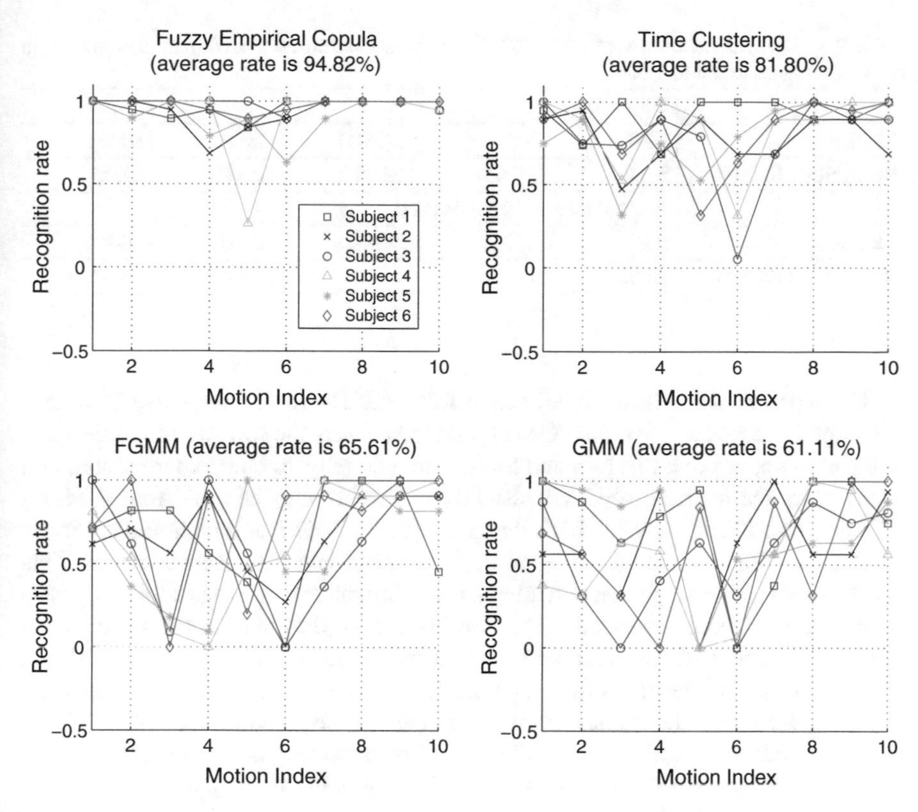

Fig. 7.10 Recognition results of different motions from different subjects on in-hand manipulation dataset with one sample from a single subject as training data

Figure 7.10 demonstrates that grasp 9 can be identified by all the proposed methods while in-hand motions 4, 5, and 6 are not straightforward to be recognised.

7.5.3 Experiment 2—Training with Six Samples from Multiple Subjects

The second experiment is to investigate how effective these methods are when they are trained with a few samples from multiple subjects and used to identify the different subjects' new samples. There are 13 grasp models and 10 in-hand models totally for all subjects. Each model is trained with 6 samples from 6 subjects. Each subject contributes one training sample. The models, combined with multiple subjects' contributions, are used to recognize other samples from all subjects. Cross testing is applied.

Table 7.3 Recognition results of FEC, TC, FGMM, GMM and HMM with six samples from multiple subjects as training data

	Grasp dataset				
Method	FEC	TC	FGMM	GMM	HMM
Recognition Rate (%)	81.90	76.65	74.89	70.17	41.45
	In-hand manipulation dataset				
Method	FEC	TC	FGMM	GMM	HMM
Recognition Rate (%)	85.83	82.19	72.10	64.82	58.42

The experiment results are tabulated in Table 7.3. The results show that FGMM is still slightly more effective than GMM. HMM still gets the lowest recognition rate although it rises from 0 to 41.45 and 58.42% individually. In both experiments 1 and 2, the proposed method, FEC, consistently outperforming the other four methods with the highest accuracy for both the grasp and in-hand manipulation datasets, is much more effective with a few samples as training data than other methods. This is a substantial advantage that can alleviate the difficulty of lacking training datasets in many applications. Detailed recognition results of different motions are shown in Figs. 7.11 and 7.12 for grasp and in-hand manipulation datasets respectively. In terms of the grasp dataset, motion 3, precisely grasping the top edge of the cup using 3 fingers, maintains the highest accuracy in the 13 grasps. Motions 2, 4, 8, 12, 13 have relative low recognition rates. Figure 7.12 presents that motion 9 remains a high recognition rate, and most of the other motions are still not easy to classify in the in-hand manipulation dataset.

7.5.4 Experiment 3—Training with Various Samples from a Single Subject

The issue addressed in this experiment is whether or not the rate of training samples affect the recognition capability and how well it enhances the performance of these five methods for a single subject. An empirical study was designed to evaluate sensitivity of these five methods in terms of change of the rate of training samples. There are 13 grasp models and 10 in-hand manipulation models for each single subject, where each subject repeated 20 times for each motion. Each model of any subject is trained with various percentage of the data ranging from 5 to 95% from the same subject. The trained model is used to test other unused data from the same subject. The experimental results with various percentage of training data are illustrated in Figs. 7.13 and 7.14 for the two datasets. It finds that the performance of these five methods are all improved with the increase of the training datasets and their accuracies stabilise with more than 30% training data. The recognition rate of HMM improves from 0% to more than 90% when training data increase from 5 to 30 % and

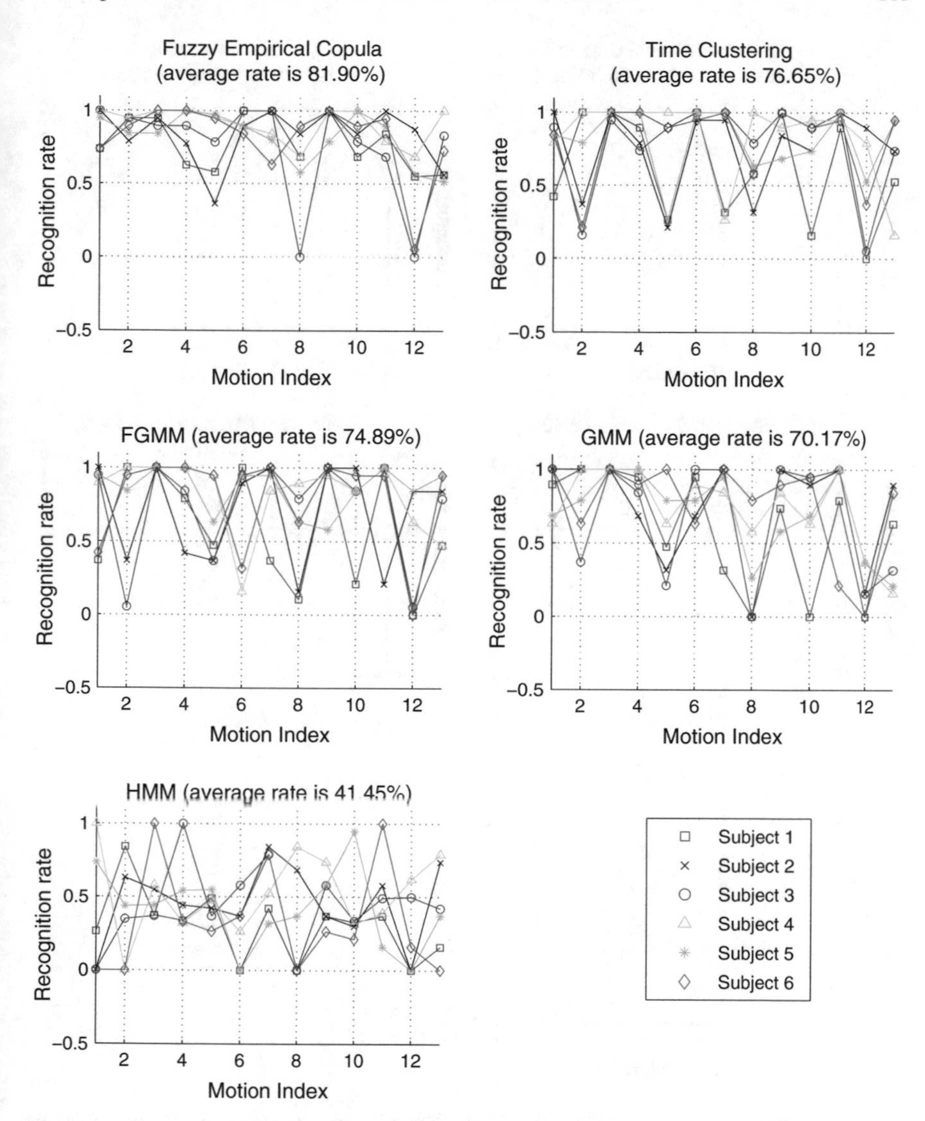

Fig. 7.11 Recognition results of different motions from different subjects on grasp dataset with six samples from multiple subjects as training data

it reaches as high as that of FEC with more than one third training data. Compared with GMM, FGMM achieves a better result with all the various sample rates. TC stays in the middle of FEC and FGMM, FEC presents the best overall performance among the five methods.

The results demonstrate that the variation of the training samples does substantially affect performance of these methods. The increase of the training data can result in about 90, 30, 30, 20 and 10 % increases of the recognition rates for HMM, GMM,

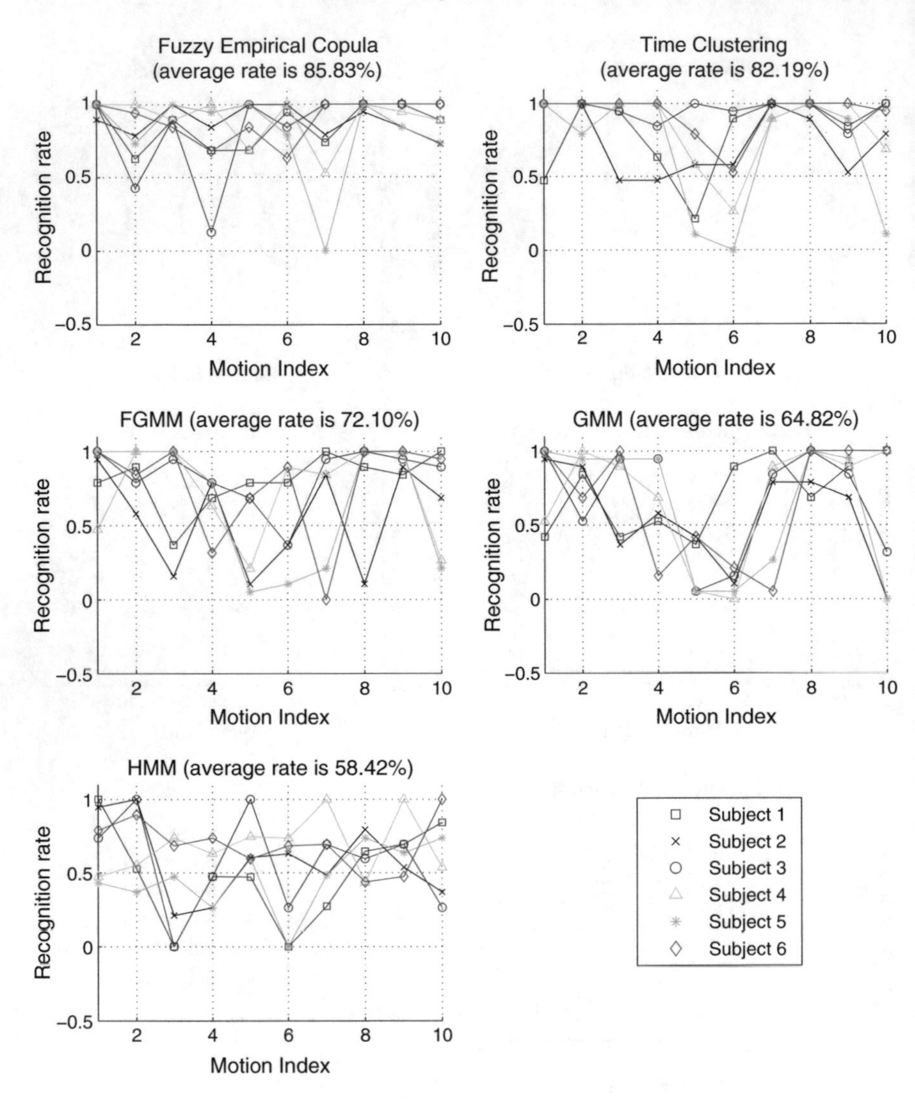

Fig. 7.12 Recognition results of different motions from different subjects on in-hand manipulation dataset with six samples from multiple subjects as training data

FGMM, TC, FEC. It presents that HMM is sensitive to the varying training data more than any of the other four methods. FEC consistently maintains high recognition rates with varying training samples, which emphasises that the proposed FEC is capable of fast extracting the intrinsic features from limited training samples and is insensitive to varying training samples. It also demonstrates that the recognition rates of the grasp dataset are always higher than that of in-hand manipulation dataset.

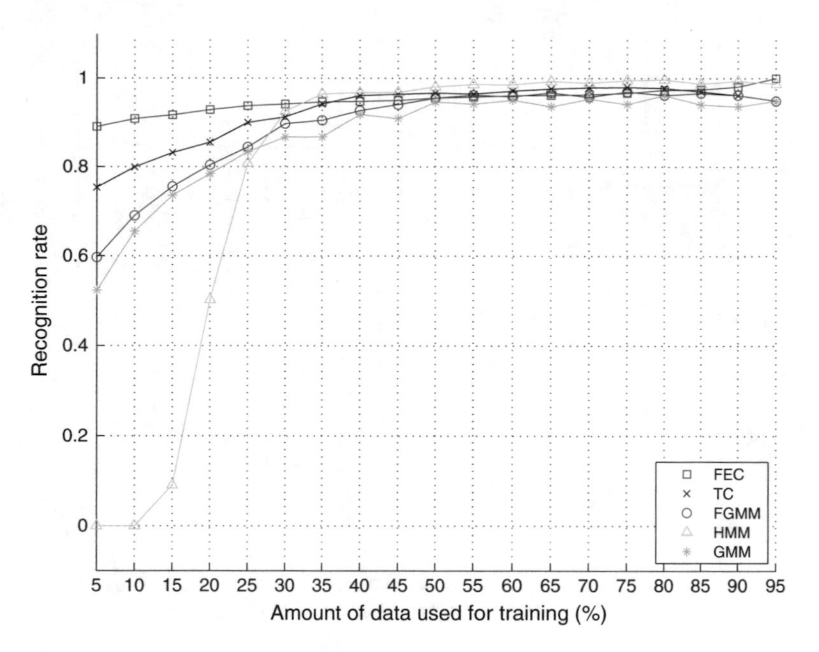

Fig. 7.13 Recognition results with various training percentages from a single subject on grasp dataset

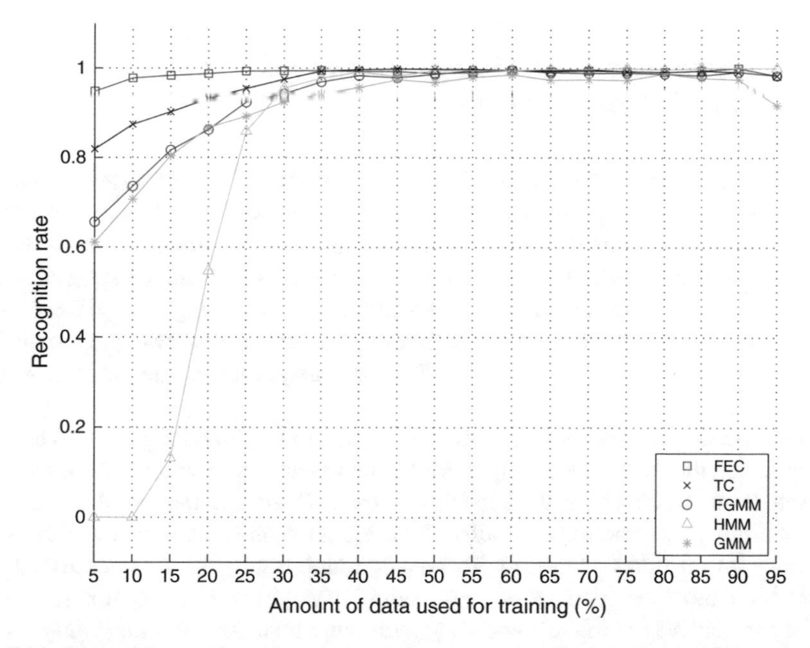

Fig. 7.14 Recognition results with various training percentages from a single subject on in-hand manipulation dataset

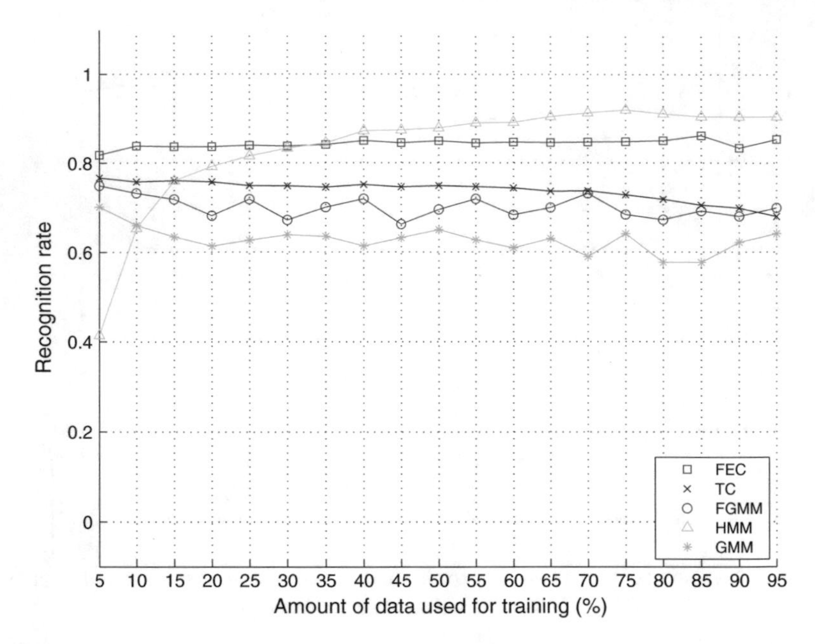

Fig. 7.15 Recognition results with various training percentages from multiple subjects on grasp dataset

7.5.5 *Experiment 4—Training with Various Samples from Multiple Subjects*

Similarly an experiment was conducted to evaluate the effects of different training rates on the classifiers' performance for multiple subjects. Regarding to experiment 2, there are 13 grasp models and 10 in-hand manipulation models for all the six subjects. Each model was trained by the mixed samples from all six subjects. Each subject contributes the same amount of samples to the training data. All the other samples which are not included in the training data are used to evaluate the models trained by multiple subjects' samples. The percentage of the training data ranges from 5 to 95%

The experimental results are provided in Figs. 7.15 and 7.16 for grasp and in-hand manipulation datasets respectively. HMM achieves an improvement of the recognition rate from less than 50 to 90% in terms of usage of the training samples. Different from the previous experiment results, there are no significant changes of recognition rates for the other methods with varying training samples. It is confirmed that HMM is the most sensitive method responding to varying training samples. Though HMM can achieve the highest accuracy with more than 40 % training samples, the proposed three methods, FEC, TC and FGMM, consistently outperform HMM and GMM when the training samples are less than one third due to the insensitiveness to the varying percentage of the training data. It should be noted that FEC achieves

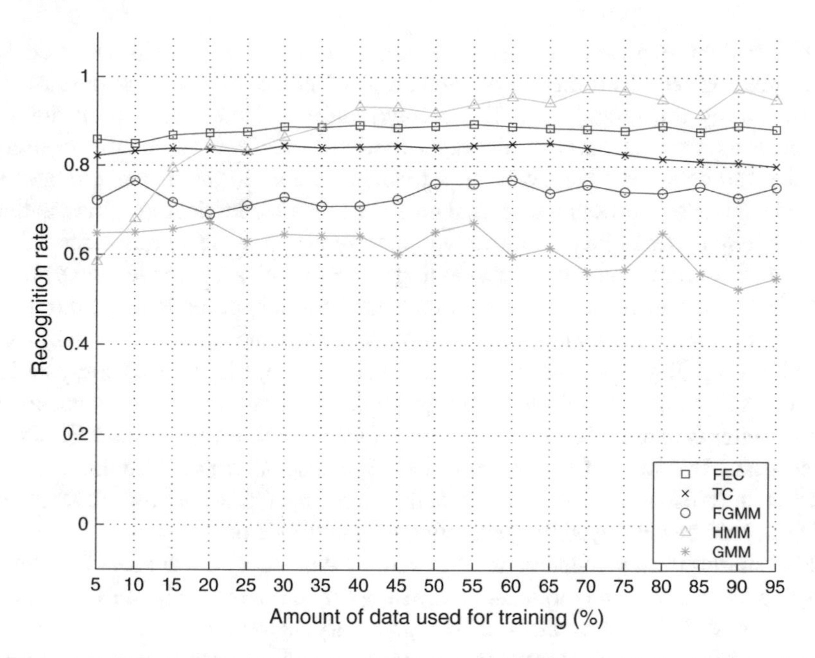

Fig. 7.16 Recognition results with various training percentages from multiple subjects on in-hand manipulation dataset

more than 80% accuracy with only 5% training data, which is far better than those of the HMM and GMM. The recognition result of TC is slightly lower than FEC, but it has the robustness of handling data sampling variations. FGMM's performance is consistently better than that GMM's thanks to the non-linear characteristic, though both recognition rates undulate intensely.

7.6 Conclusion

In this chapter, we have proposed a novel fuzzy framework of a set of recognition algorithms, TC, FGMM and FEC, in terms of numerical clustering, Gaussian patterns and dependence structure for human hand motion recognition. Experiments, covering different subjects and various scenarios, have been conducted to demonstrate their effectiveness in comparison with HMMs and GMMs under the same conditions. Two datasets consisting of 13 different types of grasps and 10 in-hand manipulations have been employed for the algorithm evaluation. Experimental results revealed a number of interesting points and the discussions of these three algorithms, which include the conditions and requirements for using different techniques within the framework, are summarised as follows.

TC is capable of generating the desired trajectory automatically since its modelling mechanism is based on the fuzzy modelling of trajectories and its output is the corresponding trajectory point. The recognition algorithm itself can instinctively identify the start point and end point of the motion. It requires low quantity of training samples and achieves a relatively high recognition rate for both single subject and multiple subjects. Low computational cost is needed for modelling. This method is applicable to motion planning directly transferred from the recognition result.

FGMM is a component-based modelling methods with a fast convergence process. With the bend principle axis, its component has the ability of modelling datasets with curve manifolds. The efficiency of convergence has been improved by the introduction of a weighting exponent on the fuzzy membership, which saves computational time of its EM algorithm. Low training samples were required, and its accuracy is always better than that of GMM, though it is lower than those of TC and FEC. Recognition algorithm was fast and the model storage space is small since the number of components are much less than points of the trajectory. Gaussian Mixture Regression (GMR) could be employed for generating desired trajectory.

FEC studies the dependence structure among the finger joint angles of the motion to recognise human hand motions. Fuzzy sampling has been adopted for a fast modelling process. FEC is the least sensitive to the varying training samples among these methods and achieves very satisfactory results for both single subject and multiple subjects with few training dataset. Both modelling and recognising were fast and model storage space with the dependence structure was small as well. It is most useful for the applications which needs fast modelling and lacks training samples.

Arranging these three algorithms in a framework provides a feasible solution to the human hand motion recognition for a wide range of hand scenarios. Experiment results have demonstrated that all the three methods in the framework are insensitive to the varying training samples to certain extent, which means that these methods have advantages over practical applications, especially when requiring a fast training model process or lacking sufficient training datasets. In the future work, the proposed fuzzy framework will be applied to applications such as behaviour based human motion recognition for prosthetic hands and human-robot skill transfer [48, 49].

References

1. A. Bicchi. Hands for dexterous manipulation and robust grasping: A difficult road toward simplicity. *IEEE Transactions on Robotics and Automation*, 16(6):652–662, 2002.
2. C.M. Light and P.H. Chappell. Development of a lightweight and adaptable multiple-axis hand prosthesis. *Medical Engineering and Physics*, 22(10):679–684, 2000.
3. R. Wei, X.H. Gao, M. Jin, Y. Liu, H. Liu, N. Seitz, R. Gruber, and G. Hirzinger. Fpga based hardware architecture for hit/dlr hand. In *IEEE/RSJ International Conference on Intelligent Robots and Systems*, pages 523–528. IEEE, 2005.
4. F. Lotti, P. Tiezzi, G. Vassura, L. Biagiotti, G. Palli, and C. Melchiorri. Development of ub hand 3: Early results. In *Proceedings of the IEEE International Conference on Robotics and Automation*, pages 4488–4493, 2005.

5. Y. Matsuoka, P. Afshar, and M. Oh. On the design of robotic hands for brain–machine interface. *Neurosurg. Focus*, 20(5):E3, 2006.
6. H. Liu. Exploring human hand capabilities into embedded multifingered object manipulation. *IEEE Transactions on Industrial Informatics*, 7(3):389–398, 2011.
7. M. Li. Dynamic grasp adaptation - from humans to robots. 2015, 2015.
8. Ravin De Souza, Sahar El-Khoury, José Santos-Victor, and Aude Billard. Recognizing the grasp intention from human demonstration. *Robotics and Autonomous Systems*, 74:108–121, 2015.
9. R. Zoliner, M. Pardowitz, S. Knoop, and R. Dillmann. Towards Cognitive Robots: Building Hierarchical Task Representations of Manipulations from Human Demonstration. *IEEE International Conference on Robotics and Automation*, pages 1535–1540, 2005.
10. A. Chella, H. Džindo, I. Infantino, and I. Macaluso. A posture sequence learning system for an anthropomorphic robotic hand. *Robotics and Autonomous Systems*, 47(2–3):143–152, 2004.
11. P. Verschure, T. Voegtlin, and R.J. Douglas. Environmentally mediated synergy between perception and behaviour in mobile robots. *Nature*, 425(6958):620–624, 2003.
12. H. Jacobsson, N. Hawes, G.J. Kruijff, and J. Wyatt. Crossmodal content binding in information-processing architectures. In *Proceedings of the ACM/IEEE International Conference on Human Robot Interaction*, pages 81–88. ACM, 2008.
13. J.H. Bae, S. Arimoto, R. Ozawa, and M. Sekimoto. Enhancement of Dexterity in Robotic Grasping Referring to Characteristics of Human Grasping. *IEEE International Conferences on Robotics and Automation*, 2:1203, 2005.
14. L. Robertsson, B. Iliev, R. Palm, and P. Wide. Perception modeling for human-like artificial sensor systems. *International Journal of Human-Computer Studies*, 65(5):446–459, 2007.
15. M.C. Carrozza, G. Cappiello, S. Micera, B.B. Edin, L. Beccai, and C. Cipriani. Design of a cybernetic hand for perception and action. *Biological cybernetics*, 95(6):629–644, 2006.
16. S. Calinon, F. Guenter, and A. Billard. On learning, representing, and generalizing a task in a humanoid robot. *IEEE Transactions on Systems Man and Cybernetics Part B*, 37(2):286, 2007.
17. S.T. Roweis and L.K. Saul. Nonlinear dimensionality reduction by locally linear embedding. *Science*, 290(5500):2323–2326, 2000.
18. S.S. Fels and G.E. Hinton. Glove-TalkII-a neural-network interface which maps gestures to parallel formant speech synthesizer controls. *IEEE Transactions on Neural Networks*, 9(1):205–212, 1998.
19. Y. Sato, M. Saito, and H. Koike. Real-time input of 3D pose and gestures of a user's hand and itsapplications for HCI. *Proceedings. IEEE Virtual Reality*, pages 79–86, 2001.
20. PRG Harding and T. Ellis. Recognizing hand gesture using Fourier descriptors. *Proceedings of the 17th International Conference on Pattern Recognition*, 3, 2004.
21. R. Zöllner, O. Rogalla, J. Zöllner, and R. Dillmann. Dynamic grasp recognition within the framework of programming by demonstration. In *IEEE International Workshop on Robot and Human Interactive Communication*, pages 418–423. Citeseer, 2001.
22. C. Li, L. Khan, and B. Prabhakaran. Real-time classification of variable length multi-attribute motions. *Knowledge and Information Systems*, 10(2):163–183, 2006.
23. D. Martinez and D. Kragic. Modeling and recognition of actions through motor primitives. In *IEEE International Conference on Robotics and Automation*, pages 1704–1709, 2008.
24. I.S. Vicente, V. Kyrki, D. Kragic, and M. Larsson. Action recognition and understanding through motor primitives. *Advanced Robotics*, 21(15):1687–1707, 2007.
25. B.C. Bedregal, A.C.R. Costa, and G.P. Dimuro. Fuzzy Rule-Based Hand Gesture Recognition. *International Federation for Information Processing-Publications-IFIP*, 217:285, 2006.
26. G. Heumer, H.B. Amor, M. Weber, and B. Jung. Grasp Recognition with Uncalibrated Data Gloves-A Comparison of Classification Methods. *IEEE Virtual Reality Conference*, pages 19–26, 2007.
27. S. Calinon and A. Billard. Incremental learning of gestures by imitation in a humanoid robot. *Proceedings of the ACM/IEEE International Conference on Human-Robot Interaction*, pages 255–262, 2007.

28. K. Kalgaonkar and B. Raj. One-handed gesture recognition using ultrasonic Doppler sonar. In *Proceedings of the 2009 IEEE International Conference on Acoustics, Speech and Signal Processing-Volume 00*, pages 1889–1892. IEEE Computer Society, 2009.
29. F.S. Chen, C.M. Fu, and C.L. Huang. Hand gesture recognition using a real-time tracking method and hidden Markov models. *Image and Vision Computing*, 21(8):745–758, 2003.
30. A. Ramamoorthy, N. Vaswani, S. Chaudhury, and S. Banerjee. Recognition of dynamic hand gestures. *Pattern Recognition*, 36(9):2069–2081, 2003.
31. K. Bernardin, K. Ogawara, K. Ikeuchi, and R. Dillmann. A sensor fusion approach for recognizing continuous human grasping sequences using hidden Markov models. *IEEE Transactions on Robotics and Automation*, 21(1):47–57, 2005.
32. Z. Ju, H. Liu, X. Zhu, and Y. Xiong. Dynamic Grasp Recognition Using Time Clustering, Gaussian Mixture Models and Hidden Markov Models. *Journal of Advanced Robotics*, 23:1359–1371, 2009.
33. A. Just and S. Marcel. A comparative study of two state-of-the-art sequence processing techniques for hand gesture recognition. *Computer Vision and Image Understanding*, 113(4):532–543, 2009.
34. N.D. Binh and T. Ejima. Real-Time hand Gesture Recognition Using Pseudo 3-D Hidden Markov Model. *IEEE International Conference on Cognitive Informatics*, 2, 2006.
35. N.D. Binh, E. Shuichi, and T. Ejima. Real-Time Hand Tracking and Gesture Recognition System. *Proceedings of International Conference on Graphics, Vision and Image Processing*, pages 362–368, 2005.
36. R. Palm, B. Iliev, and B. Kadmiry. Recognition of human grasps by time-clustering and fuzzy modeling. *Robotics and Autonomous Systems*, 57(5):484–495, 2009.
37. R. Palm, B. Kadmiry, B. Iliev, and D. Driankov. Recognition and Teaching of Robot Skills by Fuzzy Time-Modeling. In *Proceedings of IFSA World Congress and EUSFLAT Conference, Lisbon, Portugal*, pages 7–12. Citeseer, 2009.
38. A. Skoglund, B. Iliev, and R. Palm. Programming-by-demonstration of reaching motions-a next-state-planner approach. *Robotics and Autonomous Systems*, 58(5):607–621, 2010.
39. T. Takagi and M. Sugeno. Fuzzy identification of systems and its applications to modeling and control. *IEEE Transactions on Systems, Man, and Cybernetics*, 15:116–132, 1985.
40. D. E. Gustafson and W. C. Kessel. Fuzzy clustering with a fuzzy covariance matrix. *IEEE Conference on Decision and Control*, 17:761–766, 1978.
41. G.H. Golub and C.F. Van Loan. *Matrix computations*. Johns Hopkins Univ Pr, 1996.
42. A. Dasgupta and A.E. Raftery. Detecting features in spatial point processes with clutter via model-based clustering. *Journal of the American Statistical Association*, pages 294–302, 1998.
43. C. Fraley and AE Raftery. How many clusters? Which clustering method? Answers via model-based cluster analysis. *The Computer Journal*, 41(8):578–588, 1998.
44. B. Zhang, C. Zhang, and X. Yi. Active curve axis Gaussian mixture models. *Pattern Recognition*, 38(12):2351–2362, 2005.
45. Z. Ju and H. Liu. Applying fuzzy em algorithm with a fast convergence to gmms. In *Proc International Conference on Fuzzy Systems, World Congress on Computational Intelligence, Spain*, 2010.
46. Z. Ju and H. Liu. Recognizing Hand Grasp and Manipulation through Empirical Copula. *International Journal of Social Robotics*, 23:1359–1371, 2010.
47. T. Iberall. Human Prehension and Dexterous Robot Hands. *The International Journal of Robotics Research*, 16(3):285, 1997.
48. H. Liu. A fuzzy qualitative framework for connecting robot qualitative and quantitative representations. *IEEE Transactions on Fuzzy Systems*, 16(6):1522–1530, 2008.
49. H. Liu, D.J. Brown, and G.M. Coghill. Fuzzy qualitative robot kinematics. *IEEE Transactions on Fuzzy Systems*, 16(3):802–822, 2008.

Chapter 8
Human Hand Motion Analysis with Multisensory Information

8.1 Introduction

The human hand contains 27 bones with roughly 25 degrees of freedom, which are driven by 17 intrinsic muscles in the hand itself and 18 extrinsic muscles in the forearm. Such a highly articulated and complex system is capable of performing more complicated and dexterous tasks than any other existing systems, and it has been regarded as a rich source of inspiration for the engineers and scientists to design human-like robotic and prosthetic hands and to learn and model human hand motion skills. Human hand motion analysis is attracting engagement of more and more researchers in the areas of neuroscience, biomedical engineering, robotics, human-computer interaction (HCI) and Artificial Intelligence (AI).

Different properties involved in the human hand motions provide people with rich sensory information such as hand position, velocity, force and their changes with the time to build up computational models of these motions. Hand motion capturing systems can be mainly categorised into different ways: data glove based capturing, markers based capturing, vision based capturing, haptics based capturing and Electromyography (EMG) based capturing. Data glove and markers are preferred and have been widely adopted [1–4] because they use highly precise sensors to achieve hand dynamic gestures including positions, velocities and accelerations. Vision based capturing is becoming more and more popular mainly due to advances in computer vision algorithms to solve the ill-posed problem of 3D construction out of 2D images [5–7]. Compared to data gloves, using cameras to capture the hand motions is a more natural and non-contact way, which does not cling to hands and does not limit movements and volumes of the finger activities. However, the vision sensor requires quality image processing techniques and suffers from limitations of the low accuracy and high computational cost. Recently, commercially available depth cameras such as Kinect have recently been successfully employed to extract robust human body skeleton especially for games. They have also been applied to the sign language and hand gesture recognition [8, 9]. However, it is still challenging to extract the skeleton of the hand, since the hand has a small shape and high degrees of

H. Liu et al., *Human Motion Sensing and Recognition*,
Studies in Computational Intelligence 675, DOI 10.1007/978-3-662-53692-6_8

freedom compared to the main body. Haptics/tactile/force information is an important property of the human hand motions, especially for the object manipulations [10–13], e.g., Park et al. [14] designed a haptic interaction system for transferring expertise in teleoperation-based manipulation between two human users and demonstrated that the haptic knowledge is transferable both through the haptic force guidance as well as through the robotic demonstration. EMG signal is the electrical potential generated by muscle contractions and is very useful to study human movements. Different from all above sensory information capturing techniques, EMG signals for analysing human hand motions are usually taken from the human forearm instead of the hand itself, since the majority of the finger movements are driven by the extrinsic muscles in the forearm and it can be used to indirect estimate manipulation force a human hand applies [15, 16]. It is the most natural and promising way to control the prosthetic hands and attracting more and more researchers to use such signals to analyses human hand motions for prosthetic hands control [17–20], HCI [21, 22] and health recovery [23–25].

Integration of the above multiple sensory information is essential for the human hand motion analysis. Ceruti et al. [26] integrated data glove with haptics sensors to capture both the finger angles and the finger tip forces and Norman et al. [27] employed the data glove with both finger position and contact force information to analyse human in-hand manipulation. Romano et al. [28] introduced the SlipGlove providing tactile cues associated with slip between the glove and a contact surface. Such gloves with haptic-IO capability provide the vital information of human hand motion and greatly enhance the capturing of human hand skills. Yang et al. [29] proposed an on-line estimation method for the hand grasp force based on the surface EMG extracted from human forearm synchronously. More examples could be found in [30], which presents a comprehensive review of recent scientific findings and technologies including human hand analysis and synthesis, hand motion capture, recognition algorithms and applications. However, to the best of our knowledge, there are few studies, which have integrated the muscle signals with the contact forces and hand finger trajectories for human hand motion analysis.

In this chapter, we propose a framework of multiple-sensor integration of the data glove, force sensors and EMG sensors for human hand motion analysis, as shown in Fig. 8.1 based on our previous study [31]. The framework consists of four main components, i.e. motion capturing module, preprocessing module, knowledge base module, and motion identification module. The motion capturing module is to use different types of sensors to transfer the sensory information into digital signal recognisable to computers. In this chapter, three types of sensors are used, including CyberGlove joint angle sensors, FingerTPS pressure sensors and Trigno wireless EMG sensors, which capture the hand gestures, contact forces and muscle contraction signals from various hand motions respectively. The preprocessing module is to synchronise and filter the original digital data and segment them into individual tasks. Automatic synchronisation and segmentation are investigated to improve the efficiency of the acquisition of human hand tasks. The knowledge base module stores the human hand motion primitives, manipulation scenarios and correlations among the different sensory information. In the knowledge base module, we investigate the

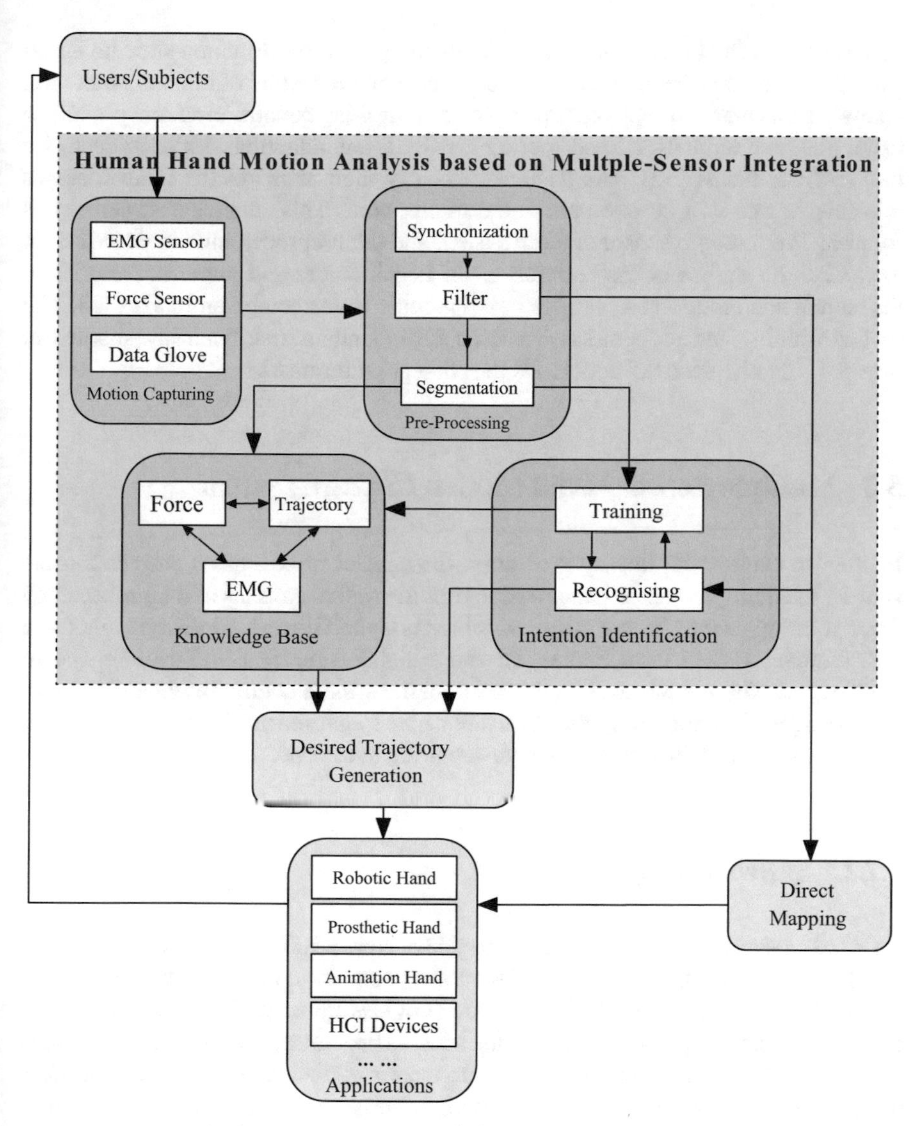

Fig. 8.1 Framework of multiple-sensor integration for human hand motion analysis

correlations of the finger trajectories, contact forces and EMG signals, using FEC from Chap. 6. The identification module is to use the clustering and machine learning methods to train the motion models using the preprocessed datasets and to recognise new or testing data from one or more types of sensors. Different recognition methods including FGMMs introduced in Chap. 5, GMMs and Support Vector Machine (SVM) are investigated for the EMG based manipulation identification and comparison results are analysed. In addition, the desired trajectory generation module can be regarded as the link between the motion analysis framework and specific applications. It generates the desired trajectory based on the results of the human motion analysis framework for applications such as controlling robotic hand and prosthetic hand, planning animation hand motion and HCI, e.g. adapting motion trajectories into artificial hands [32]. The desired trajectory module is not the main focus of this chapter and will be one part of the future work. This chapter is structured as follows: The multiple sensor capture system and the preprocessing are described in Sect. 8.2. The studies of the correlations of EMG, forces and finger trajectories in the human hand motion knowledge based module have been given in Sect. 8.3. The motion training and recognition based on EMG signals have been investigated in Sect. 8.4. Finally, Sect. 8.5 concludes the chapter with remarks.

8.2 Multiple-Sensor Hand Motion Capture System

In order to capture the finger trajectories, the contact force signals and the muscle signals, the multiple-sensor hand motion capture system consists of a high-accuracy finger joint-angle measurement system which is CyberGlove, a wireless tactile force measurement system from FingerTPS and a high-frequency EMG capture system with Trigno Wireless Sensors. In this section, the system configuration and the preprocessing module including the hardware based synchronisation and segmentation will be described, followed by the data capturing at the end.

8.2.1 System Configuration

As a fully instrumented glove, the CyberGlove shown in Fig. 8.2a uses proprietary resistive bend-sensing technology to accurately transform hand and finger motions into real-time digital joint-angle data and provides up to 22 high-accuracy joint-angle measurements, where three flexion sensors per finger, four abduction sensors, a palm-arch sensor, and sensors to measure flexion and abduction. Tracking sensors are also included in the CyberGlove system to measure the position and orientation of the forearm in space. The sensor resolution is 0.5°, the repeatability is one degree, and the sampling rate is 150 Hz. Highly sensitive capacitive-based pressure sensors of FingerTPS shown in Fig. 8.2b have also been utilised to reliably quantify forces applied by the human hand. Calibration is achieved by using a reference force sensor. The system has 6 comfortable capacitive sensors per hand, wirelessly connected to the computer. The sensors have a data rate of 40 Hz. The full scale range is 10–50 lbs and the sensitivity is 0.01 lbs. Video images can be captured and displayed in real-time, synchronised with tactile data. In addition, the EMG capture system employs Trigno Wireless Sensors shown in Fig. 8.2c and has 16 EMG channels and

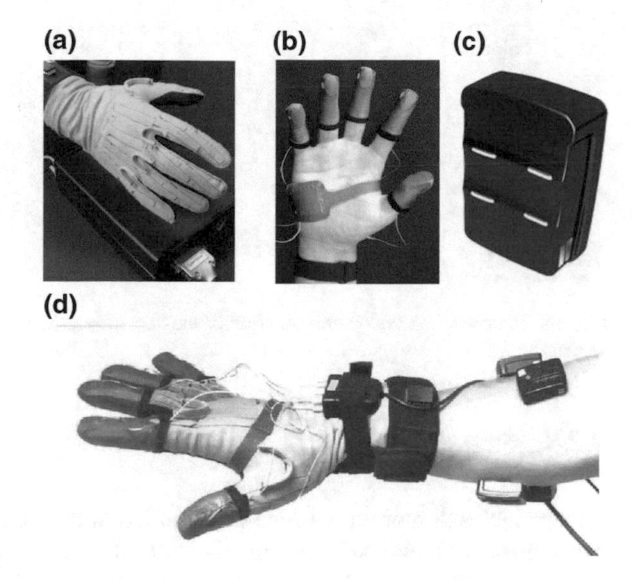

Fig. 8.2 Sensors: **a** CyberGlove; **b** FingerTPS force sensor; **c** Trigno wireless EMG sensor; **d** Multiple sensors for one hand

48 accelerometer channels. The resolution is 16 bits and the sampling rate is 4000 Hz. Its size is 37 mm × 26 mm × 15 mm and the range of its guaranteed performance is 40 m. It has 64 channels of real-time analog output for motion capture integration. Triaxial accelerometers can be used to resolve orientation with respect to the normal force, as well as capture dynamic movements and impacts. These data are captured simultaneously with the EMG data.

8.2.2 Synchronization

The integration of the data glove, the force sensors and the EMG sensors requires a high-speed digital signal processor (DSP) to acquire, process and send raw synchronised information digitally to a PC for analysis shown in Fig. 8.3. The CPU speed of the DSP is greater than 10 MHz for a faster and efficient data acquisition. The force signals from FingerTPS and the EMG signals from the Trigno system are sampled simultaneously at 4 K samples per second by the onboard Analog to Digital Converter (ADC). The data captured from the CyberGlove can be gathered via a universal asynchronous receiver/transmitter (UART) and transmit back to a computer along with other data. The interface connection between the DSP and the computer is USB whose maximum data transfer rate is 10 megabits per second. The three devices are sampled simultaneously and the resolution is 16 bits.

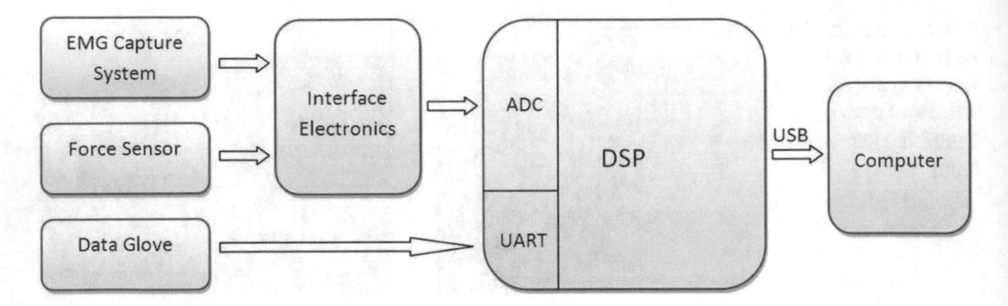

Fig. 8.3 Hardware based synchronization of the data glove, the force sensor and the EMG capture system

8.2.3 Segmentation

To separate each motion with the next motion in the same type, before moving the hand, the hand should be in intermediate state that is a flat hand with no strength. It is assumed that the motion begins when the finger angles change from the intermediate state, and ends when the finger angles change to the intermediate state. In this way, the computer can identify both the start point and end point of each motion using threshold based methods, shown in Fig. 8.4. Five-quick-grasp is to do five power grasps quickly and it enables the muscle to quickly continuously contract five times, and enables the sensory signals with five prominent maximum in their trajectories, simultaneously, shown in Fig. 8.5. These five contractions and maximum are very easy to be identified using peak-detection algorithms. To automatically segment the motions, five-quick-grasp is utilised in the experiments when one type of the motions is finished and the participants are performing the next type of the motions. During the experiment, even trained participants may have difficulties in fulfilling all the tasks at one go. Usually, the motions required in the experiment cannot be finished properly; for example, the cup is dropped by accident. In this case, the data recorded previously is invalid and the experiment needs to be repeated. To solve this problem, we design a protocol that if the motion fails, the participants need to do a four-quick-

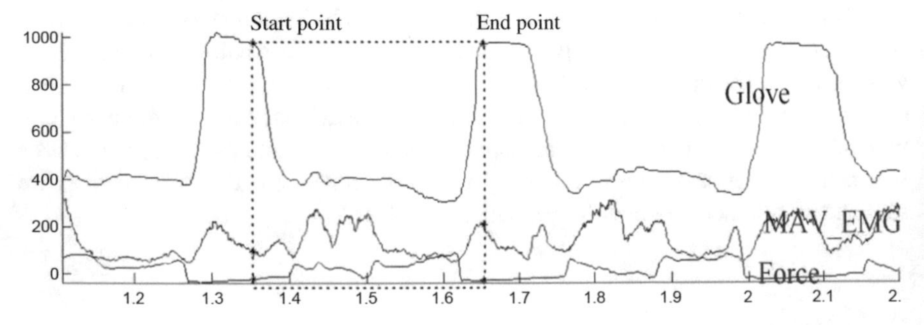

Fig. 8.4 Start and end point for the motion separation

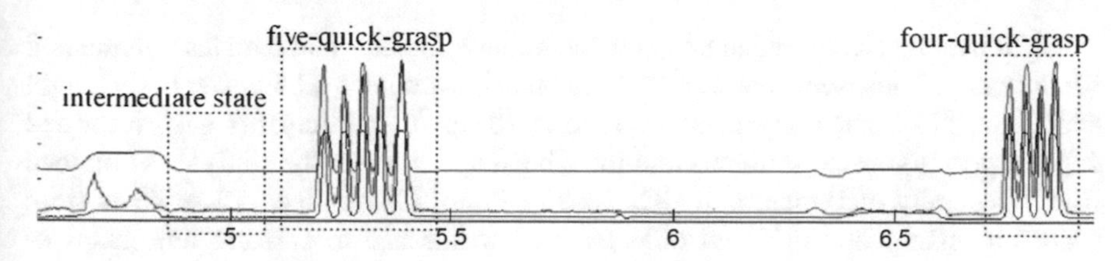

Fig. 8.5 Intermediate state, five-quick-grasp and four-quick-grasp

grasp that indicates the motion recorded before is invalid. In the separation process, the motion before the 'four-quick-grasp' will be deleted automatically.

8.2.4 Data Capturing

The sEMG of 5 forearm muscles shown in Fig. 8.6, i.e. flexor carpi radialis, flexor carpi ulnaris, flexor pollicis longus, flexor digitorum profundus and extensor digitorum, were measured. To obtain clearer signals, subjects were scrubbed with alcohol and shaved if necessary and then electrodes were applied over the body using the die cut medical grade double-sided adhesive tape. Electrodes locations were selected according to the musculoskeletal of these five muscles and confirmed by muscle specific contractions, which include manually resisted finger flexion, extension and abduction. The real time sEMG signals were visualised on a computer screen giving participants feedback to choose the positions of electrodes with stronger sEMG signals. During the experiments, the accelerometers of the EMG systems and the position tracking of the CyberGlove have not been used.

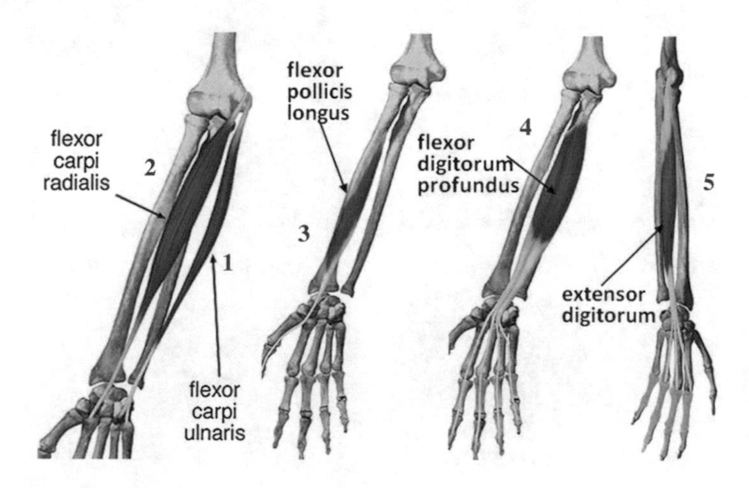

Fig. 8.6 Muscle positions

Eight healthy right-handed subjects including 2 females and 6 males volunteered for the study. Their ages range from 23 to 40 and average is 32.5 years; body height average is 175.5 cm; body mass average is 70 kg. All participants gave informed consent prior to the experiments and the ethical approval for the study was obtained from University of Portsmouth CCI Faculty Ethics Committee. All subjects were trained to manipulate different objects. Participants had to perform ten grasps or in-hand manipulations which are shown in Fig. 8.7 and the motions are listed as following

1. Grasp and lift a book using five fingers with the thumb abduction.
2. Grasp and lift a can full of rice using thumb, index finger and middle finger only.
3. Grasp and lift a can full of rice using five fingers with the thumb abduction.
4. Grasp and lift a big ball using five fingers.
5. Grasp and lift a disc container using thumb and index finger only.
6. Uncap and cap a marker pen using thumb, index finger and middle finger.
7. Open and close a pen box using five fingers.
8. Pick up a pencil using five fingers, flip it and place it on the table.
9. Hold and lift a dumbbell.
10. Grasp and lift a cup using thumb, index finger and middle finger.

The way to grasp or manipulate objects had been shown to the participants in the demonstration before they performed and every motion lasted about 2 to 4 s. Each motion was repeated 10 times. Between every two repetitions, participants had to relax the hands for 2 s in the intermediate state that is opening hand naturally without any muscle contraction. These intermediate states were used to segment the motions.

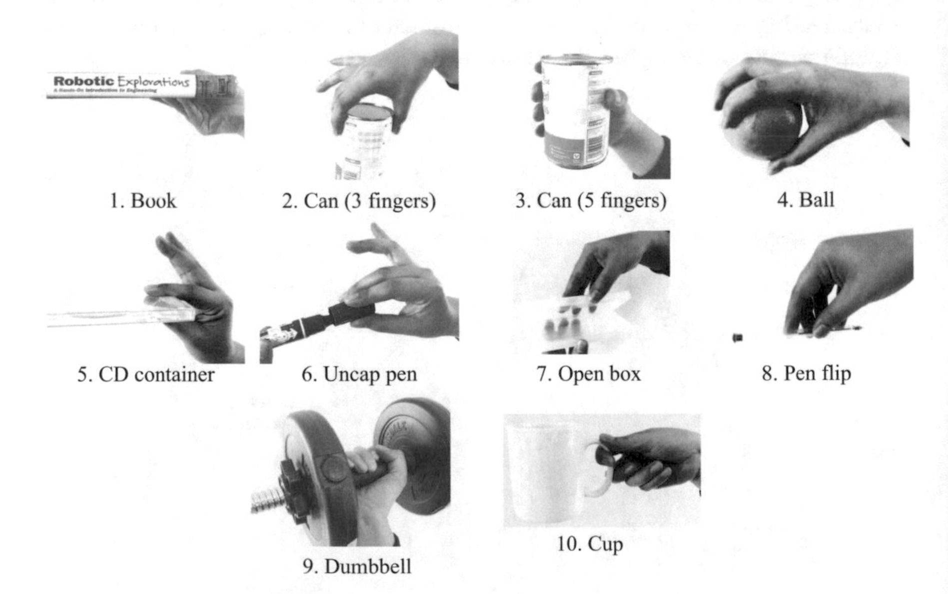

Fig. 8.7 Hand motions including grasps and in-hand manipulations

Once one motion with ten repetitions was finished, participants had to relax the hand for 2 min before the next motion started. This was designed to overcome the effects of muscle fatigue.

8.3 Correlations of Finger Trajectories, Contact Forces and EMG Signals

The hand motion capture system integrating multiple sensors provides us with the ability to study the correlations among the finger force, the finger trajectory and the muscle signal. Spearman's rank correlation coefficient or Spearman's rho is a non-parametric measure of statistical dependence between two variables, and it assesses how well the relationship between two variables can be described using a monotonic function [33]. In this section, we will concentrate on Spearman's rho which is best-known in economic and social statistics. Copula is a popular statistical tool to model and estimate the distribution of random vectors by estimating marginals and to describe the dependence between random variables. The copula of a random vector can capture the properties of the joint distribution which are invariant under transformations of the univariate margins [34], so it is natural to consider the dependence measure, Spearman's rho, which is based on the distribution's copula [35]. Sklar's theorem is fundamental to the theory of copula and underlies most applications of the copula. It elucidates the role that copula plays in the relationship between multivariate distribution functions and their univariate margins. More details of the copula definition and Sklar's theorem can be found in [36, 37]. In this section, we will investigate correlations of the sensory information using the Spearman's rho based on the FEC, whose algorithm can be found in Chap. 6.

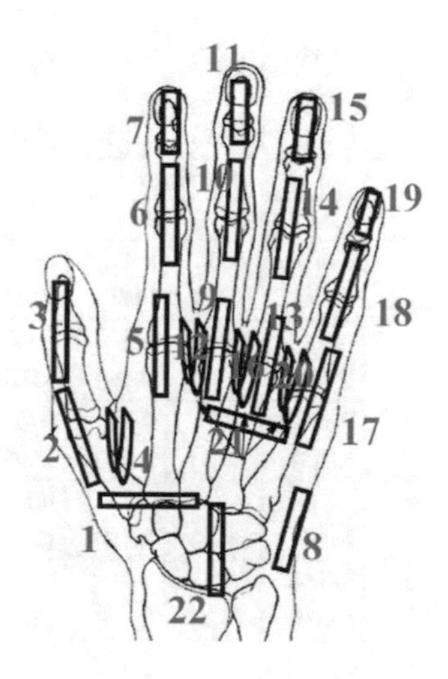

Fig. 8.8 Sensor positions in CyberGlove

Correlations among Mean Absolute Value (MAV) of sEMG signal, finger force and finger angle trajectory have accessed by the Spearman's rho calculated based on the copula. A sliding window with size of 300ms and increment of 50ms has been used to compute MAV. Figures 8.6 and 8.8 present position indexes of the EMG sensors and CyberGlove sensors respectively. The index of the force sensors are one to six for thumb, index finger, middle finger, ring finger, small finger and palm respectively, shown in Fig. 8.2b. Table 8.1 shows the average correlation coefficients (Spearman's rho) and their variances in the bracket for the relations between the muscle contractions and the finger angle trajectories of the motion eight, which is picking up a pencil and flipping it by eight different subjects. From the table, it is clear that there are significant relationships between the muscle signals and the finger angle trajectories: the first muscle signal captured from the flexor carpi ulnaris has strong positive relationships with the little finger and ring finger movements; for the second muscle signal, the contraction from the flexor carpi radialis has the strongest positive relationship with the index finger movements and the second strongest the

Table 8.1 Relation between the muscle contraction and the finger angle trajectories of the motion eight; the row indexes 1–5 represent the different EMG sensors shown in Fig. 8.6; the column indexes 1–22 represent the different CyberGlove sensors shown in Fig. 8.8; the relative strong relations for each EMG sensor are highlighted

	Thumb finger				Index finger				Middle finger		
	1	2	3	4	5	6	7	8	9	10	11
1	0.14 (0.06)	0.23 (0.07)	−0.31 (0.06)	−0.25 (0.04)	0.11 (0.06)	−0.10 (0.06)	0.17 (0.04)	0.24 (0.04)	0.25 (0.06)	0.39 (0.06)	0.29 (0.07)
2	0.42 (0.09)	−0.34 (0.07)	0.04 (0.06)	−0.32 (0.07)	**0.55** **(0.10)**	**0.41** **(0.05)**	**0.49** **(0.05)**	0.21 (0.04)	**0.81** **(0.09)**	**0.48** **(0.07)**	**0.47** **(0.08)**
3	**0.73** **(0.04)**	**0.68** **(0.12)**	**0.71** **(0.07)**	−0.52 (0.04)	0.33 (0.03)	0.55 (0.10)	0.48 (0.09)	0.05 (0.03)	0.43 (0.07)	0.39 (0.08)	0.31 (0.04)
4	0.00 (0.05)	0.17 (0.07)	0.28 (0.08)	0.11 (0.06)	**0.68** **(0.05)**	**0.45** **(0.06)**	**0.36** **(0.08)**	0.44 (0.05)	**0.52** **(0.09)**	**0.44** **(0.05)**	**0.31** **(0.04)**
5	−0.19 (0.11)	0.13 (0.07)	−0.10 (0.08)	−0.12 (0.09)	**−0.42** **(0.11)**	**−0.31** **(0.05)**	**−0.25** **(0.06)**	−0.13 (0.07)	**−0.43** **(0.09)**	**−0.21** **(0.10)**	**−0.14** **(0.12)**
		Ring finger				Little finger					
	12	13	14	15	16	17	18	19	20	21	22
1	−0.18 (0.06)	**0.63** **(0.06)**	**0.51** **(0.06)**	**0.40** **(0.06)**	0.23 (0.05)	**0.83** **(0.06)**	**0.49** **(0.06)**	**0.48** **(0.07)**	0.22 (0.04)	−00.06 (0.08)	−0.37 (0.08)
2	0.32 (0.08)	−0.07 (0.06)	−0.37 (0.09)	−0.38 (0.05)	0.26 (0.06)	−0.21 (0.08)	−0.36 (0.09)	−0.25 (0.07)	−0.17 (0.05)	−0.11 (0.08)	−0.43 (0.07)
3	**0.53** **(0.05)**	0.18 (0.08)	0.43 (0.03)	0.51 (0.05)	0.51 (0.05)	0.21 (0.06)	0.48 (0.03)	0.52 (0.08)	0.31 (0.07)	0.19 (0.10)	0.10 (0.10)
4	0.21 (0.06)	0.30 (0.08)	0.01 (0.05)	0.19 (0.05)	−00.06 (0.06)	0.32 (0.09)	0.16 (0.05)	0.25 (0.07)	0.24 (0.08)	0.28 (0.08)	−0.16 (0.06)
5	−0.08 (0.10)	0.25 (0.06)	−0.02 (0.13)	0.04 (0.09)	−0.21 (0.09)	0.22 (0.10)	−0.14 (0.11)	−0.13 (0.09)	−0.24 (0.06)	−0.11 (0.08)	0.16 (0.08)

Table 8.2 Relation between the muscle contraction and the finger tip force of the motion eight; the row indexes 1–5 represent the different EMG sensors shown in Fig. 8.6; the column indexes 1–5 represent the different force sensors; the relative strong relations for each EMG sensor are highlighted

	thumb	index	middle	ring	little	palm
1	0.32 (0.08)	0.26 (0.09)	0.28 (0.10)	**0.56 (0.07)**	**0.62 (0.09)**	−0.14 (0.06)
2	0.34 (0.09)	**0.41 (0.08)**	**0.56 (0.06)**	0.16 (0.07)	0.17 (0.07)	0.16 (0.08)
3	**0.73 (0.09)**	0.45 (0.10)	0.32 (0.09)	0.22 (0.06)	0.33 (0.11)	−0.12 (0.06)
4	0.33 (0.09)	**0.45 (0.07)**	**0.42 (0.06)**	0.21 (0.05)	0.32 (0.07)	0.02 (0.11)
5	0.15 (0.10)	**−0.42 (0.08)**	**−0.46 (0.08)**	−0.16 (0.07)	−0.25 (0.10)	0.09 (0.09)

middle finger movements; the third muscle signal from the flexor pollicis longus has the strongest positive relationship with the thumb movement; similar to the second muscle signal, the fourth muscle signal from the flexor digitorum profundus has the strongest relationship with the index finger and the middle finger; the fifth muscle signal from extensor digitorum has strong inverse relationship with the index finger and the middle finger. For each muscle signal, the strongest correlations are highlighted in the table. In addition, from the correlation coefficients for the sensor number 4, 8, 12, 16, 20 and 22 which are measuring the finger adduction, abduction and the angles on the palm and wrist, it is hard to see their obvious relationships with the muscle signals compared to the flexion and extension angles such as thumb fingers in the table. Though the third muscle signal has 'large' relationship with the sensor number 4, 12 and 16, their spearman's rho values are much smaller than those between the thumb finger and the third muscle signal.

On the other hand, the relationships between the muscle signals and the finger tip forces have also been studied and the results are shown in the Table 8.2. From the table, the results are consistent with the results in the Table 8.1 except that all the correlation coefficient averages and variances are smaller than those in the Table 8.1.

8.4 Motion Recognition via EMG Intention

As one of the most active research areas, hand motion recognition, in general, consists of neural network approaches [38], support vector machines (SVM) methods [39, 40], rule-based reasoning approaches [41] and probabilistic graphical models [3, 42]. Neural networks are more efficient and achieve better results for complex applications with a huge amount of data. SVMs are very popular because the optimisation problem has a unique solution, but the choice of the kernel function has a significant effect on its performance and the best choice is application dependent.

The rule-based reasoning approaches are easy to implement but their performance highly depends on the applications. Probabilistic graphical models such as hidden Markov models (HMM) and Gaussian Mixture Models (GMMs) have demonstrated their high potential in hand gesture recognition, since they are very rich in mathematical structure and hence their theoretical basis can be adopted for a wide range of applications. Gaussian Mixture Models (GMMs) are one of the most statistically mature methods in pattern recognition and machine learning [42, 43], and has been successfully implemented to identify the high frequency signals such as in speech and EMG recognition [44, 45]. In our previous Chap. 5, Fuzzy Gaussian Mixture Models (FGMMs) are proposed with a better fitting performance and a faster convergence speed than conventional GMMs. In this chapter, we will employ FGMMs to recognise hand motions including different hand grasps and in-hand manipulations via the EMG based signals. In this section, we will firstly revisit the Expectation-Maximization (EM) algorithm for the FGMMs and then propose the novel training and recognising methods. The comparative experimental results and discussions are given in the end.

8.4.1 Training Models via FGMMs

Root Mean Square (RMS), modelled as amplitude modulated Gaussian random process, relates to the constant force and non-fatiguing contraction. Suppose the EMG signal is $f(t)$, where $1 \le t \le N$, N is the number of the sample points, then the RMS is given by

$$f_{rms}(t) = \sqrt{\frac{1}{2w+1} \sum_{i=t-w}^{t+w} f^2(i)} \tag{8.1}$$

where $2w + 1$ denotes the length of the signal window, and $1 \le i \le N$. In this chapter, we use RMS as the EMG feature, since RMS shows powerful performance in robust noise tolerance than other time domain features such as integrated EMG, simple square integral, mean absolute value, mean absolute value slop, and variance [46]. The RMS feature is used both for the FGMMs learning and recognising.

The inputs of the FGMMs include the RMS feature which in our case is a five dimensional time series, number of components k, degree of fuzziness m and threshold ε. Then EM algorithm for FGMMs has been utilised to find the optimised centres of the components μ, their covariances Σ and the control parameters C, T, Q, which are the outputs of the FGMMs and will be used in the recognition process. For the details to implement the EM algorithm of FGMMs, please refer to the algorithm appendix in [47]. An example of the model trained by the introduced FGMMs with six components on the extracted RMS is shown in Fig. 8.9.

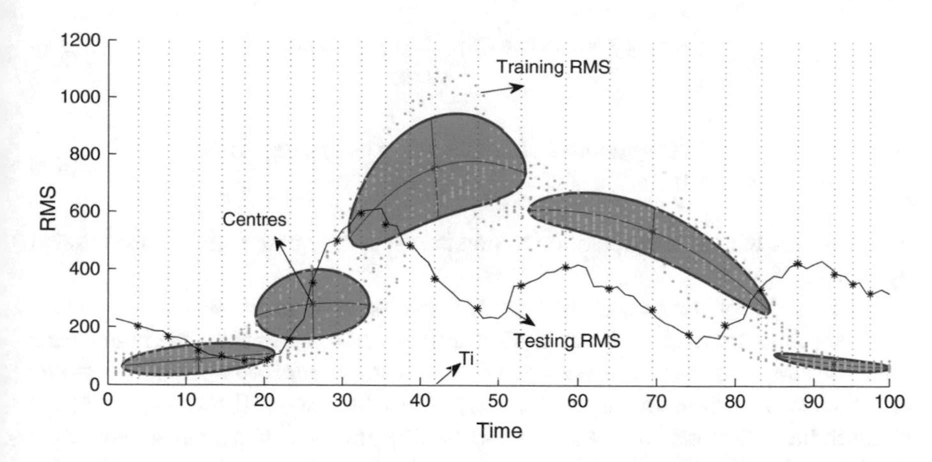

Fig. 8.9 An example of the FGMMs trained result with six components (*blue*) on the RMS feature (*greed dots*); The *black line* is the RMS from the testing motion; *black dots* are the re-sampled points for the testing RMS at T_i time instance

8.4.2 Recognition

Supposing there are k components in the FGMMs trained result. To recognise the testing motion, similarity function is proposed in Eq. 8.2. The similarity of the testing motion and the trained model of FGMMs is defined by the normalised log-likelihood between the re-sampled testing points and the FGMMs components as:

$$Si = \frac{1}{5k}\sum_{j=1}^{5k}\log\left(\sum_{i=1}^{k}\alpha_i p_i(x_{T_j}|\theta_i)\right) \tag{8.2}$$

where α_i is the mixing coefficient of the ith component, if the component's curvature parameter $a_i < \varepsilon$, the $p(x|\theta)$ will be calculated by:

$$p(x|\theta) = \frac{1}{(2\pi)^{\frac{d}{2}}\sqrt{|\Sigma|}}\exp\left(-\frac{(x-\mu)^T\Sigma^{-1}(x-\mu)}{2}\right) \tag{8.3}$$

if its curvature parameter $a_i \geq \varepsilon$, $p(x|\theta)$ is achieved by Eq. 5.37. x_{T_j} is the selected points from the testing data $\mathbf{x}$ at the time instance of T_j, which can be achieved by Eq. 8.4.

$$T_j = \mu_f - \frac{\eta_j(\mu_f - \mu_{f-1})}{3} + \frac{\gamma_j(\mu_{f+1} - \mu_f)}{3}; \tag{8.4}$$

where T_j is the time sampling points for the testing data; $j \in (1, \ldots, 5\cdot k)$;μ_f is the time label of the fth component centre; $f = \lfloor (j-1)/5 \rfloor$; the parameters, η and γ, are achieved by Eqs. 8.5 and 8.6:

$$\eta_j = \begin{cases} 2 - [(j-1)mod \quad 5] & \text{if}\,[(j-1)mod \quad 5] < 2 \\ 0 & \text{else} \end{cases} \tag{8.5}$$

$$\gamma_j = \begin{cases} [(j-1)mod \quad 5] - 2 & \text{if}\,[(j-1)mod \quad 5] > 2 \\ 0 & \text{else} \end{cases} \tag{8.6}$$

where *mod* is the modulo operation to find the remainder of division of one number by another.

An example of the re-sampling process can be seen in Fig. 8.9 where the re-sampled points are marked by the black star points from the testing data in the black line. The idea of resampling testing points is to reduce the computation cost and at the same time to maintain the accuracy of the likelihood. If the number of the points in the testing data $n >> k$, the re-sampling process can manage to reduce the number of the testing points to $5k$, which can alleviate the computational burden for the recognition. Additionally, the re-sampling process selects the points according to the distribution of the components, which guarantees the re-sampled points cover the major distribution for the testing process.

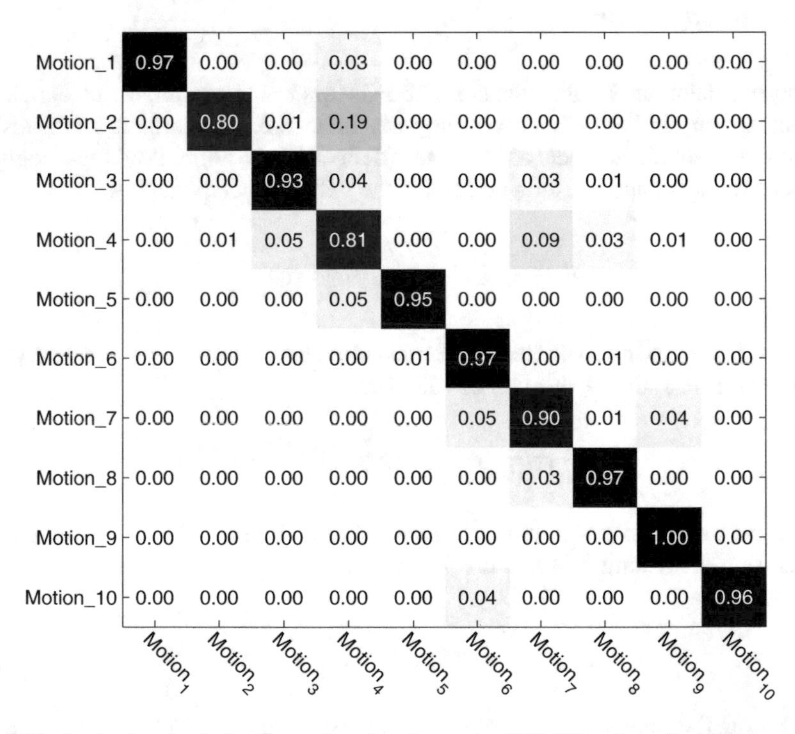

Fig. 8.10 Confusion matrix for the ten hand motions using FGMMs, where the total accuracy is 92.75%

8.4.3 Experimental Results

To evaluate the performance of FGMMs for classifying the EMG signals, FGMMs are compared with both traditional Gaussian Mixture Models (GMMs) and Support Vector Machine (SVM). To have a fare comparison, the recognition process for GMMs is the same as FGMMs which have been proposed in Sect. 8.4.2. The parameter for GMMs and FGMMs is the number of the component ranging from 2 to 20 with increments of one. As another popular machine learning method for classification [48], SVM use a kernel function to implicitly map the input vector into a high-dimensional space, and to maximise the margin between classes based on computational statistical theory. In this chapter, radial basis function for the SVM classifier has been employed, which has been demonstrated with satisfactory performance in pattern recognition tasks [40, 49]. We used the one-against-all multi-class method for the multi-label classification, where for each label it builds a binary-class problem so instances associated with that label are in one class and the rest are in another class. Thc parameters for SVM are the kernel parameter ranging from 1 to 10 with increments of one and penalty cost whose range is from 1 to 501 with increments of 50 achieved by using LIBSVM [50] package. These parameters are selected with their best performance.

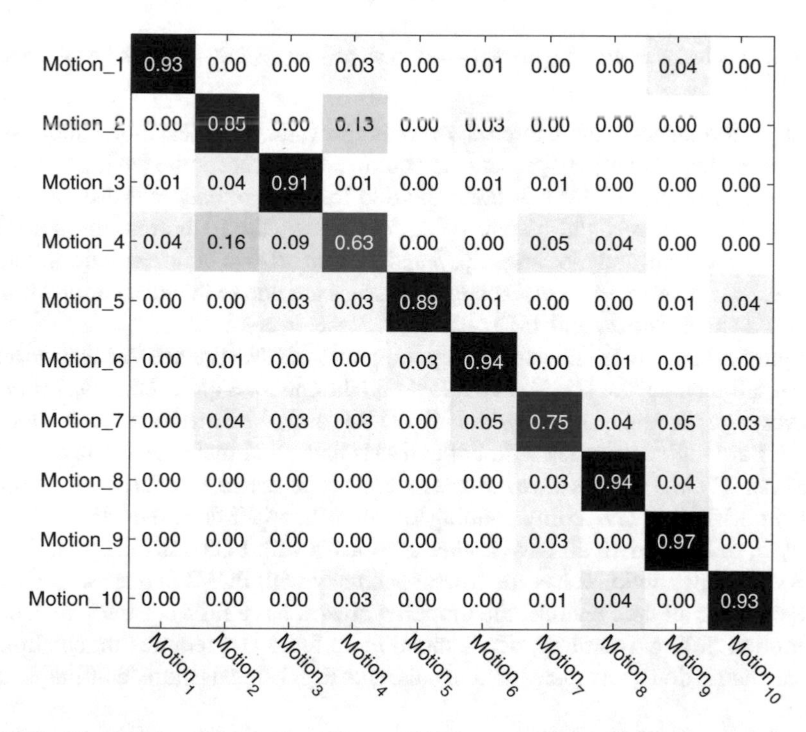

Fig. 8.11 Confusion matrix for the ten hand motions using GMM, where the total accuracy is 87.25%

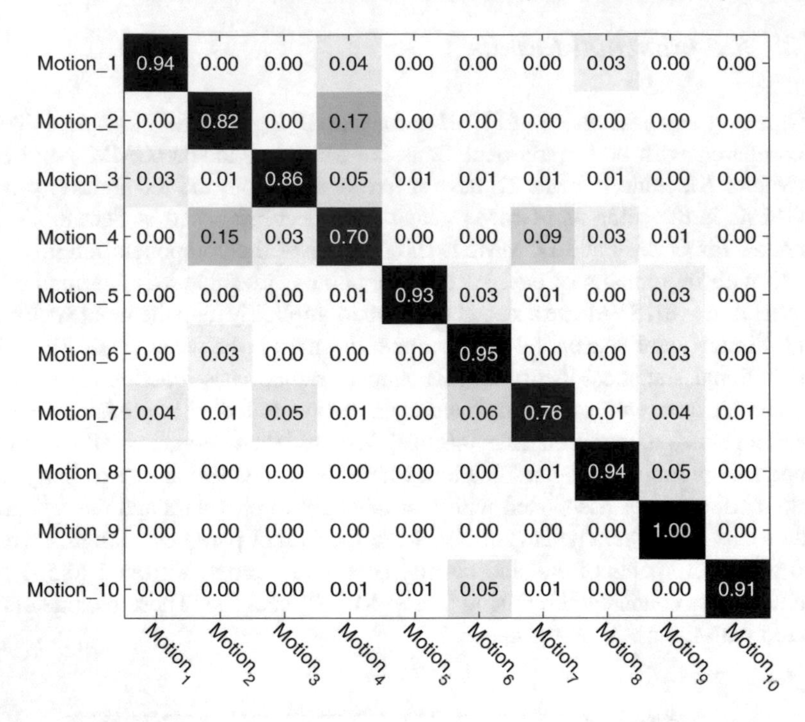

Fig. 8.12 Confusion matrix for the ten hand motions using SVM, where the total accuracy is 88.13 %

The performances of these three algorithms are evaluated by leave-one-subject-out cross-validation. Figure 8.10 presents the confusion matrix for the ten hand motions in Sect. 8.2.4 using FGMMs. Among the 800 testing motions, FGMMs obtain 58 errors, and the total recognition rate is 92.75%. Among the 10 motions, motion 9 has the full correct rate, and motions 1, 6, 8 and 10 also receive high recognition rate of above 95%. However, the worst recognition rates are due to the misclassification of motions 2 and 4 with 20 and 19% error rates respectively.

Figures 8.11 and 8.12 show the recognition rates of GMMs and SVM respectively. The overall accuracy for GMMs is 87.25% and the one for SVM is 88.13%. Motion 9 achieves high recognition rates with both GMMs and SVM methods, while motions 4 and 7 have high error rates with both GMMs and SVM methods. Compared with GMMs and SVM, FGMMs have the best overall performance, which reduces the error rate from 12.75 to 7.25%, corresponding to a more than 40% error reduction. Motions 1, 3, 4, 5, 6, 7, 8 and 10 all have higher accuracies with FGMMs than with GMMs and SVM. Only motion 2 has the lowest accuracy with FGMMs than with GMMs and SVM. For all the methods, the motion 9 always has a high accuracy rate, since this motion, lifting a cambelt, needs much more force and requires much stronger muscle contraction than others, which makes the EMG signals more identifiable than others.

On the other hand, Fig. 8.13 shows the average recognition rates and their variances of the eight subjects using different methods. For different subjects performing all

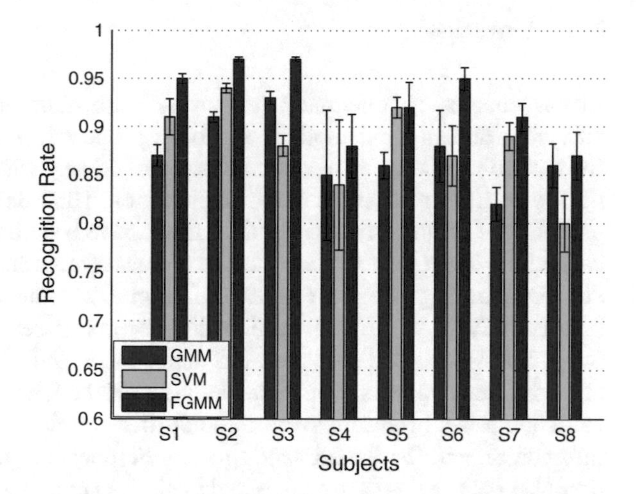

Fig. 8.13 Recognition results with means and variances of different subjects using different methods

motions, the recognition rates range from 80 to 97%. The 2nd subjects has the highest average accuracy rate of 94%, while the subject eight has the lowest average rate of 84.33%. From both the average accuracies and variances shown in Fig. 8.13, it concludes that FGMMs can reduce the error rates for all the subjects except the 5th subject, for whom the SVM has the highest accuracy while FGMMs have the second one. When achieving relatively high accuracy, e.g. subjects 1, 2 and 3, FGMMs have very small variances which are smaller than 0.02. For the 6th subject, FGMMs have only five motions misclassified and it manages to improve the recognition rates of GMMs and SVM from around 87–95%. Figure 8.14 presents the box plot of different classifiers for the different subjects. Generally, FGMMs outperform GMMs and SVM for all the eight subjects in terms of the performance and the latter two have similar performances with each other.

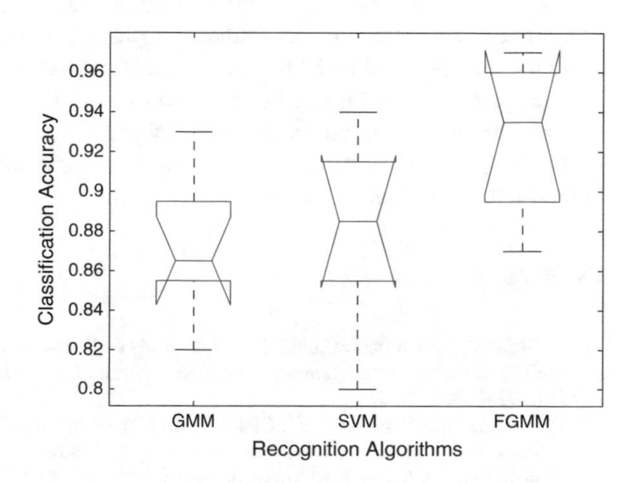

Fig. 8.14 Box plot results for the different classifiers for all subjects

8.5 Conclusion

In this chapter, an integrated framework with multiple sensory information for analysing human hand motions has been proposed, providing effective solutions for motion capturing, data synchronisation and segmentation, correlation study of the sensory information and motion recognition. Three devices, i.e. CyberGlove, FingerTPS and Trigno wireless EMG sensors, have been integrated to simultaneously capture the finger angle trajectories, the contact forces and the forearm EMG signals at a fast sampling rate. An effective solution to automatically segment the manipulation primitives of different motions has been proposed using five-quick-grasp and four-quick-grasp protocols. Ten different grasps and in-hand manipulations from eight different subjects have been analysed. In the knowledge base module, empirical copula has been employed to study the correlations of the different sensory information and the experimental results demonstrate there exist significant relationships between muscle signals and finger trajectories and between muscle signals and contact forces. In the motion recognition module, FGMMs have been used to recognise these ten motions from eight different subjects based on RMS features of the EMG signals and it achieved an overall 92.75% recognition rate, which is much higher than those of GMMs and SVM due to its nonlinear fitting capability. In terms of different subjects, FGMMs still outperformed the other two with improved accuracies.

The proposed framework integrates the state-of-the-art sensor technology, mature machine learning methods and signal processing algorithms. It provides a versatile and adaptable platform for researchers in robotics, biomedical engineering, AI, and HCI to analyse human hand motions. The strong correlations among the signals indicate that the muscle signals can be potentially used to estimate the gesture or force of the hand motions. The main application of this framework is to control prosthetic hand via EMG signals. The proposed classification algorithms can effectively identify the amputees intended movements. The next step for controlling the prosthetic hand is to generate corresponding finger movements including the gestures and forces. The studied correlations in our framework provide a good reference for the further study to generate the desired trajectories. The future work is targeted to extend the knowledge base with human hand motion regressed primitives from the training motions and manipulation scenarios e.g object shape and contact points [51], and further to apply the framework into automatically controlling prosthetic hands such as the i-LIMB hand from Touch Bionics.

References

1. L. Dipietro, A.M. Sabatini, and P. Dario. A survey of glove-based systems and their applications. *IEEE Transactions on Systems, Man, and Cybernetics, Part C: Applications and Reviews*, 38(4):461–482, 2008.
2. C.D. Metcalf, S.V. Notley, P.H. Chappell, J.H. Burridge, and V.T. Yule. Validation and application of a computational model for wrist and hand movements using surface markers. *IEEE Transactions on Biomedical Engineering*, 55(3):1199–1210, 2008.

3. Z. Ju and H. Liu. A unified fuzzy framework for human hand motion recognition. *IEEE Transactions on Fuzzy Systems*, 19(5):901–913, 2011.
4. Kyung-Won Kim, Mi-So Lee, Bo-Ram Soon, Mun-Ho Ryu, and Je-Nam Kim. Recognition of sign language with an inertial sensor-based data glove. *Technology and Health Care*, 24(s1):S223–S230, 2015.
5. T.B. Moeslund and E. Granum. A survey of computer vision-based human motion capture. *Computer Vision and Image Understanding*, 81(3):231–268, 2001.
6. A. Malinowski and H. Yu. Comparison of embedded system design for industrial applications. *IEEE Transactions on Industrial Informatics*, 7(2):244–254, 2011.
7. X. Ji and H. Liu. Advances in View-Invariant Human Motion Analysis: A Review. *IEEE Transactions on Systems, Man and Cybernetics Part C*, 40(1):13–24, 2010.
8. Z. Ren, J. Meng, J. Yuan, and Z. Zhang. Robust hand gesture recognition with kinect sensor. In *Proceedings of the 19th ACM international conference on Multimedia*, pages 759–760. ACM, 2011.
9. Z. Zafrulla, H. Brashear, T. Starner, H. Hamilton, and P. Presti. American sign language recognition with the kinect. In *Proceedings of the 13th international conference on multimodal interfaces*, pages 279–286. ACM, 2011.
10. S. Calinon, P. Evrard, E. Gribovskaya, A. Billard, and A. Kheddar. Learning collaborative manipulation tasks by demonstration using a haptic interface. In *International Conference on Advanced Robotics*, pages 1–6. IEEE, 2009.
11. S. Saliceti, J. Ortiz, A. Cardellino, L. Rossi, and J.G. Fontaine. Fusion of tactile sensing and haptic feedback for unknown object identification aimed to tele-manipulation. In *IEEE Conference on Multisensor Fusion and Integration for Intelligent Systems*, pages 205–210. IEEE, 2010.
12. T. Takaki and T. Omata. High-performance anthropomorphic robot hand with grasping-force-magnification mechanism. *IEEE/ASME Transactions on Mechatronics*, 16(3):583, 2011.
13. M. Ferre, I. Galiana, R. Wirz, and N. Tuttle. Haptic device for capturing and simulating hand manipulation rehabilitation. *IEEE/ASME Transactions on Mechatronics*, 16(5):808–815, 2011.
14. C.H. Park, J.W. Yoo, and A.M. Howard. Transfer of skills between human operators through haptic training with robot coordination. In *IEEE International Conference on Robotics and Automation*, pages 229–235. IEEE, 2010.
15. T.S. Saponas, D.S. Tan, D. Morris, R. Balakrishnan, J. Turner, and J.A. Landay. Enabling always-available input with muscle-computer interfaces. In *Proceedings of the 22nd Annual ACM Symposium on User Interface Software and Technology*, pages 167–176. ACM, 2009.
16. X. Tang, Y. Liu, C. Lv, and D. Sun. Hand motion classification using a multi-channel surface electromyography sensor. *Sensors*, 12(2):1130–1147, 2012.
17. C. Cipriani, F. Zaccone, S. Micera, and M.C. Carrozza. On the shared control of an emg-controlled prosthetic hand: analysis of user–prosthesis interaction. *IEEE Transactions on Robotics*, 24(1):170–184, 2008.
18. C. Castellini and P. van der Smagt. Surface emg in advanced hand prosthetics. *Biological cybernetics*, 100(1):35–47, 2009.
19. D. Zhang, X. Chen, S. Li, P. Hu, and X. Zhu. Emg controlled multifunctional prosthetic hand: Preliminary clinical study and experimental demonstration. In *IEEE International Conference on Robotics and Automation*, pages 4670–4675. IEEE, 2011.
20. Daniele Leonardis, Michele Barsotti, Claudio Loconsole, Massimiliano Solazzi, Marco Troncossi, Claudio Mazzotti, Vincenzo Parenti Castelli, Caterina Procopio, Giuseppe Lamola, Carmelo Chisari, et al. An emg-controlled robotic hand exoskeleton for bilateral rehabilitation. *IEEE transactions on haptics*, 8(2):140–151, 2015.
21. M.R. Williams and R.F. Kirsch. Evaluation of head orientation and neck muscle emg signals as command inputs to a human–computer interface for individuals with high tetraplegia. *IEEE Transactions on Neural Systems and Rehabilitation Engineering*, 16(5):485–496, 2008.
22. A.B. Usakli, S. Gurkan, F. Aloise, G. Vecchiato, and F. Babiloni. On the use of electrooculogram for efficient human computer interfaces. *Computational Intelligence and Neuroscience*, 2010:1–5, 2010.

23. P. Langhorne, F. Coupar, and A. Pollock. Motor recovery after stroke: a systematic review. *The Lancet Neurology*, 8(8):741–754, 2009.
24. N. Ho, K. Tong, X. Hu, K. Fung, X. Wei, W. Rong, and E.A. Susanto. An emg-driven exoskeleton hand robotic training device on chronic stroke subjects: Task training system for stroke rehabilitation. In *IEEE International Conference on Rehabilitation Robotics*, pages 1–5. IEEE, 2011.
25. Aisha Al-Mansoori, Alkhzami Al-Harami, Rawda AlMesallam, Uvais Qidwai, and Abbes Amira. An intelligent sensing system tor healthcare applications using real-time emg and gaze fusion. In *SAI Intelligent Systems Conference (IntelliSys), 2015*, pages 491–498. IEEE, 2015.
26. M.G. Ceruti, V.V. Dinh, N.X. Tran, H. Van Phan, L.R.T. Duffy, T.A. Ton, G. Leonard, E. Medina, O. Amezcua, S. Fugate, et al. Wireless communication glove apparatus for motion tracking, gesture recognition, data transmission, and reception in extreme environments. In *Proceedings of the 2009 ACM symposium on Applied Computing*, pages 172–176. ACM, 2009.
27. N. Hendrich, D. Klimentjew, and J. Zhang. Multi-sensor based segmentation of human manipulation tasks. In *IEEE Conference on Multisensor Fusion and Integration for Intelligent Systems*, pages 223–229. IEEE, 2010.
28. J.M. Romano, S.R. Gray, N.T. Jacobs, and K.J. Kuchenbecker. Toward tactilely transparent gloves: Collocated slip sensing and bibrotactile actuation. In *Proceedings of the World Haptics - Third Joint EuroHaptics conference and Symposium on Haptic Interfaces for Virtual Environment and Teleoperator Systems*, pages 279–284, Washington, DC, USA, 2009. IEEE Computer Society.
29. D. Yang, J. Zhao, Y. Gu, L. Jiang, and H. Liu. Estimation of hand grasp force based on forearm surface emg. In *International Conference on Mechatronics and Automation*, pages 1795–1799. IEEE, 2009.
30. H. Liu. Exploring human hand capabilities into embedded multifingered object manipulation. *IEEE Transactions on Industrial Informatics*, 7(3):389–398, 2011.
31. Z. Ju and H. Liu. A generalised framework for analysing human hand motions based on multisensor information. In *IEEE International Conference on Fuzzy Systems*, pages 1 –6, june 2012.
32. T. Yoshikawa. Multifingered robot hands: Control for grasping and manipulation. *Annual Reviews in Control*, 34(2):199–208, 2010.
33. C. Spearman. The proof and measurement of association between two things. *The American Journal of Psychology*, 15(1):72–101, 1904.
34. E.F. Wolff. N-dimensional measures of dependence. *Stochastica: revista de matemática pura y aplicada*, 4(3):175–188, 1980.
35. F. Schmid and R. Schmidt. Multivariate conditional versions of spearman's rho and related measures of tail dependence. *Journal of Multivariate Analysis*, 98(6):1123–1140, 2007.
36. R.B. Nelsen. *An introduction to copulas*. Springer Verlag, 2006.
37. A. Sklar. Fonctions de répartition a n dimensions et leurs marges. *Publ Inst Statist Univ Paris*, 8:229–231, 1959.
38. M.S.K. Yewale and M.R.P.K. Bharne. Artificial neural network approach for hand gesture recognition. *International Journal of Engineering Science and Technology*, 3(4):2603–2608, 2011.
39. D. Martinez and D. Kragic. Modeling and recognition of actions through motor primitives. In *IEEE International Conference on Robotics and Automation*, pages 1704–1709, 2008.
40. X. Chen, X. Zhu, and D. Zhang. A discriminant bispectrum feature for surface electromyogram signal classification. *Medical Engineering & Physics*, 32(2):126–135, 2010.
41. B.C. Bedregal, A.C.R. Costa, and G.P. Dimuro. Fuzzy Rule-Based Hand Gesture Recognition. *International Federation for Information Processing-Publications-IFIP*, 217:285, 2006.
42. Z. Ju, H. Liu, X. Zhu, and Y. Xiong. Dynamic Grasp Recognition Using Time Clustering, Gaussian Mixture Models and Hidden Markov Models. *Journal of Advanced Robotics*, 23:1359–1371, 2009.
43. S.O. Ba and J. M. Odobez. Recognizing visual focus of attention from head pose in natural meetings. *IEEE Transactions on Systems, Man, and Cybernetics, Part B: Cybernetics*, 39(1):16–33, Feb. 2009.

44. Y. Huang, K.B. Englehart, B. Hudgins, and A.D.C. Chan. A gaussian mixture model based classification scheme for myoelectric control of powered upper limb prostheses. *IEEE Transactions on Biomedical Engineering*, 52(11):1801–1811, 2005.
45. L. Lu, A. Ghoshal, and S. Renals. Regularized subspace gaussian mixture models for speech recognition. *IEEE Signal Processing Letters*, 18:419–422, 2011.
46. A. Phinyomark, C. Limsakul, and P. Phukpattaranont. Emg feature extraction for tolerance of 50 hz interference. In *Proc. of PSU-UNS Inter. Conf. on Engineering Technologies, ICET*, pages 289–293, 2009.
47. Z. Ju and H. Liu. Fuzzy gaussian mixture models. *Pattern Recognition*, 45(3):1146–1158, 2012.
48. V.N. Vapnik. *The nature of statistical learning theory*. Springer Verlag, 2000.
49. I. Steinwart and A. Christmann. *Support vector machines*. Springer Verlag, 2008.
50. C.C. Chang and C.J. Lin. Libsvm: A library for support vector machines. *ACM Transactions on Intelligent Systems and Technology*, 2:27:1–27:27, May 2011.
51. Q.V. Le, D. Kamm, A.F. Kara, and A.Y. Ng. Learning to grasp objects with multiple contact points. In *IEEE International Conference on Robotics and Automation*, pages 5062–5069, 2010.

Chapter 9
A Novel Approach to Extract Hand Gesture Feature in Depth Images

9.1 Introduction

Recently, the problem of acquisition and recognition of human hand gestures from RGB-Depth (RGB-D) sensors, such as Microsofts Kinect, is an important subject in the area of the computer vision and pattern analysis. In order to extract and recognise hand gestures from RGB-D data, many researchers conducted significant contribution, including the hand gesture extracting, tracking, recognising and so on [1–3]. These achievements are of much importance for research in areas of human-computer interaction(HCI). Researchers largely welcome the Kinect developed by Microsoft Corporation, as it can simultaneously acquire data of RGB image and depth map of the scene by its IR emitter and camera sensors. Its broad applications cover 3D reconstruction [4, 5], image processing [6, 7], human-machine interface [2, 8–13], robotics [14, 15], object recognition [16, 17], just to name a few [18–20]. However, there are many problems such as distortion and disaccord of depth and RGB images in corresponding pixels, especially the limitations in extracting of correct human hand gestures [20]. Due to the noises and holes in the RGB-D data, precisely segmenting the hand gestures is still a challenging task.

Kinect is the Microsoft's somatic human-computer interaction device developed for its game console XBOX360. Actually it is a three-dimension camera that can provide colour image stream, depth image stream, skeleton image stream and voice stream in real time. Thus it is able to catch the movement of players in three-dimension. Then we can use simple gestures and voice to do the human-computer interaction. Kinect hardware includes a colour camera, a depth camera witch consist of an infrared emitter and a infrared receiver, a set of microphone sensors, a acceleration sensor and a servo motor that drive the pitching action.

Traditional three-dimension measurement uses structured light or time of flight (ToF) [1]. The illumination is dispersing in time and space. While Kinect depth camera is based on a new technology called Light-coding, in which continuous illumination is used. In this way, general CMOS sensor can be used to reduce cost. In addition, being a game peripheral, Kinect has strong real-time computing power, especially for its skeleton stream witch is calculated using depth information. Due to these advantages, Kinect sensor works as a low-cost and high-performance hardware

H. Liu et al., *Human Motion Sensing and Recognition*,
Studies in Computational Intelligence 675, DOI 10.1007/978-3-662-53692-6_9

platform in many research fields besides computer games, such as virtual trying on clothes [21], three-dimension scanner [22], robot navigation [23] and so on [20].

Kinect is designed as a game peripheral, so despite its powerful real-time human-computer interaction function, there are some weaknesses in respects that is of no importance for game user experience due to cost cutting. And these become obstacles for the application of Kinect in scientific research. Specifically, depth image data and skeleton data are usually mixed up with lots of noise, because they are created by embedded computer using its inner algorithm, and this will be a problem for some high-robust-needed applications; the depth data has a limitation of distance measurement which is designed for computer game players, Kinect cannot provide reliable distance information to exceed this limitation [24]. So many projects using Kinect as a hardware platform are trying to find the way to make up these weaknesses. They include the calibration of the Kinect's colour image data and depth image data [25]; algorithm for calculation of skeleton data using raw data [26]; some filtering algorithm of Kinect data streams [27], and so on.

Hand is the most flexible connector between human and the outside world, so do human-computer interaction based on physical sensory data. The research of hand gesture recognition based on the Kinect has become very important and also has gained wide attention. To recognise hand gestures, there are three aspects of important work that need be discussed. First of all, hand detection is the basis of all other analysis algorithm. Hand detection base on two kinds of data, they are RGB pixels and depth data. For hand detection from depth data, segmentation is considered as a depth clustering problem, where the pixels are divided into different depth levels in [28]. The critical part is to determine a threshold, indicating at which depth level the hand is located. In [29], Lianget comes up with a concise method, using a clustering algorithm followed by morphological constraints, he detect hands on the depth images, then a distance transform to the segmented hand contour is performed to estimate the palm and its center. Secondly, hand pose estimation is the key of most hand gesture recognition research. Finger-Earth Mover's distance is proposed in [30] to measure the dissimilarities between different hand shapes. Skeleton-based approaches deduce the hand pose based on the configuration of the hand skeleton. The skeleton generation is the key to these methods. A pioneering work [31] is the random decision forests (RDF)-based hand skeleton tracking, which can be seen as a hand-specific version of Shotton's human-body skeleton tracking. It performs per-pixel classification by means of the RDF technique and assigns each pixel a hand part. Then, the mean shift algorithm is applied to estimate the centres of hand parts to form a hand skeleton. Finally, hand gesture classification makes it possible to understand the meanings of hand gestures [15]. Tang et al. [32] establish a real-time system that is able to identify simple hand gestures such as grasping and pointing. In this system, a person's hand is estimated based on a skeletal tracker. Next, local shape-related features, such as radial histogram and modified speeded-up robust features (SURF) [33], are extracted from the image. A support vector machine (SVM) classifier (with a radial basis function) is used to distinguish the hand gestures. In this chapter, a real-time hand gesture feature extraction method is proposed by

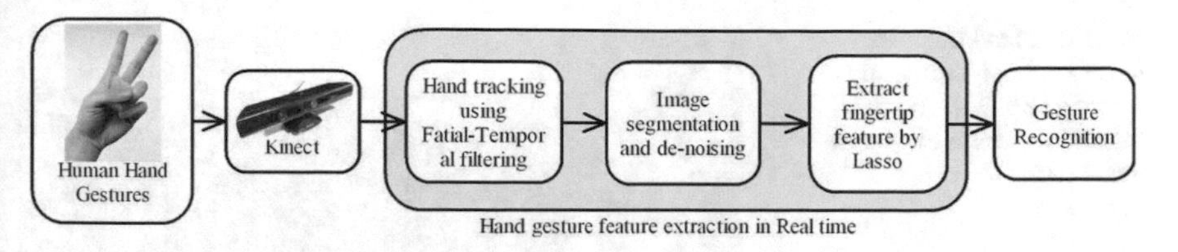

Fig. 9.1 The proposed approach for hand gesture feature extraction in real time

a novel finger earth mover's distance algorithm and Lasso algorithm to locate the palm and effectively extract the hand contour curve. The diagram of the proposed approach is shown in Fig. 9.1.

9.2 Image Segmentation and De-noising Combining with Depth Information

9.2.1 Image Segmentation

Kinect sensor [34] can output colour information and depth information flow, and with the combination of the two information flows, the required palm image can be drawn out from a complex background. Colour information flow can be set with several grades of resolution ratio and different formats. High-resolution ratio can provide more information, while the update rate is low. It is opposite to the low-resolution ratio, users can select according to specific applications. To provide depth information ratio is one of the characteristics of Kinect, and is also the foundation of the realization of basic physical sensory. The value of each frame of image pixel of depth information flow represents the minimum distance between the point on the image and the real object. Then it is possible to track the action of human or erase the background object by the depth information. Besides, Kinect can also provide the inner processor and meaningful information for users, such as player-separated data. Separated data is provided with bitmap format, each value of image pixel is corresponding to the number of the player who is the nearest to the camera at the position of the pixel in the field of view. The palm will be roughly segmented based only on the depth information. Firstly, Spatial-temporal filtering (STF) [21, 22] is employed to track the hand position and based on the tracking result the hand initial depth can be automatically chosen, as shown in Fig. 9.2. Then, the initial depth is used to determine the depth threshold value, based on which the palm images are segmented. In addition, the initial edge of the hand gesture can be achieved using Sobel method [23].

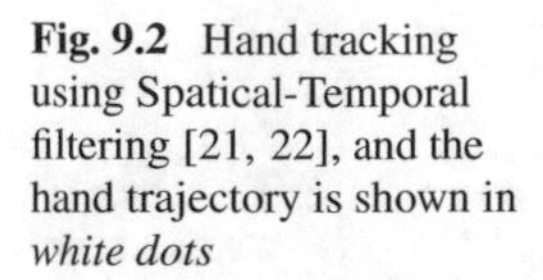

Fig. 9.2 Hand tracking using Spatical-Temporal filtering [21, 22], and the hand trajectory is shown in *white dots*

9.2.2 Select the Edge Contour of Image

With the above image segment result, it is much simple to select the edge contour of image. Traversal the pixel of four directions can get the contour of all eight directions. The pseudo-code shall be found in the Appendix E.1.

9.2.3 De-noise by Contour Length Information

The palm image information extracted by method above is of severe noise. Because the depth information provided by Kinect sensor has a low-resolution ratio. With such ratio it is difficult to recognise the object like palm, which is with abundant details. Also, the continuity of player separation data can not be fully ensured among the frames, it is with some breaks in times. All the above reasons make the palm image segmented by only Kinect depth information very hard and player separation data is not used in recognition algorithm, which is of high robustness. In Fig. 9.3, there are some holes in the segmented palm contour. These holes are wrongly taken as pixel that is not belonging to the palm.

By experiment, there are mainly 2 kinds of noises caused by image segmentation: one kind of noisy point is connected with the background, because of its existence, the palm image is separated into several parts, such as the No.3 black part in Fig. 9.3; another kind of noisy point is inside the palm image, around which are the recognized palm image pixels. The existence of such noise will lead to the whole structure. Such as No.1 and 2 in Fig. 9.3.

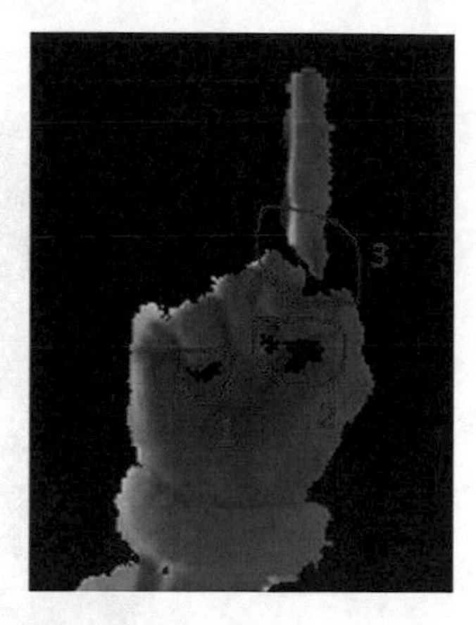

Fig. 9.3 Problems of noise of image segmentation

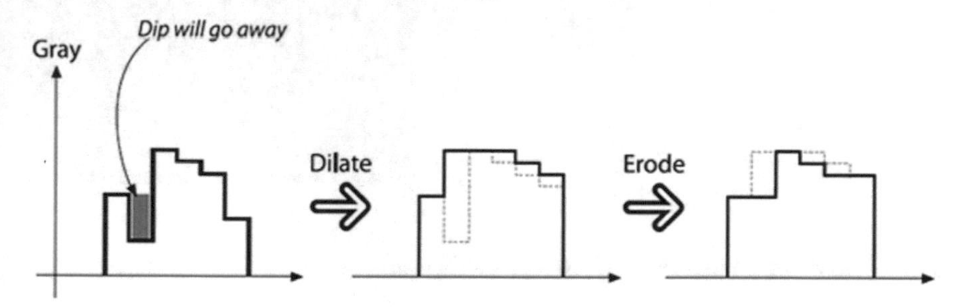

Fig. 9.4 Closed operation of morphology

In order to filtrate the first kind of noise, we use closed operation that is widely used in image morphology, and firstly expansion and then corrosion should be undertaken in Fig. 9.4. The principle shall be found as follows:

The operation of morphology can also filtrate the second kind of noise, but we consider a method that can totally eliminate the second noise. It is a method that can use the contour length information. The principle of the method is: Suppose the shape of palm and the noise hollow are all the edges, extract the respective contour point, divide the eight connected contour point as a group, you will get several groups of contour chain. Calculate the length of each group of contour chain, and find out the longest group. The noise hole is relatively smaller compared with the real size of palm, so we can consider the longest group of contour chain as the real contour of palm; other short chains are formed because of the second kind of noise. Figure 9.5 indicates that such method can effectively recognise and filtrate the second kind of noise.

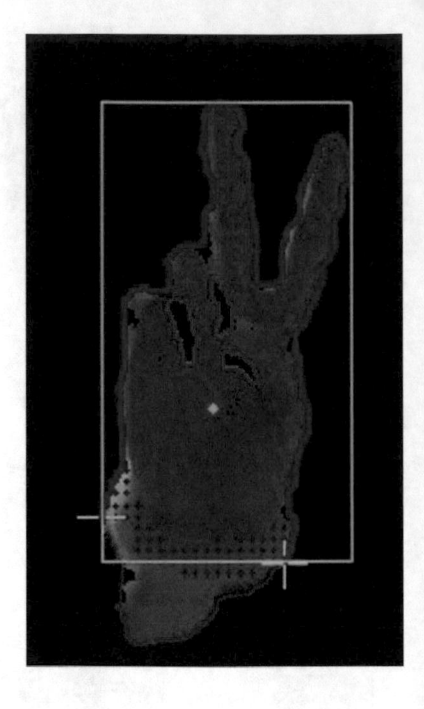

Fig. 9.5 Use long contour information to filtrate the noise

9.3 Extract Fingertip Feature by Lasso

In this section, the earth mover's distance is introduced and modified, based on which the Lasso algorithm is proposed.

9.3.1 Earth Mover's Distance

Earth Mover's Distance (EMD) is defined as measure criterion compared with histogram. The problem it concerned is how to change the histogram from one shape into another one by certain movement (including move part of the histogram or whole to a new position). Generally, people always construct the information of histogram by statistical methods with the application of EMD, then match the different histograms of different images according to the algorithm to distinguish different images. Zhou Ren raised a new method to construct histogram, which is FEMD (Finger Earth Mover's Distance [35]). In this method, it is the direct conversion from gesture contour into histogram instead of the image calculation histogram information applied in EMD calculation. The detailed steps of the construction of finger contour histogram:

1. Segment the complete image of palm.
2. Find out the central point of palm image by certain method.

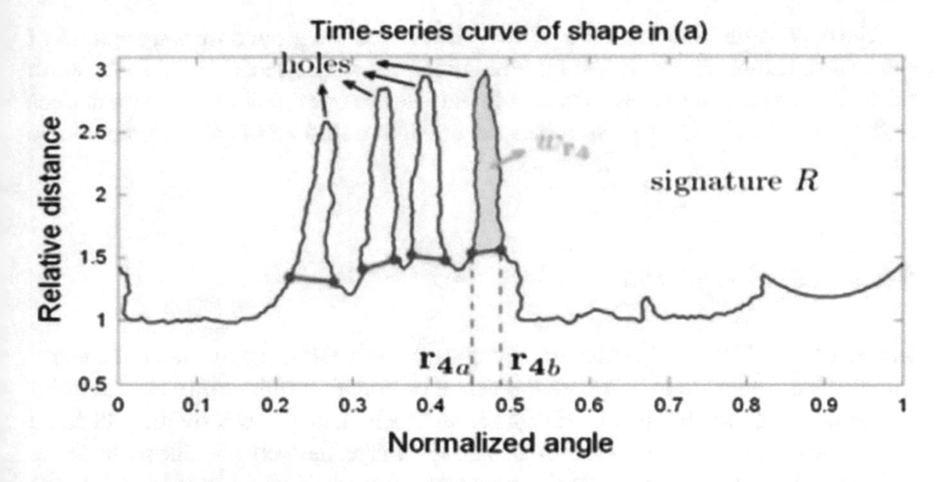

Fig. 9.6 FEDM gesture recognition algorithm

3. Start from the left contour of wrist; traverse the whole contour of palm until the right of wrist. Record the distance from the central point to each contour point and the angel from the start around the palm.
4. Suppose the angle as X-coordinate and the distance data as Y-coordinate, form the curve graph.

Find the corresponding position of each finger in curve graph, construct the histogram data required by EMD algorithm with the area by its width and the X-coordinate, as shown in Fig. 9.6.

r_{4a} and r_{4b} are the two identified points for the fourth finger; holes mean the identified peaks. To compute the FEMD, we need to compute the value of F. F is defined by minimising the work needed to move all the earth piles:

$$F = \arg\min WORK(R, T, F) = \arg\min \sum_{i=1}^{m} \sum_{j=1}^{n} d_{ij} f_{ij} \tag{9.1}$$

so that

$$\begin{cases} f_{ij} \geq 0 & 1 \leq i \leq m, 1 \leq j \leq n, \\ \sum_{j=1}^{n} f_{ij} \leq \omega_{r_i} & 1 \leq i \leq m, \\ \sum_{i=1}^{m} f_{ij} \leq \omega_{t_j} & 1 \leq j \leq m, \\ \sum_{i=1}^{m} \sum_{j=1}^{n} = \min\left(\sum_{i=1}^{m} \omega_{r_i}, \sum_{j=1}^{n} \omega_{t_j}\right) \end{cases} \tag{9.2}$$

Matric f is the flow matrix in EMD, matric d is the ground distance matric of the tow signature, R and T. We find the minimum work needed to move the earth piles. The first constraint restricts the moving flow to one direction: from earth piles to the holes. The last constraint forces the maximum amount of earth possible to be moved.

9.3.2 Lasso Algorithm

In order to use FEMD algorithm, the position coordinates of palm need to be determined. One method is to calculate the centre of gravity position of the whole palm as the palm required by the FEMD algorithm. Such method is clearly thought and a simple algorithm, but there is an obvious disadvantage: theoretically, the palm centre in the FEMD algorithm shall be the centre of palm centre image not including the finger part. Calculating the centre position of the whole image without any distinction will definitely lead into deviation. Another simple and available method is applied in this chapter, which is to find out all small regions nearest to the edge contour point by image corrosion. Then the position is the centre of region similar with the palm centre. The last rest region is a single pixel, and we can get the centre of circle for palm contour image by such method under the most ideal circumstance. Such method can also judge the threshold value by setting the size of corrosion and adjusting the area of the rest region, also limiting the deviation into a reasonable range. The program flow chart shall be found as Fig. 9.7. There are two loops: one is for the condition of square and the other is for the condition of the threshold. After processing, change of the algorithm is correct and available, the robustness and instantaneity are perfect.

Besides the calculation of palm centre, the extraction of fingertip feature is another important step to construct palm contour curve by EMD algorithm histogram. The fingertip feature required by FEMD algorithm consists of two kinds of data, which are the X-coordinate of finger root and the areas between the fingers. In [35], two kinds of methods for extraction of features of fingertip were mentioned, threshold and Near-Convex Decomposition. Near-Convex Decomposition can segment the feature of fingertip in palm image accurately, and is with well robustness. The disadvantage is also obvious, such as complicated algorithm and awful real-time. The threshold applies the simplest strategy, by given a certain value of threshold and all parts that higher than the threshold value shall be considered as the feature of fingertip. This method has high requirement of palm contour curve, and is easily effected by noise. Another new algorithm to extract the features of fingertips, called Lasso, is proposed with both fast computation and high robustness. By imitating the looping into the pillar, take the fingertip as a pillar. The width of these pillars is almost the same. Extraction of the features of these pillars is like the process of looping the pillars with the lasso whose diameter is similar with the width of the pillars. When the lasso goes down to the bottom or it is tightened, the extraction can be seen as finished. The process of the above algorithm shall be found in the following Fig. 9.8. The algorithm of the Lasso can be found in Appendix E.2.

Fig. 9.7 Palm extract algorithm

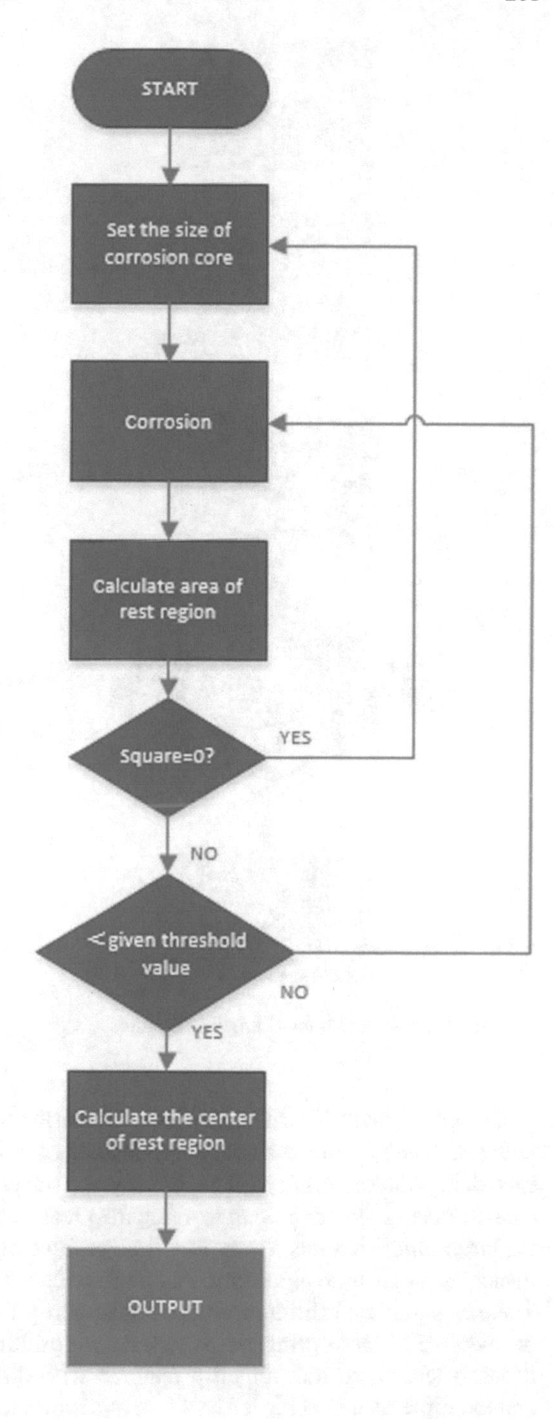

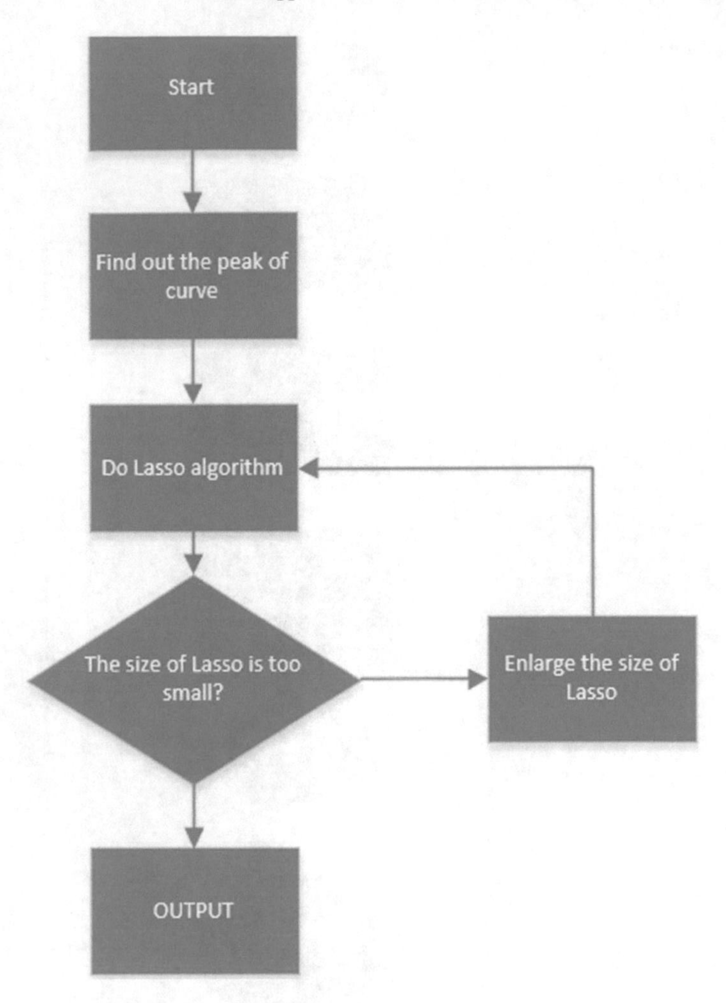

Fig. 9.8 Process of extract of fingertip features

Compared with Near-Convex Decomposition algorithm, Lasso algorithm deals with a simple two-dimensional curve instead of a three-dimensional image. And for each data point of the curve, it needs only one traverse in the worst case. In fact, those data points that do not belong to a fingertip feature will be ignored automatically. So the lasso algorithm has a very low time complexity and shows outstanding performance as a real-time algorithm. Also, lasso algorithm shows pretty good reliability. There is significant difference between fingertip features and noise signal from the perspective of lasso principle. And lasso algorithm supports dynamic modification of lasso length, so that fingertip features with different size can be adapted. Several example results achieved by lasso algorithm to extract the fingertip features are shown in Fig. 9.9.

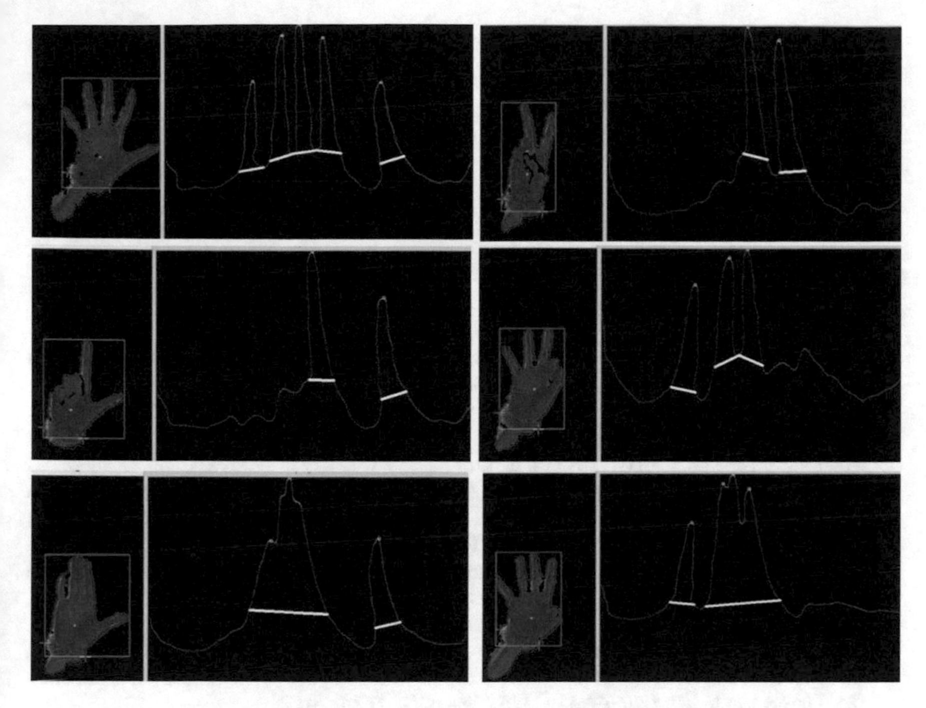

Fig. 9.9 Result of lasso algorithm

9.4 Conclusion

In this chapter, we investigated how to extract hand gesture features from depth data in real-time via a Kinect sensor. Firstly, we described the technical principles and features of the Kinect camera. And on this basis, we analysis the difficulties and key content on Kinect study. A real-time approach was proposed to extract the hand gesture feature by using image segmentation, Earth Mover's Distance and Lasso algorithms. In the step of palm image segmentation, we use a contour length information based de-noise method, which is simple but practical. The hand gesture image has a regular and stable contour, while the noise signal in the depth data is scattered and unstable, so the hand gesture feature always has a longer contour than the noise signal. Therefore, significant effect can be achieved by setting a contour length threshold value. Then FEMD algorithm was proposed with new methods in the calculation of central point of palm image and the extraction of fingertip feature. Especially the proposed lasso algorithm can extract the fingertip feature from a hand contour curve correctly with excellent real-time performance. This work will provide a fundamental and distinctive feature for hand gesture analysis and recognition, especially in real-time human-computer interactions. Due to the feature extracted by this method is based on a single RGBD image, such finger features from continuous images can potentially form a dynamic and 3 dimensional feature, which could be very useful for dynamic gesture analysis and recognition. Future research will be focused on application in effective dynamic hand gestures and their recognition via the proposed hand features [36–38].

References

1. K. Khoshelham and S.O. Elberink. Accuracy and resolution of kinect depth data for indoor mapping applications. *Sensors*, 12(2):1437–1454, 2012.
2. J. Shotton, T. Sharp, A. Kipman, A. Fitzgibbon, M. Finocchio, A. Blake, M. Cook, and R. Moore. Real-time human pose recognition in parts from single depth images. *Communications of the ACM*, 56(1):116–124, 2013.
3. C. Herrera, J. Kannala, et al. Joint depth and color camera calibration with distortion correction. *IEEE Transactions on Pattern Analysis and Machine Intelligence*, 34(10):2058–2064, 2012.
4. R.A. Newcombe, A.J. Davison, S. Izadi, P. Kohli, O. Hilliges, J. Shotton, D. Molyneaux, S. Hodges, D. Kim, and A. Fitzgibbon. Kinectfusion: Real-time dense surface mapping and tracking. In *IEEE International Symposium on Mixed and Augmented Reality*, pages 127–136. IEEE, 2011.
5. S. Izadi, R. A Newcombe, D. Kim, O. Hilliges, D. Molyneaux, S. Hodges, P. Kohli, J. Shotton, A.J. Davison, and A. Fitzgibbon. Kinectfusion: real-time dynamic 3d surface reconstruction and interaction. In *ACM SIGGRAPH 2011*, page 23. ACM, 2011.
6. N. Silberman and R. Fergus. Indoor scene segmentation using a structured light sensor. In *IEEE International Conference on Computer Vision Workshops*, pages 601–608. IEEE, 2011.
7. Yue Gao, You Yang, Yi Zhen, and Qionghai Dai. Depth error elimination for rgb-d cameras. *ACM Transactions on Intelligent Systems and Technology*, 6(2):13, 2015.
8. J. Preis, M. Kessel, M. Werner, and C. Linnhoff-Popien. Gait recognition with kinect. In *1st International Workshop on Kinect in Pervasive Computing*, pages 1–6, 2012.
9. J.L. Raheja, A. Chaudhary, and K. Singal. Tracking of fingertips and centers of palm using kinect. In *Third International Conference on Computational Intelligence Modelling and Simulation*, pages 248–252. IEEE, 2011.
10. W. Xu and E.J. Lee. Gesture recognition based on 2d and 3d feature by using kinect device. In *International Conference on Information and Security Assurance*, 2012.
11. L. Shao, L. Ji, Y. Liu, and J. Zhang. Human action segmentation and recognition via motion and shape analysis. *Pattern Recognition Letters*, 33(4):438–445, 2012.
12. Jianfeng Li and Shigang Li. Eye-model-based gaze estimation by rgb-d camera. In *Proceedings of the IEEE Conference on Computer Vision and Pattern Recognition Workshops*, pages 592–596, 2014.
13. Zhenbao Liu, Jinxin Huang, Junwei Han, Shuhui Bu, and Jianfeng Lv. (in press) human motion tracking by multiple rgbd cameras. *IEEE Transactions on Circuits and Systems for Video Technology*, 2016.
14. J. Sturm, S. Magnenat, N. Engelhard, F. Pomerleau, F. Colas, W. Burgard, D. Cremers, and R. Siegwart. Towards a benchmark for rgb-d slam evaluation. In *Proc. of the RGB-D Workshop on Advanced Reasoning with Depth Cameras at Robotics: Science and Systems Conference*, volume 2, page 3, 2011.
15. C. Li, H. Ma, C. Yang, and M. Fu. Teleoperation of a virtual icub robot under framework of parallel system via hand gesture recognition. In *IEEE International Conference on Fuzzy Systems*, pages 1469–1474. IEEE, 2014.
16. L. Bo, X. Ren, and D. Fox. Unsupervised feature learning for rgb-d based object recognition. *Experimental Robotics*, pages 387–402, 2013.

17. L. Spinello and K.O. Arras. People detection in rgb-d data. In *IEEE/RSJ International Conference on Intelligent Robots and Systems*, pages 3838–3843. IEEE, 2011.
18. Z. Zhang. Microsoft kinect sensor and its effect. *Multimedia, IEEE*, 19(2):4–10, 2012.
19. L. Cruz, D. Lucio, and L. Velho. Kinect and rgbd images: Challenges and applications. In *SIBGRAPI Conference on Graphics, Patterns and Images Tutorials*, pages 36–49. IEEE, 2012.
20. Jungong Han, Ling Shao, Dong Xu, and Jamie Shotton. Enhanced computer vision with microsoft kinect sensor: A review. *Cybernetics, IEEE Transactions on*, 43(5):1318–1334, 2013.
21. P. Dollár, V. Rabaud, G. Cottrell, and S. Belongie. Behavior recognition via sparse spatio-temporal features. In *Proceeding of Joint IEEE International Workshop on Visual Surveillance and Performance Evaluation of Tracking and Surveillance*, pages 65–72. Beijing, 2005.
22. H. Zhu and C. Pun. Hand gesture recognition with motion tracking on spatial-temporal filtering. In *Proceedings of the 10th International Conference on Virtual Reality Continuum and Its Applications in Industry*, pages 273–278. ACM, 2011.
23. H. Farid and E.P. Simoncelli. Optimally rotation-equivariant directional derivative kernels. In *Computer Analysis of Images and Patterns*, pages 207–214. Springer, 1997.
24. Z. Ren, J. Yuan, C. Li, and W. Liu. Minimum near-convex decomposition for robust shape representation. In *IEEE International Conference on Computer Vision*, pages 303–310. IEEE, 2011.
25. Z. Ju, Y. Wang, W. Zeng, S. Chen, and H. Liu. Depth and rgb image alignment for hand gesture segmentation using kinect. In *International Conference on Machine Learning and Cybernetics*, volume 2, pages 913–919. IEEE, 2013.
26. D.S. Alexiadis, P. Kelly, P. Daras, N.E. O'Connor, T. Boubekeur, and M.B. Moussa. Evaluating a dancer's performance using kinect-based skeleton tracking. In *Proceedings of the 19th ACM international conference on Multimedia*, pages 659–662. ACM, 2011.
27. S. Matyunin, D. Vatolin, Y. Berdnikov, and M. Smirnov. Temporal filtering for depth maps generated by kinect depth camera. In *3DTV Conference: The True Vision-Capture, Transmission and Display of 3D Video*, pages 1–4. IEEE, 2011.
28. R. Tara, P. Santosa, and T. Adji. Hand segmentation from depth image using anthropometric approach in natural interface development. *International Journal of Scientific and Engineering Research*, 3(5):1–4, 2012.
29. H. Liang, J. Yuan, and D. Thalmann. 3d fingertip and palm tracking in depth image sequences. In *Proceedings of the 20th ACM international conference on Multimedia*, pages 785–788. ACM, 2012.
30. Z. Ren, J. Meng, J. Yuan, and Z. Zhang. Robust hand gesture recognition with kinect sensor. In *Proceedings of the 19th ACM international conference on Multimedia*, pages 759–760. ACM, 2011.
31. C. Keskin, F. Kıraç, Y.E. Kara, and L. Akarun. Real time hand pose estimation using depth sensors. In *Consumer Depth Cameras for Computer Vision*, pages 119–137. Springer, 2013.
32. M. Tang. Recognizing hand gestures with microsoft's kinect. *Palo Alto: Department of Electrical Engineering of Stanford University:[sn]*, 2011.
33. H. Bay, T. Tuytelaars, and L. Van Gool. Surf: Speeded up robust features. In *Computer vision–ECCV 2006*, pages 404–417. Springer, 2006.
34. A. Ferreira, W.C. Celeste, F.A. Cheein, T.F. Bastos-Filho, M. Sarcinelli-Filho, and R. Carelli. Human-machine interfaces based on emg and eeg applied to robotic systems. *Journal of NeuroEngineering and Rehabilitation*, 5(1):1, 2008.
35. Zhou Ren, Junsong Yuan, Jingjing Meng, and Zhengyou Zhang. Robust part-based hand gesture recognition using kinect sensor. *IEEE transactions on multimedia*, 15(5):1110–1120, 2013.
36. Z. Ju and H. Liu. Fuzzy gaussian mixture models. *Pattern Recognition*, 45(3):1146–1158, 2012.
37. Z. Ju and H. Liu. A unified fuzzy framework for human hand motion recognition. *IEEE Transactions on Fuzzy Systems*, 19(5):901–913, 2011.
38. Z. Ju, C. Yang, Z. Li, L. Cheng, and H. Ma. Teleoperation of humanoid baxter robot using haptic feedback. In *International Conference on Multisensor Fusion and Information Integration for Intelligent Systems*, pages 1–6. IEEE, 2014.

Chapter 10
Recognizing Constrained 3D Human Motion: An Inference Approach

10.1 Introduction

Motion behaviour understanding is usually performed by comparing observations to models inferred from examples using different learning algorithms under enormous uncertainties from complex human kinematic structure, occlusion and environmental factors [1, 2]. Motion recognition appeared in the mid nineties as a secondary topic mainly linked to computer vision based experiments. The relatively recent advent of 3d motion capture technology and subsequent spread of view invariant representation models contributed to its emergence as a field of research of its own more concerned with the qualitative analysis of motion itself, than with the extraction of a human pose estimation from videos or images. The rise of gesture based human interface devices in the entertainment industry, the need for automated detection of abnormal behaviour in security surveillance and health care, and the existence of a growing market for computer assisted sport performance analysis are some of the catalysts that accelerate the present growth of the field. However, due to the relative novelty of 3d motion capture technology, the creation of learning data sets for different types of motions, even when using off-the-shelf easy to use depth data systems introduced by Microsoft kinect [3–9], is still computationally expensive and labour intensive. This calls in a timely fashion for Machine Learning methods that can classify motions from learning samples of reduced size.

General representation systems in use are template matching [10] which seeks to match features of a search to template images or vectors, state-spaces approaches [11] as dynamical discrete systems, or semantic description [12]. In order to balance computational cost and accuracy, this study focuses on a state space approach that can differentiate subtle motion changes with a view invariant 3D skeletal representation. The machine learning techniques used in motion recognition can be presented as belonging to the following paradigms: probabilistic graphical models, sequential Monte Carlo methods, Kernel based methods, connectionist approaches, syntactic techniques, and hybrid approaches including soft computing. The probabilistic generation of graphical models is the most widely spread paradigm. It includes methods such as Hidden Markov Models (HMM), Bayes Net, Finite State Machine (FSM), Conditional Random Fields (CRF), and Maximum Entropy Markov Models (MEMM). HMM [13] is a double stochastic process that associate an underlying

H. Liu et al., *Human Motion Sensing and Recognition*,
Studies in Computational Intelligence 675, DOI 10.1007/978-3-662-53692-6_10

Markov chain with a finite number of states to a set of random functions, each associated with one state. Bayes Net [14] is another popular method aiming at producing directed acyclic graph where nodes represent variables and edges represent conditional dependencies. FSM [15] can model gestures as ordered sequences of states in a spatio-temporal configuration space. CRF [16] generates undirected graph that can model independence relationships between labels but not between observations. It was recently used for activity recognition from sensor data [17]. MEMM [18] combines HMM and maximum entropy estimation to produces hierarchical probabilistic graphical models that are applicable to the classification of human activities. Sequential Monte Carlo methods (SMC) such as Kalman filter [19], particle filters, and condensation algorithms [20] are mainly used for motion recognition associated with tracking and pose estimation. Kernel based methods such as PCA [21], Support Vector Machines (SVM) [22], and Relevance Vector Machines (RVM) [23] are computationally efficient and give good results in high dimensional space. Connectionist approaches such as Multi-layer Perceptrons [24], Time Delay Neural Network (TDNN) [25], Radial Basis Function networks [26] are also very popular. Syntactic techniques [27] use grammar parsing to recognise sequences of discrete behaviours on top of a low level standard independent probabilistic temporal behaviour detector. This approach smooths the output of uncertain low-level detection by including a priori knowledge about the structure of temporal behaviours in a given domain. Instance-based learning methods such as Nearest Neighbours classifiers [28] or Dynamic Time Warping [14] in the temporal domain have the advantages of being fast and cope well with small learning samples. Finally, hybrid approaches [29] and soft computing methods [30] have recently seen an explosion of research interest. The present work fits into this category.

Existing machine learning methods are facing specific problems in the context of action understanding applied to 3d motion capture data. The generation of probabilistic graphical models through methods such as HMM, or the training of neural networks require learning sample of significant size to work efficiently, which is not ideal for 3D motion capture data sets where learning samples are of sub-optimal size. Instance based classifiers like Nearest Neighbours generally suffer from a high sensitivity to noise. Kernel based methods present a delicate and computationally expensive hyper-parameter tuning process. This situation creates a functional niche for a Machine Learning method that could satisfy the following functional requirements: the ability to classify from learning samples of sub-optimal size, a low sensitivity to noise, and simplicity regarding the parameter tuning process. From a functional point of view, the present framework has been developed to address these issues while taking into account the need for a context-aware and time-sensitive recognition of occluded motions. From a theoretical prospective, one can observe two phenomenas: the omnipresence of probabilistic methods in Machine Learning and the emergence of fuzzy set theory as described in [31]. Probabilistic methods express degrees of belief in a non-compositional way while fuzzy set theory introduces degrees of truth which have the advantage of allowing overlapping classes. Despite their differences, probability theory and fuzzy logic can be seen as complementary rather than competitive [32]. This complementarity has remained relatively unexplored when facing the fundamental problem of Fuzzy Membership Function generation from data in the sense that there seem to be no design methodology that allows the direct mapping from a fuzzy representation to a probability distribution. The predominance of methods such as Fuzzy Clustering, Inductive Reasoning, Neural Networks, Evolutionary Algorithms, and some work based on the optimization of Fuzzy Entropy [33] seems to confirm that fact. One notable exception is possibility theory [34] that allows the transformation of probability distributions to possibility distributions. However, if possibility theory does extend to probability theory to deal with uncertainty, it does not confer it compositionality [32]. Another practical attempt mapping a Fuzzy Membership Function to a Normal Distribution is hinted at by Frantti [35] in the context of mobile network engineering. Unfortunately, this work has its shortcomings as the minimum and maximum of the observed data are the absolute limits of the membership function. Such a system does ignore motions which are over the extrema of the range of the learning sample. One possible way to overcome this problem would be to introduce a function that maps a degree of membership to the probability that values fall within a given cumulative probabilistic distribution.

As a consequence, there seems to be a theoretical and functional gap that leaves open the advent of a motion classification framework based on the three following components. First, a standalone classifier using Fuzzy Quantile Generation (FQG), a novel method that generates Fuzzy Membership Functions using metrics derived from the probabilistic quantile function. Secondly, a Genetic Programming filter that produces time-dependent and context-aware rules to smooth the qualitative outputs of the classifier. Finally, a module that deals with occlusion by implementing new procedures for feature selection and feature reconstruction. The remainder of this chapter is organised as follows. The human skeletal representation in use is described in Sect. 10.2. The three components of the proposed inference framework, i.e., fuzzy quantile generation, the context-aware filter and occlusion are detailed in Sect. 10.3. Experiments have been conducted to evaluate the components of the framework in Sect. 10.4, concluding remarks are given in Sect. 10.5.

10.2 Human Skeletal Representation

A wide variety of ways can be chosen to represent the human body according to application contextual requirement. For instance, either precise 3-D joint data, force and torques of a movement or their combination have been applied for modelling the human body with either kinematic analysis or analytical dynamics in kinesiology or biomechanics [36]. It is evident that these data capture methods either are unsuitable for complex human motion analysis in practice [37] or bring intractable computational burden for data processing [38]. Thanks to recently booming gaming and animation, BVH, that stands for Biovision hierarchical data [39], has been intensively developed for motion capture device. BVH represents a human skeleton as a hierarchical skeletal limbs linked by rotational joints. Not only does it inherit the

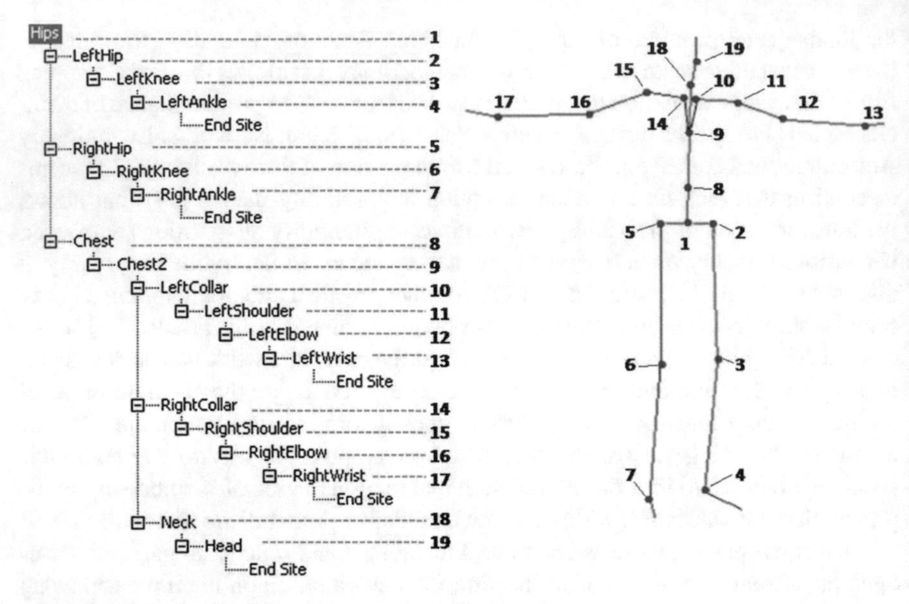

Fig. 10.1 BVH motion capture format encoding human skeletal representation with 19 joints

advantages of the existing representations, but it also has the capability of keeping track of trajectories of a multiple degrees-of-freedom joint such as the elbow joint.

Motion capture device and BVH representation are employed for this study as shown in Fig. 10.1, the body is represented in terms of nineteen main joints which is the standard number of joints used to animate human representations in consistence with human anatomy model. Each joint has three degrees of freedom, the rotations are represented by three rotation angles of Euler ZXY in order of axis Z, X and Y, and the sampling rate is 120 frames per second. In practice, for every frame, the observed data is represented by a nineteen-by-three matrix containing ZXY Euler Angles for all nineteen joints in a human skeleton. It is assumed that this number is sufficient to characterise and capture the general motions of a human skeleton performing boxing combinations in this study.

10.3 The 3D Motion Recognition Framework

The inference based framework is proposed to recognise 3D spatial-temporal human motions while taking into account contextual information and occlusion. The framework consists of a central component presenting a Fuzzy Quantile Generation (FQG) based classifier, an occlusion layer, and a context-aware filter as shown in Fig. 10.2.

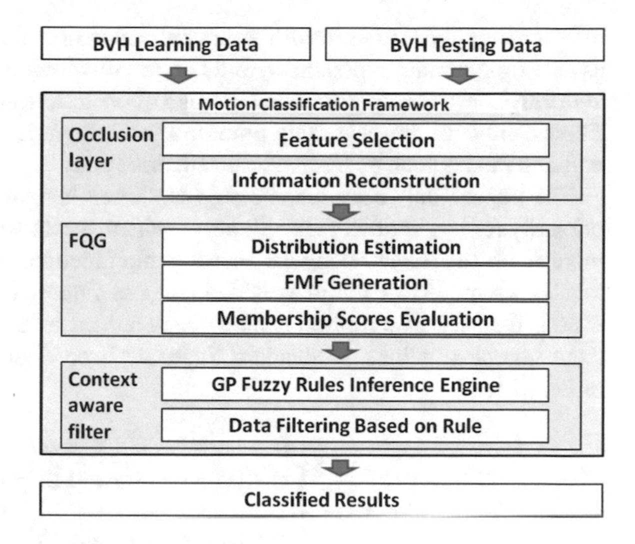

Fig. 10.2 A block diagram of the motion classification framework

10.3.1 Fuzzy Quantile Generation

FQG is a method that builds and maps Fuzzy Membership Functions to probability distributions. It builds a function that maps a degree of membership to the probability that values fall within a given cumulative probabilistic distribution. This function is constructed in four steps. First, probabilistic and fuzzy models in use must be identified. Secondly, the upper and lower bases of a Fuzzy Membership Function are initially estimated. Thirdly, the FMF shape is modified in order to follow a shift of the probability distribution inferred from the mean of the learning sample. Finally, the resulting model is used to classify instances by evaluating their membership scores.

10.3.1.1 Choosing the Type of Probability Distribution and Fuzzy Membership Function

Human motions are non-trivial to model because nobody moves exactly the same way, and even the same person, when repeating the same motion, does not perform it in a strictly identical fashion. Despite of this, we seem to be able to recognise a motion when performed by different actors at various speeds and in very different conditions. From this observation, and the existence of the Central Limit Theorem, it is assumed that motions and by extension joints rotations, are somehow distributed. The proposed method does not automate the choice of a distribution, so it is left to the user to decide which distribution best represents the domain the sample is extracted from. The normal distribution being commonly encountered in practice is used extensively throughout statistics as a simple model for complex phenomena. In the context of this experiment, it is chosen as the most likely base case and is used as a proof of concept. However, there are times when other distributions might offer a better alternative for different application domains. That is why others such as the Log-Normal, Exponential, and Laplace distributions are presented as possible alternatives in order to give a wider range of choices from a theoretical perspective. These can be used as a starting point to later map more complex distributions with multiple modes such as Beta and Sine distributions.

The trapezoidal fuzzy membership is chosen here as a mean of representation primarily for its simplicity and efficiency with respect to computability. It offers a bit more flexibility than a triangular membership function, and set a membership score to a known move to 1 for an extended range of values. This makes it more practical for this data set, as a motion can be characterized by a range representing a set of fixed angular rotations. A standard trapezoid fuzzy-four-tuple (a, b, α, β) is given in Eq. 10.1.

$$\mu(x) = \begin{cases} 0 & x < a - \alpha \\ \alpha^{-1}(x - a + \alpha) & x \in [a - \alpha \quad a] \\ 1 & x \in [a \quad b] \\ \beta^{-1}(b + \beta - x) & x \in [b \quad b + \beta] \\ 0 & x > b + \beta \end{cases} \tag{10.1}$$

A measure between a fuzzy membership and a probability distribution can be constructed via inverse cumulative distribution function or quantile function. Intuitively, the quantile function is employed to evaluate the shape of a fuzzy membership function using metrics expressed in numbers of standard deviations.

10.3.1.2 Building the Upper/Lower Base of the Fuzzy Membership Function

In order to estimate the upper and lower bases of a fuzzy membership, a function is derived in Eq. 10.2 where n_{min} and n_{max} be the minimum and maximum in number of standard deviations (here, n_{min} and n_{max} are typically set to -4 and 4 to cover most of the distribution); $s \in [0, 1]$ be a parameter describing the proportion of the population whose values are in the interval $[a, b]$, with a and b being respectively the min and max values of the learning sample; $CDF^{-1}(x)$ be the quantile function, $CDF^{-1}(0.5)$ be the position of the median of the distribution; a, the lowest value of the learning sample, defined by $a = CDF^{-1}(0.5 - (s/2))$ and b, the highest value, defined by $b = CDF^{-1}(0.5 + (s/2))$ in numbers of standard deviations. It is illustrated in Fig. 10.3 for a normal distribution.

$$\begin{cases} \alpha = |CDF^{-1}(0.5 - (s/2)) - n_{min}| \\ \beta = |n_{max} - CDF^{-1}(0.5 + (s/2))| \end{cases} \tag{10.2}$$

Normal, Laplace, Exponential and Log-Normal distributions are presented in Fig. 10.4. For each of these distributions, the corresponding quantile function CDF^{-1} is provided: for a normal distribution, $CDF^{-1} = \{x : CDF^{-1}(x|\mu, \sigma) = p\}$, where

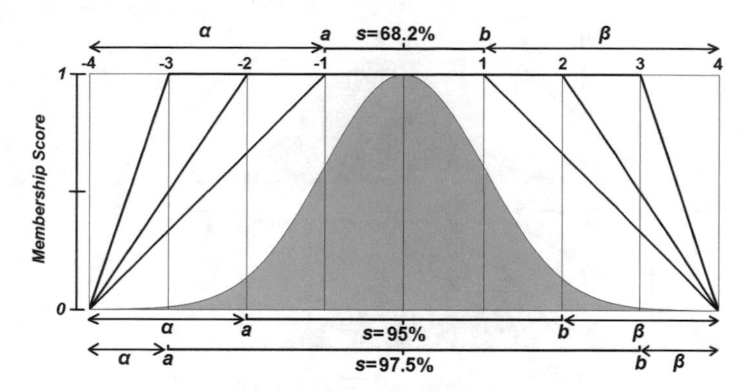

Fig. 10.3 Influence of s parameter (expressed in %) on the shape of a normal distribution

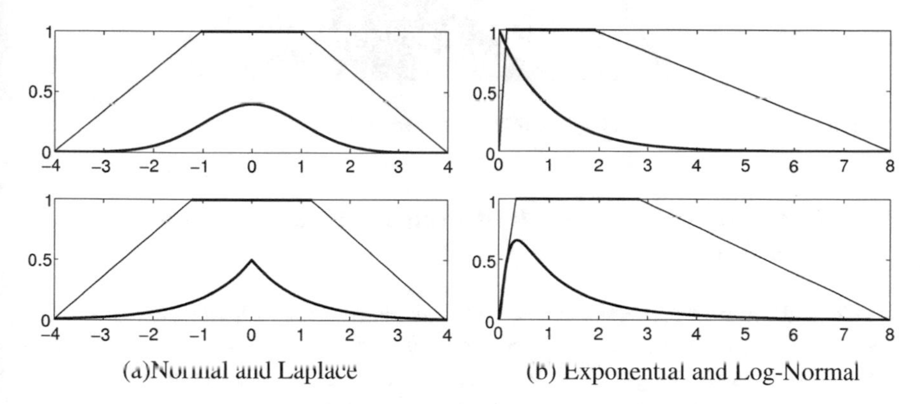

(a)Normal and Laplace (b) Exponential and Log-Normal

Fig. 10.4 Mapping from selected distributions to a fuzzy membership when $s = 0.7$

$p = F(x|\mu, \sigma) = \frac{1}{\sigma\sqrt{2\pi}} \int_{-\infty}^{x} \exp(\frac{-(t-\mu)^2}{2\sigma^2})dt$. The result, x, is the solution of the integral equation above where you supply the desired probability, p; for a Laplace distribution, of mean μ and standard deviation σ, we have:

$$CDF^{-1} = \begin{cases} x = \mu + \sigma \cdot log(2p) & \text{if } p \leq 0.5 \\ x = \mu - \sigma \cdot log(2(1-p)) & \text{otherwise} \end{cases} \tag{10.3}$$

For an exponential distribution, of parameter λ, we have $CDF^{-1}(p, \lambda) = \frac{-\ln(1-p)}{\lambda}$; for a log-normal distribution, of mean parameter μ and standard deviation σ, we have $CDF^{-1}(p, \sigma) = \exp(\sigma\Phi^{-1}(p))$, where Φ is the quantile function of the normal distribution.

This analysis is valid for distributions with only one mode i.e. this excludes saddle shaped distributions. The ratio between upper and lower bounds of the fuzzy membership is identified, its initial characteristics are set.

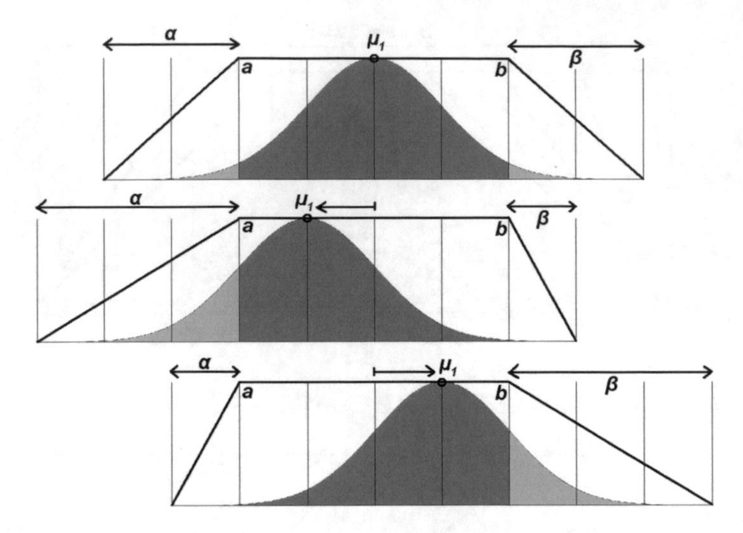

Fig. 10.5 Effect of shifting on fuzzy membership generation

10.3.1.3 Deforming the Fuzzy Membership Function to Follow a Shift of Distribution

The Central Limit Theorem states that sampling distributions drawn from a uniformly distributed population will tend to form a nearly perfect normal distribution when the sample size is large enough. This guarantees that a mean based on a randomly chosen sample of sufficiently large size will be remarkably close to the true mean of the population. Therefore, the mean of the theoretical underlying distribution is shifted to overlap with the mean of the learning sample. The shape of the corresponding Fuzzy Membership Function deformed in kind (see Fig. 10.5). This shift of the distribution δ_μ is a ratio expressing how far the mean of the learning sample is from its midpoint which represents the mean of the underlying distribution in number of standard deviations.

The FMF is deformed to follow the shift δ_μ in different ways depending on the type of distribution in use (see Fig. 10.6). Let μ_0 be the initial mean of the distribution and μ_1 be the mean of the learning sample respectively expressed in number of standard deviations by $CDF^{-1}(\mu_0)$ and $CDF^{-1}(\mu_1)$. Let a and b be the extrema of the learning sample such that: $a = CDF^{-1}(0.5 - (s/2))$, and $b = CDF^{-1}(0.5 + (s/2))$ and α_0, and β_0 be the initial values of α, and β. If the chosen distribution has only one mode and is central symmetric, then δ_μ, and the new values of α and β can be computed as follows:

$$\begin{cases} \delta_\mu = (CDF^{-1}(\mu_1) - a)/(b - a) \\ \alpha = (1 - \delta_\mu)(\alpha_0 + \beta_0) \\ \beta = (\delta_\mu)(\alpha_0 + \beta_0) \end{cases} \tag{10.4}$$

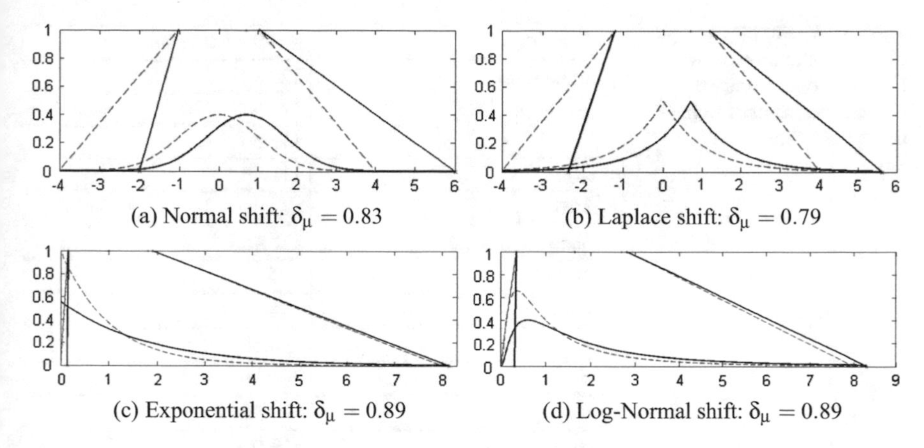

(a) Normal shift: $\delta_\mu = 0.83$ (b) Laplace shift: $\delta_\mu = 0.79$

(c) Exponential shift: $\delta_\mu = 0.89$ (d) Log-Normal shift: $\delta_\mu = 0.89$

Fig. 10.6 Shifting distributions with $s = 0.7$ depending on δ_μ (*plain lines* are the shifted distributions)

For right-skewed distributions of the type Exponential or Log-Normal, δ_μ, and the new values of α and β can be computed as follows:

$$\begin{cases} \delta_\mu = (CDF^{-1}(\mu_1) - CDF^{-1}(\mu_0))/(b - CDF^{-1}(\mu_0)) \\ \alpha = \alpha_0 - (\delta_\mu \cdot \alpha_0) \\ \beta = \beta_0 + (\delta_\mu \cdot \alpha_0) \end{cases} \tag{10.5}$$

10.3.1.4 Membership Score Evaluation

Instances of moves are classified by evaluating their membership scores to the Fuzzy Membership Functions generated by FQG. A posture being defined by a set of 57 Euler angles is modelled by 57 Fuzzy Membership Functions. Having no initial prior knowledge about the eventual predominance of some of these joints, the overall membership of a test instance to a known move is computed by calculating the average of all the 57 membership scores of the Euler Angles. This approach could probably be improved in the near future by introducing weighted average for certain joints (for example, the position of the elbow might be more important than the position of the knee when in guard). A parameter t expresses the membership threshold to a move. In practice, all frames with a membership score equal or greater than t are classified as belonging to that move. The lower the value of t, the lower is the selectivity of the classifier, and the higher the value of t, the more difficult it becomes for a move to be given a membership score of 1. When the same frame has a high membership score for several fuzzy sets representing different moves, an order of preference of these sets can be established by comparing the Euclidean distance of the observed data to the centroid of each fuzzy set. The existence of t allows the introduction of a convenient a-posteriori way to fine tune parameters in order to tailor the precision of

Table 10.1 Simple example of FMF generation using FQG based on a normal distribution inferred from three input data

Variable	Associated value
Input data	122
	121.9
	111.56
Min	111.56
Max	122
Sample Average	118.486
nmin	−4
nmax	4
s	0.68
a	−0.994
b	0.994
α_0	3.006
β_0	3.006
δ_μ	0.663
α	2.023
β	3.988

every model to the quantity of available learning data for each move. If results show that the relative size s of a learning sample for given move was over-estimated, the membership scoresmi can be re-scaled by fine-tuning the thresholds t_i linked to each move i, effectively "truncating" the upper part of the Fuzzy Membership Functions without having to recalculate the model (see Eq. 10.6).

$$m_i = (m_i - t_i) \, / \, (1 - t_i) \tag{10.6}$$

A simplified example of Fuzzy Membership Function generation using FQG is presented from three input data in Table 10.1.

10.3.2 The Context-Aware Filter

The context-aware filter has the capability to label image frames by computing membership scores to known motions. The classification is conducted frame by frame and is not time-dependent. It does not take into account the previous or following motions. Hence, it is necessary to smooth the qualitative output using knowledge base such as context-aware fuzzy-rules. These rules might combine very specific operators such as logical functions, measurements of speed, and multiple input labels ranked as first, second or third best choices for a motion. No initial prior knowledge about how motion can be smoothed is assumed. The high dimensionality and specificity of the problem where results might also form nested logical structures of arbitrary

Table 10.2 Grammar rules defining the individual motions produced with Strongly Typed GP

Parent nodes	Associated children nodes	
	Function nodes	Terminal nodes
Root	If then replace X by Y	Empty
If then replace X by Y	Membership_1	Is_Short
	Membership_2	Is_Medium
	Membership_3	Is_long
	Left_1, Left_2, Left_3	Mvt Type 1, Mvt Type 2, Mvt Type 3
	Right_1, Right_2, Right_3	Mvt Type 4, Mvt Type 5, Mvt Type 6
	And, Or, Not	Mvt Type 7
Membership_1	Empty	Mvt Type 1, Mvt Type 2, Mvt Type 3
Membership_2		Mvt Type 4, Mvt Type 5, Mvt Type 6
Membership_3 Left_1, Left_2, Left_3Right_1, Right_2, Right_3		Mvt Type 7
And Or	Membership_1	Is_Short
	Membership_2	Is_Medium
	Membership_3 Left_1, Left_2, Left_3 Right_1, Right_2, Right_3 And, Or, Not	Is_long
Not	Membership_1	Is_Short
	Membership_2	Is_Medium
	Membership_3 Left_1, Left_2, Left_3 Right_1, Right_2, Right_3 And, Or	Is_long

complexity make standard approaches like the Takagi-Sugeno-Kang (TSK) or Mamdani models unsuitable. Similarly, mainstream inference methods that are designed to obtain such models through techniques such as Fuzzy Neural Networks [40] or standard Evolutionary Algorithms [41] might also be difficult to reuse in this context. Rules are therefore inferred using a specifically adapted variation of genetic programming. For the purpose of this research, a Strongly-Typed Genetic Programming open-source distribution was built [42]. The GP system evolves logic rules that transform the qualitative output of each frame. The detailed GP terminal and function sets are provided in Table 10.2. Some operators return the groups of frames that are identical for a given duration, e.g. *is_short* expresses a duration of less than 5 frames. Other operators return the groups of frames that belong to the first, second and third best membership scores of a motion, e.g. *membership_2(left_hook)* returns groups of frames with the second best membership score for a motion as a left hook. Others return the groups of frames that have specific previous motions, e.g. *left_2(guard,jab)* returns groups of frames preceded in order by a guard and then a jab motion. Logical operators are present, e.g., and, or, not. Logical structures of the type *If Then Replace X by Y* use the previous operators to identify groups of frames and replace their best motion membership score by a different one, e.g., *If (membership_2(guard)) Then Replace (jab) by (Cross)*.

A tree structure consists of four interconnected *If Then Replace X by Y* rules. The GP parameters are: (a) a population size of 1000 individuals with a maximum number of 400 generations per evaluation; (b) tournament selection of size seven and selection probability 0.8; (c) the probabilities of crossover, mutation, and reproduction are respectively 0.5, 0.49 and 0.01. The fitness function simply sums the number of frames that have a different qualitative output from the frames present in a "perfect" sequence as defined by a human observer. So if n is the total number of frames observed, Δ is one unit that expresses a difference of classification on one frame between FQG and the human observer, then the fitness F is $F = \sum_{r=1}^{n} \Delta_r$. The Koza operators such as the tree building "ramped half and half" algorithm, and operators such as crossover, mutation are all modified with a Strongly-Typed flavour. In practice, this means that the structure of all the individuals generated will be defined by a set of rules. These rules associate for each parent node an ordered set of children nodes. Each parent node maps to a list of possible children nodes which can be either function nodes or terminal nodes.

10.3.3 Dealing with Occluded Motion

Occlusion is a de-facto standard problem when dealing with the classification of real human motion data. In order to effectively deal with occlusion, feature selection and feature reconstruction are introduced in this section. The former deals with the optimisation of the feature selection phase via the introduction of an improved measure of similarity. The latter is to reconstruct plausible rotational data from occluded joints using a modified version of fuzzy robot kinematics.

Regarding to feature selection, fuzzy rough feature selection [43] is chosen that allows real-valued noisy data to be reduced without the need of user supplied information. The quality of the results relies on the estimation of dependency degrees between attributes which itself is based on the measure of how similar two objects x and y can be for a feature a. That is to say, the fuzzy similarity relation between two objects for a given attribute can determine the efficiency of the whole attribute selection process. In this chapter, several fuzzy similarity relations are presented and compared to the fuzzy Laplace Similarity relation. It is evident that reducing the features to a reduction subset with minimal information loss provides a way to estimate which joints are the most important in the chapter context. If none of these essential joints are occluded, the uncertainty of the classification can be lowered. The importance of these joints, when computing a distance between membership scores of motions, can be taken into account. Fuzzy rough feature selection is based on fuzzy lower and upper approximations $\mu_{\underline{R_P}}(x)$ and $\mu_{\overline{R_P}}(x)$, where a T-transitive fuzzy similarity relation approximates a fuzzy concept X.

$$\mu_{\underline{R_P}}(x) = \inf_{y \in \mathbb{U}} I(\mu_{R_P}(x, y), \mu_X(y)) \tag{10.7}$$

$$\mu_{\overline{R_P}}(x) = \sup_{y \in \mathbb{U}} T(\mu_{R_P}(x, y), \mu_X(y)) \tag{10.8}$$

where, I is a fuzzy implicator and T a t-norm. R_P is the fuzzy similarity relation induced by the subset of features P.

$$\mu_{R_P}(x, y) = \bigcap_{a \in P} \{\mu R_a(x, y)\} \tag{10.9}$$

where $\mu R_a(x, y)$ expresses the degree to which objects x and y are similar for a feature a. The fuzzy positive region can be defined as:

$$\mu_{POS_{R_P(Q)}}(x) = \sup_{X \in \mathbb{U}/Q} \mu_{\underline{R_P X}}(x) \tag{10.10}$$

The resulting degree of dependency is:

$$\gamma' P(Q) = \frac{\sum_{x \in \mathbb{U}} \mu_{POS_{R_P(Q)}(x)}}{|\mathbb{U}|} \tag{10.11}$$

Core features can be determined by the change in dependency of the full set of conditional features when individual attributes are removed:

$$Core(\mathbb{C}) = \left\{a \in \mathbb{C} | \gamma'_{\mathbb{C}-\{a\}}(Q) < \gamma'_{\mathbb{C}}(Q)\right\} \tag{10.12}$$

The degree of similarity $\mu R_a(x, y)$ can be expressed by different fuzzy similarity relations. Depending on which relation is used, a different subset of attributes will be selected, influencing in turn directly the performance of a classifier. The fuzzy Laplace similarity relation can be expressed in function of two parameters: the location parameter of the Laplace Distribution μ_0 and the scale parameter λ. Equation 10.13 shows the general case while Eq. 10.14 represents the most used case when $\mu_0 = 0$ and $\lambda = 0.5$.

$$\mu_{R_a} = \min\left(\left[\frac{\exp\left(\frac{-||a(x)-a(y)|-\mu_0|}{\lambda}\right)}{2\lambda}\right], 1\right) \tag{10.13}$$

$$\mu_{R_a} = \min(\exp(-2 \cdot |a(x) - a(y)|), 1) \tag{10.14}$$

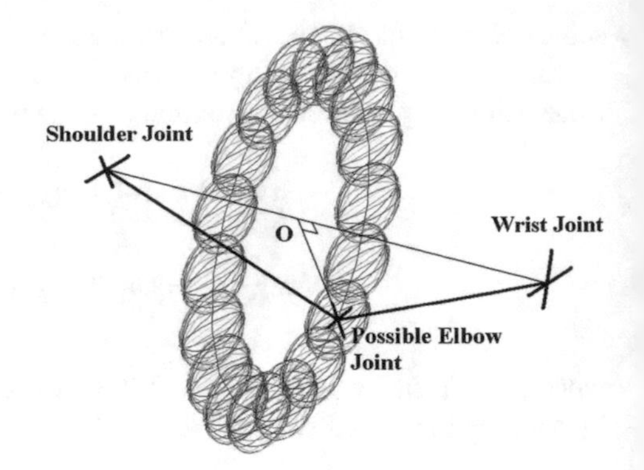

Fig. 10.7 Finding possible qualitative locations of an occluded elbow joint

On the other hand, for feature reconstruction, the granularity of the search space is increased by using fuzzy qualitative Euler angles. Fuzzy qualitative trigonometry is adapted here [44], first to infer possible fuzzy qualitative states representing plausible rotations, then to rank possible solutions using existing and reconstructed rotational data. Since the skeletal representation is normalised, the distances to the occluded joints are known. That is to say that the position of the occluded joint lies along a circle of centre O defined by these distances, as shown in Fig. 10.7 for an example of occluded elbow joint, where the circle is divided by $n = 16$ possible spheres where each sphere represents the volume of a possible position. Considering 8 different cameras placed evenly placed around the stick-figure at belt height, the occlusion of all the spheres is reconstructed in 3D for all 8 points of views. Only spheres which are occluded by other limbs are considered as possible joints positions. This significantly reduces the search space. The coordinates of the centre of each occluded sphere are extracted. For each centre, one crisp joint position is extracted and then converted into all possible corresponding Fuzzy Qualitative Euler Angles, giving a limited number of plausible suggestions for the Euler Angles of the hidden joint. This conversion of one Crisp Euler angle rotation into the set of all equivalent fuzzy qualitative Euler angles is done via a pre-computed mapping table. Let the unit circle be divided into $n = 16$ fuzzy qualitative angles shown in Fig. 10.8. Each of the 3 fuzzy qualitative Euler angle expressing a rotation will have n possible discrete states. Therefore, there exist $n!/(n-3)!$ possible fuzzy qualitative Euler angles combinations. Each combination will correspond to a resulting fuzzy qualitative surface on the unit sphere in Fig. 10.8.

When dealing with insufficient data, one generally has to first make the best of the precious little information available in order to then venture into making educated guesses. This idea is translated by implementing the ranking strategy in two steps. First, an initial evaluation is conducted from the available (or non-occluded) data. An overall average of the membership scores of visible joints to fuzzy membership functions that model the rotations of corresponding joints in known stances is produced. Let $j = 1, \ldots, 19$ be one of the 19 joints and M_i with $i = 1, \ldots, 6$ be one of the six possible moves: Guard, Jab, Cross, Left Hook, Right Hook, and Lower Left Hook.

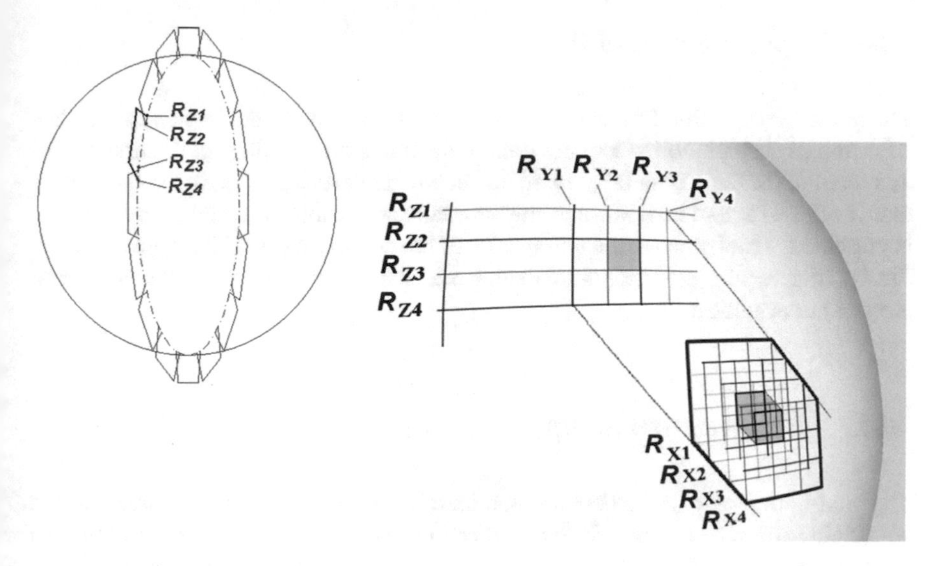

Fig. 10.8 Mapping from the fuzzy qualitative circle to fuzzy qualitative euler angles positions

Let $S_{j_{M_i}}$ be the membership score of a non-occluded joint j to a trapezoid fuzzy membership function that models the rotation of that joint for a known move M_i. We end up with a set of six average membership scores: $\langle S_{M_1}, S_{M_2}, S_{M_3}, S_{M_4}, S_{M_5}, S_{M_6} \rangle$. If there is very little occlusion, the information from visible joints conveys more certainty. Inversely, if there are many occluded joints, the information from these visible joints conveys a much greater uncertainty. If most of the joints (around 80%) are non-occluded, then the average membership scores of the visible joints is the most important information, while the plausible rotations estimated for the occluded joints is just here to give possible interpretations. In the inverse situation where 80% of joints would be occluded, the plausible estimations of rotations would have more weight on the classification. In this experiment, as around 80% the joints are visible at any given time, the move with the highest average membership scores from non-occluded joints tends to reflect the most likely estimation. Therefore, the best move from visible data satisfies the constraint: $M_{V_{1st}} = \max \langle S_{M_1}, S_{M_2}, S_{M_3}, S_{M_4}, S_{M_5}, S_{M_6} \rangle$. In this case, solutions generated by plausible joints are simply here to confirm or infirm this initial assessment. In the second step, for every occluded joint, plausible fuzzy qualitative Euler Angles are ranked by geometrical distance to the fuzzy membership functions of known moves. The three closest moves are selected and ranked as possible solutions: $\langle M_{O_{1st}}, M_{O_{2nd}}, M_{O_{3rd}} \rangle$. If one of the most plausible solutions from occluded joints corresponds to the move estimated as most likely from visible data, the system confirms that the occluded frame can be classified as the latter. In other words, if: $M_{V_{1st}} \in \langle M_{O_{1st}}, M_{O_{2nd}}, M_{O_{3rd}} \rangle$ then $M_{V_{1st}}$ is considered the most plausible suggestion for the occluded frame.

10.4 Experiments and Results

The motion recognition framework is evaluated in an experiment involving the classification of real 3d motion capture data in the context of boxing. In the first part of this section, the experimental method and setup are described. Secondly, the performance of FQG as a standalone learning paradigm applicable to behaviour recognition is presented. Thirdly, experimental results of the context-aware GP filter are shown. Finally, the feature selection and feature reconstruction aspects of the occlusion module are evaluated.

10.4.1 Experimental Setup

The boxing motion capture data are obtained from a Vicon Motion Capture Studio with eight infra-red cameras. Each motion capture suit is set with a total of 49 optical passive reflective markers. The motion recognition is implemented in MATLAB on a standard PC with 2 Gigs of RAM. An additional MATLAB toolbox presented in [45] is also used for extracting Euler angles from BVH files. The relative positions of the cameras and the subject are shown in Fig. 10.9. Three male subjects, aged between 18 and 21, of light to medium-average size (167–178 cm) and weight (59–79kgs), all practicing boxing in competition at the national level. None of them presented any abnormal gait. Optical Markers were placed in a similar way on each subject to ensure a consistent motion capture shown in Fig. 10.10.

Participants are required to perform modern boxing moves on their own, in a preordained and controlled fashion. The motion capture data is obtained from several subjects performing each boxing combination four times. There are twenty-one different boxing combinations, each separated by a guard stance. These are performed at two different speeds (medium-slow and medium fast). The boxing combinations

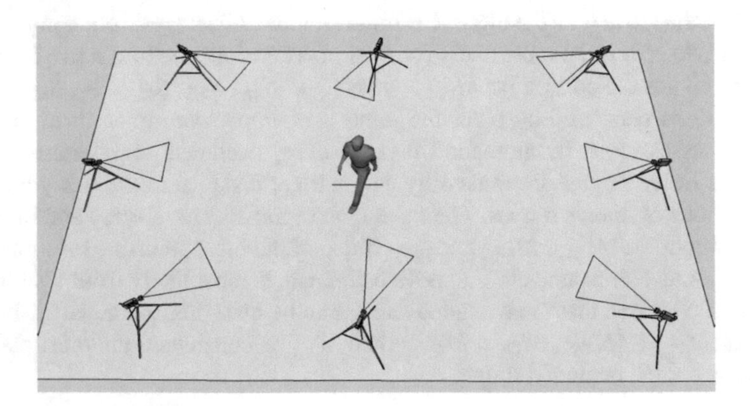

Fig. 10.9 3d motion capture studio floor plan

Fig. 10.10 Optical markers placement—front view

are ordered sets of basic boxing stances. There are in total six precisely defined basic stances: Guard, Jab, Cross, Left Hook, Right Hook, Lower Left Hook. The level of precision needed to identify such motions is non-negligible. These are well-known boxing stances and are accurately described in [46]. Each participant goes through the following steps: on arrival, after a 10 min introduction, the participant is briefed on the experiment and given consent and information forms to read and sign. He is then putting the suit and captors on during the next 10 min. The next step which is the set up and calibration of the motion capture equipment takes around one hour. The data capture of different motions follows and is spread over 20 min. The participant then goes to the changing room and leaves.

A set of fuzzy membership functions corresponding to a specific stance is extracted from various samples. First, all three participants are employed to learn and to test how well the system recognises the stances. Then, an evaluation is conducted to see how the system cope to learn from only two participants, and test how well it recognise stances from a third different participant. Analysis is initially focused on classifying only one move and then to six moves at the same time. The inputs for each given time frame are the six membership scores of each known move.

10.4.2 Evaluating Fuzzy Quantile Generation

FQG performance is evaluated in this section as a standalone classifier using Receiver Operating Characteristic (ROC) analysis. It involves the recognition of multiple moves from a 3d motion capture boxing data set. FQG classifier is evaluated by comparing its performance with a human observer.

FQG is compared to algorithms that are implemented using the WEKA Machine Learning package presented in [47]. Initial comparison was started with a time-dependent Hidden Markov Model (HMM) classifier. However, due to the particularities of the boxing data set, artificially created classes representing boxing stances could not be obtained via standard clustering methods and would otherwise give singular matrices with HMM. Another fact was that HMM seemed to need considerably larger learning samples to be able to perform. It was then decided to focus the study on a comparison between FQG and sixteen time-invariant classification algorithms aiming at recognising boxing stances. These classifiers take exactly the same input data as FQG in order to provide a fair comparison. All these algorithms are implemented using the WEKA Machine Learning package presented in [47]. Several of these techniques correspond to some of the methods used for motion recognition presented in the introduction. Bayes Net and Naive Bayes fit into the Probabilistic graphical models category. Similarly, Radial Basis Function networks and Multilayer Perceptrons are part of the connectionist approach. SMO can be classified as a kernel based method, while Hyperpipes and Voting Feature Intervals are part of the voting strategies paradigm. IB1 is an instance based classifier, and Fuzzy Nearest Neighbours and Fuzzy Rough Nearest Neighbours can fit into hybrid methods. The WEKA classifiers are used with reasonably adjusted parameters mostly based on default settings. In order to keep the comparison relatively fair, FQG parameter s described in Sect. 10.3.1.4 was set by default to 0.7 through the whole comparative experiment. While some of these classifiers might perform substantially better with deeply optimised parameters, the purpose of this study, is not to focus on their optimisation but is rather to give a comparison ground with FQG and commonly-used techniques representing different paradigms on the same boxing 3d Motion Capture data sample. These algorithms display high computational efficiency and are used in state of the art research. It is difficult to generalise comparisons between machine learning methods due to the particulars of the topology of the search space of each application domain. However, it is possible to offer a comprehensive and indicative study that does not necessarily have to deal in absolutes to say if a method performs roughly at the same level of efficiency as the best methods available.

Classifiers performances are evaluated using accuracy, and F-Score. Accuracy is defined as:

$$accuracy = \frac{tp + tn}{tp + fp + fn + tn} \tag{10.15}$$

where tp represents true positive rate, tn true negative rate, fp false positive rate, and fn false negative rate. It represents the proportion of true results (both true positives

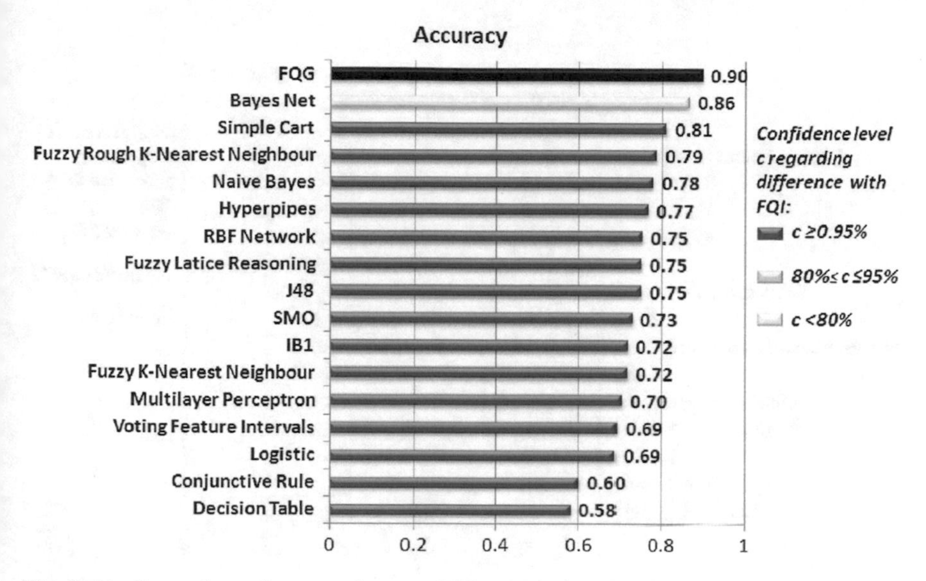

Fig. 10.11 Comparison of accuracy between FQI and 16 other classifiers

and true negatives) in the population. It is the most used empirical measure in the area of machine learning classifiers. At a first glance (see Fig. 10.11), results seem quite encouraging as FQI shows the best average accuracy of 90% over sixteen high performance machine learning classifying methods (Fig. 10.11). This difference in accuracy with FQI is significant at the level 0.05 on a two-tailed t-test for fifteen of these methods at the exception of Bayes Net (86%) which performs similarly and does not show a significant difference, even at the level 0.20. So far, if exclusively focusing on the accuracy, FQG and Bayes Net seem to be in first position, while the next three best methods are: SimpleCart (81%), Fuzzy Rough K-Nearest neighbours (79%) and Naive Bayes (78%). Beside FQI, the most accurate methods belong broadly to Bayes, decision trees and Fuzzy Rough Nearest Neighbours types. The worst performances belong to rule based classifiers—Conjunctive Rule (60%) and Decision Table (58%). They present a lost of accuracy of nearly 10% compared to the next best classifier. This gap could be explained by the difficulty for the bottom-up design to generate the sheer quantity of rules that could deal with real numbers in a noisy and biologically imprecise data set. When using accuracy alone, results seem to be in favor of FQG and Bayes Net. However, using only accuracy to measure the performance of the FQG classifier would only show a part of the whole picture.

F-Measure or balanced F-Score (Eq. 10.16) is a measure that combines Precision and Recall. It is the harmonic mean of precision and recall (recall measures the proportion of actual positives which are correctly identified as such and precision is the proportion of the true positives against all the positive results).

$$F - score = \frac{Precision \cdot Recall}{Precision + Recall} \tag{10.16}$$

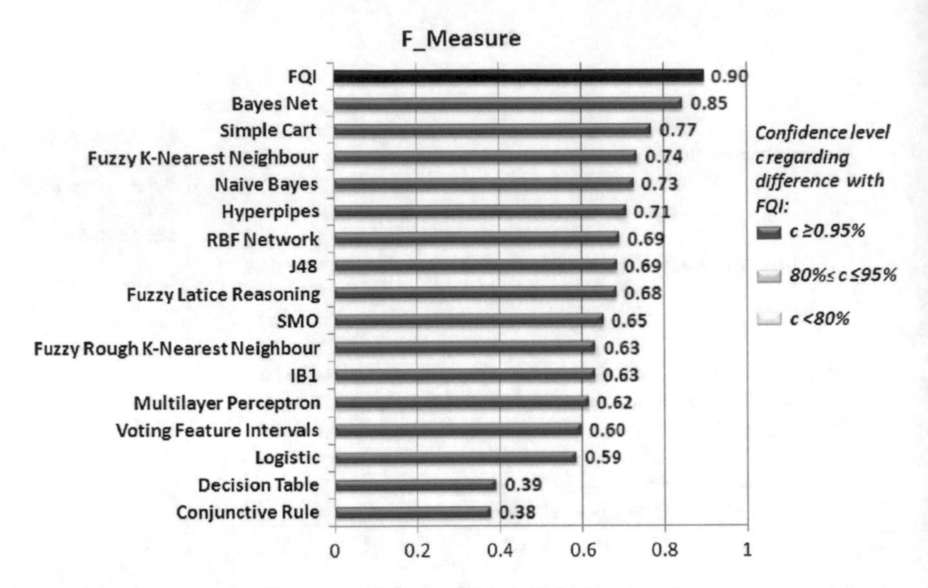

Fig. 10.12 Comparison of F-Measure between FQI and 16 other classifiers

FQG shows the best average F-score of 90% over the sixteen other classifiers (Fig. 10.11). This difference in F-score with FQG is significant at the level 0.05 on a two-tailed t-test for all of these methods. The next four best classifiers are Bayes Net (85%), SimpleCart (77%), Fuzzy Rough K-Nearest neighbours (74%) and Naive Bayes (73%). These results confirm the overall dominance of FQI so far.

FQG seems to compare favourably with the considered time invariant techniques when classifying static stances. Results show that FQG ends up with the best Accuracy, and F-Score, and suggest that FQG present certain advantages such as: the ability to decouple conditional feature distributions into one-dimensional sets, the capacity to link these features by combining their memberships, the ability to deal with imprecision and overlapping classes scores, and the capacity to deal with partial knowledge by taking into account the degree of incompleteness of the learning sample when mapping it to an underlaying distribution (Fig. 10.12).

10.4.3 Evaluating Occlusion Module

Feature selection and feature reconstruction are being tested in experiments in order to evaluate its capability of handling occlusion. For feature selection, the performance of Laplace similarity is compared with Gaussian and triangle-2 fuzzy similarity relations. Each fuzzy similarity relation is measured by taking into account both the reduction of the number of attributes and the resulting accuracy of a naive Bayesian classifier. Compression is measured with the ratio of attributes selected over the

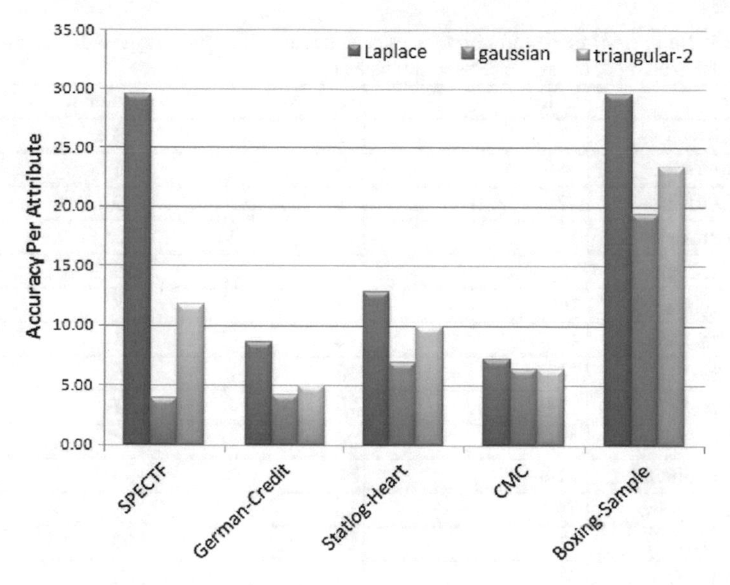

Fig. 10.13 Accuracy/Attribute comparisons for laplace, gaussian and triangular-2 similarity relations on different datasets

total number of attributes. Accuracy is obtained with the average over a 10 fold evaluation of the accuracy of the Bayesian classifier. Results are compared over four standard datasets from the UCI Machine Learning Repository and a boxing motion sample coming from 3d motion capture data. For all datasets, fuzzy Laplace similarity displays the highest compression rate for the first to second best accuracy. When combining both compression and accuracy by evaluating the percentage of accuracy gained per attribute, it becomes possible to compare the performance of each measure of similarity as shown in Fig. 10.13. In all cases, when looking at the ratio of compression over accuracy, the fuzzy Laplace similarity relation systematically outperforms the Gaussian and the Triangular-2 relations. These results are promising so that the Laplace similarity relation is employed in the context of attribute selection when developing the occluded human motion classification. In the case of the motion capture boxing sample, triangular 2 similarity gave a smaller reduct set including 3 numbered features 45, 48, 56 than Gaussian similarity consisting 4 features 7, 44, 45, 48 (see Table 10.3). They both produced reduction sets with similar dataset consistency around 0.91. Laplace similarity gives the smallest subset of all with only 2 attributes 3, 31 and a data consistency around 0.95. All these ended up with an acceptable accuracy when classifying the motions. Triangular-2 selects the head right shoulder and right elbow rotations to differentiate boxing motions. Gaussian selects the left knee, right shoulder, and right elbow to differentiate motions. Interestingly, Laplace similarity uses the left shoulder and the hip to differentiate boxing motions. This is actually closer to what boxing experts observed as they take into account both the upper body movements and the hips rotations.

Table 10.3 Numbered key features of joint euler angles identified by fuzzy rough feature selection (gaussian labelled by $*$, triangular-2 by $\bullet$ and laplace by $\circ$)

Joint name	Euler Z	Euler X	Euler Y
ROOT Hips	1	2	3 $\circ$
JOINT LeftHip	4	5	6
JOINT LeftKnee	7 $*$	8	9
JOINT LeftAnkle	10	11	12
JOINT RightHip	13	14	15
JOINT RightKnee	16	17	18
JOINT RightAnkle	19	20	21
JOINT Chest	22	23	24
JOINT Chest2	25	26	27
JOINT LeftCollar	28	29	30
JOINT LeftShoulder	31 $\circ$	32	33
JOINT LeftElbow	34	35	36
JOINT LeftWrist	37	38	39
JOINT RightCollar	40	41	42
JOINT RightShoulder	43	44$*$	45$*\bullet$
JOINT RightElbow	46	47	48$*\bullet$
JOINT RightWrist	49	50	51
JOINT Neck	52	53	54
JOINT Head	55	56 $\bullet$	57

Experiment results presented in Fig. 10.13 show that the fuzzy Laplace similarity relation outperforms other known fuzzy similarity relations such as Gaussian and Triangular-2 measures in the context of fuzzy rough feature selection. The proposed fuzzy similarity relation is integrated into the motion classification framework in order to identify essential motion joints, e.g. here shoulders and hips. The feature selection algorithm estimates uncertainty to be around 0.95 if all these essential joint are occluded, 0.316 if only one of them is occluded, and 0.05 if none is occluded. Left or right shoulders are considered only as the hips are always visible in the experimental data set. The case, when only one essential joint is occluded, is of interest as it appears about 60% of the time in the present boxing motions data samples. If the Euler Angles rotations of these joints can be accurately reconstructed, it becomes possible to reduce uncertainty and suggest plausible end results for the classification.

The ability to infer correct plausible moves from occluded data is evaluated on samples with simulated occlusion from 8 different points of views. Occlusion is reconstructed using Matlab by computing which joints are masked by a 3 dimensional mesh composed of cylinders representing the limbs and portions of the human body. This multiplies by 8 the size of the data set. The process of generation of these occluded data is computationally expensive and slow. Therefore, considering the time

Table 10.4 Classification of motions with one occluded shoulder joint

Joint	Nb frames	Correct guesses	Rates (%)
Right shoulder	3389	2604	76.83
Left shoulder	3096	2343	75.67
Total	6490	4947	76.22

constraints, one sample of reasonable complexity is used for testing: a combination of guards and jabs. A binary mask expressing the state of occlusion for all 19 joints is produced at every frame. It seems that in more than half of the cases (58.35% of all the frames when combining the two samples), only one of the essential joints that were previously determined through feature selection is occluded. In more than 99% of the cases, this is a shoulder joint. This represents a data set of around 6490 frames with one shoulder joint occluded. As shown in Table 10.4, the most plausible rotation chosen for occluded shoulder joints arrives at the correct classification of the moves for 4947 of these 6490 frames. For instance, when looking at one participant performing moves from different points of views, on cases when one shoulder joint is occluded, e.g. 58% of all frames, the classifier infers the correct moves 76.2% of the time. The occluded data classification is substantially improved when applying the GP filter on top of these results.

The experiment shows that FQG can be extended to deal with occlusion reasonably well. Initially all occluded data are intractable. After using the system described above on the given sample, around 44% of all the occlusions scenarios are correctly classified. One drawback of this approach is that it is computationally expensive. The number of fuzzy qualitative angles generated to express one plausible rotations often exceeds 160. Having 3 plausible rotations for one occluded joint is not rare. The system has to compute geometrical distances to seven fuzzy membership functions of known moves, i.e., $3 \times 160 \times 7 = 3360$ operations. Then, it requires sorting all these in order to find the three closest moves estimated by all these plausible rotations. These numerous operations have to be computed for one frame of average occlusion complexity. Our current work is focused on improvements for conducting occluded motion recognition in near real-time conditions.

10.5 Conclusion

This chapter presents a three-component motion classification framework. First, fuzzy quantile generation is proposed to generate fuzzy membership functions using metrics derived from the probabilistic quantile function; secondly, a genetic programming based filter is developed to produce time-dependent and context-aware rules to smooth the qualitative outputs of fuzzy quantile generation; finally, motion occlusion is effectively handed by introduced feature selection and feature reconstruction.

The framework has demonstrated its effectiveness on motion capture data from real boxers in terms of fuzzy membership generation, context-aware rule generation, motion occlusion. Note that the motion capture data presents challenges for classification problem in general: real biologically "noisy" data, cross-gait differentials from one individual to another, relatively high dimensionality of a skeletal representation that has 57 degrees of freedom, and large number of learning samples of suboptimal size.

References

1. S. Mitra and T. Acharya. Gesture Recognition: A Survey. *IEEE Transactions on systems, man and cybernetics. Part C, Applications and reviews*, 37(3):311–324, 2007.
2. T.B. Moeslund, A. Hilton, and V. Krüger. A survey of advances in vision-based human motion capture and analysis. *Computer Vision and Image Understanding*, 104(2–3):90–126, 2006.
3. T. Wan, Y. Wang, and J. Li. Hand gesture recognition system using depth data. In *International Conference on Consumer Electronics, Communications and Networks*, pages 1063–1066, April 2012.
4. A. Fernndez-Baena, A. Susin, and X. Lligadas. Biomechanical validation of upper-body and lower-body joint movements of kinect motion capture data for rehabilitation treatments. In *International Conference on Intelligent Networking and Collaborative Systems*, pages 656–661, Sept. 2012.
5. J. Shotton, A. Fitzgibbon, M. Cook, T. Sharp, M. Finocchio, R. Moore, A. Kipman, and A. Blake. Real-time human pose recognition in parts from single depth images. In *IEEE Conference on Computer Vision and Pattern Recognition*, pages 1297–1304. IEEE, 2011.
6. Z. Ju, Y. Wang, W. Zeng, S. Chen, and H. Liu. Depth and rgb image alignment for hand gesture segmentation using kinect. In *International Conference on Machine Learning and Cybernetics*, volume 2, pages 913–919. IEEE, 2013.
7. Zhou Ren, Junsong Yuan, Jingjing Meng, and Zhengyou Zhang. Robust part-based hand gesture recognition using kinect sensor. *IEEE transactions on multimedia*, 15(5):1110–1120, 2013.
8. Nianfeng Li, Yinfei Dai, Rongquan Wang, and Yanhui Shao. Study on action recognition based on kinect and its application in rehabilitation training. In *Big Data and Cloud Computing (BDCloud), 2015 IEEE Fifth International Conference on*, pages 265–269. IEEE, 2015.
9. Hesham Alabbasi, Alex Gradinaru, Florica Moldoveanu, and Alin Moldoveanu. Human motion tracking & evaluation using kinect v2 sensor. In *E-Health and Bioengineering Conference (EHB), 2015*, pages 1–4. IEEE, 2015.
10. Z. Ren, J. Meng, J. Yuan, and Z. Zhang. Robust hand gesture recognition with kinect sensor. In *Proceedings of the 19th ACM international conference on Multimedia*, pages 759–760. ACM, 2011.
11. A.F. Bobick and A.D. Wilson. A state-based approach to the representation and recognition of gesture. *IEEE Transactions on Pattern Analysis and Machine Intelligence*, 19(12):1325–1337, 1997.
12. P. Remagnino, T. Tan, and K. Baker. Agent orientated annotation in model based visual surveillance. In *Proceeding of International Conference on Computer Vision*, pages 857–862, 1998.
13. J. Yamato, J. Ohya, and K. Ishii. Recognizing human action in time-sequential images using hidden markov model. In *Proceedings of the 1992 IEEE Computer Society Conference on Computer Vision and Pattern Recognition*, pages 379–385, 1992.
14. K. Takahashi, S. Seki, E. Kojima, and R.E. Oka. Recognition of dexterous manipulations from time-varying images. In *Proceedings of the 1994 IEEE Workshop on Motion of Non-Rigid and Articulated Objects*, pages 23–28, 1994.

15. J. Davis and M. Shah. Visual gesture recognition. *IEEE Proceedings-Vision, Image and Signal Processing*, 141(2):101–106, 1994.
16. John L., Andrew M., and F. Pereira. Conditional random fields: Probabilistic models for segmenting and labeling sequence data. In *Proc. 18th International Conf. on Machine Learning*, pages 282–289. Morgan Kaufmann, San Francisco, CA, 2001.
17. D.L. Vail. *Conditional Random Fields for Activity Recognition*. PhD thesis, CMU School of Computer Science, 2008.
18. J. Sung, C. Ponce, B. Selman, and A. Saxena. Unstructured human activity detection from rgbd images. In *IEEE International Conference on Robotics and Automation*, pages 842–849. IEEE, 2012.
19. G. Welch and G. Bishop. An introduction to the kalman filter. Technical Report TR 95-041, University of North Carolina at Chapel Hill, Department of Computer Science, Chapel Hill, NC 27599-3175, 1995.
20. M. Isard and A. Blake. Contour tracking by stochastic propagation of conditional density. *European Conference on Computer Vision*, 1(2), 1996.
21. F. Endres, J. Hess, and W. Burgard. Graph-based action models for human motion classification. In *German Conference on Robotics, ROBOTIK*, pages 1–6, May 2012.
22. B.E. Boser, I.M. Guyon, and V.N. Vapnik. A training algorithm for optimal margin classifiers. In David Haussler, editor, *Proceedings of the 5th Annual ACM Workshop on Computational Learning Theory*, pages 144–152, Pittsburgh, PA, July 1992. ACM Press.
23. F. Guo and G. Qian. Dance posture recognition using wide-baseline orthogonal stereo cameras. In *International Conference on Automatic Face and Gesture Recognition*, pages 481–486, 2–6 2006.
24. T. Jan, M. Piccardi, and T. Hintz. Neural network classifiers for automated video surveillance. *13th IEEE Workshop on Neural Networks for Signal Processing*, pages 729–738, 2003.
25. M.H. Yang and N. Ahuja. Extraction and classification of visual motion patterns for hand gesture recognition. In *in Proc. IEEE Conf. Computer Vision and Pattern Recognition*, pages 892–897, June 1998.
26. K. Symeonidis. Hand gesture recognition using neural networks. *Neural Networks*, 13.5.1, 1996.
27. Y.A. Ivanov and A.F. Bobick. Recognition of visual activities and interactions by stochastic parsing. *IEEE Transactions on Pattern Analysis and Machine Intelligence*, 22(8):852–872, August 2000.
28. A.F. Bobick and J.W. Davis. The recognition of human movement using temporal templates. *IEEE Transactions on Pattern Analysis and Machine Intelligence*, 23:257–267, 2001.
29. H. Bourlard and C.J. Wellekens. Links between markov models and multilayer perceptrons. *IEEE Transactions on Pattern Analysis and Machine Intelligence*, 12(12):1167–1178, December 1990.
30. S.K. Korde and K.C. Jondhale. Hand gesture recognition system using standard fuzzy c-means algorithm for recognizing hand gesture with angle variations for unsupervised users. In *International Conference on Emerging Trends in Engineering and Technology*, pages 681–685, 16–18 2008.
31. L.A. Zadeh. Toward human level machine intelligence - is it achievable? the need for a paradigm shift. *Computational Intelligence Magazine, IEEE*, Volume 3(3):11–22, August 2008.
32. D. Dubois and H. Prade. Possibility theory, probability theory and multiple-valued logics: A clarification. *Annals of Mathematics and Artificial Intelligence*, 32:35–66, 2001.
33. G. Nieradka and B.S. Butkiewicz. A method for automatic membership function estimation based on fuzzy measures. In *Lecture Notes in Computer Science*, volume 4529, pages 451–460. Springer, 2007.
34. L.A. Zadeh. Fuzzy sets as a basis for a theory of possibility. *Fuzzy Sets and Systems*, 1:3–28, 1978.
35. Tapio F. Timing of fuzzy membership functions from data. Academic Dissertation - University of Oulu - Finland, July 2001.

36. International Society of Biomechanics. Isb recommendation on definitions of joint coordinate system of various joints for the reporting of human joint motion–part i: ankle, hip, and spine. *Journal of Biomechanics*, 35(4):543–548, 2002.
37. J. Favre, R. Aissaoui, B.M. Jolles, O. Siegrist, J.A. de Guise, and K. Aminian. 3d joint rotation measurement using mems inertial sensors: Application to the knee joint. In *ISB-3D: 3-D Analysis of Human Movement, Valenciennes, France.*, 2006.
38. C.S. Chan, H. Liu, and D.J. Brown. Recognition of human motion from qualitative normalised templates. *Journal of Intelligent and Robotic Systems*, 48(1):79–95, January 2007.
39. M. Meredith and S. Maddock. Motion capture file formats explained. Technical report, Computer Science Department, University of Sheffield, 2000.
40. A.F. Gobi and W. Pedrycz. Fuzzy modelling through logic optimization. *International Journal of Approximate Reasoning*, 45(3):488–510, 2007.
41. T. Yu, D. Wilkinson, and D. Xie. A hybrid GP-fuzzy approach for resevoir characterization. In R.L. Riolo and B. Worzel, editors, *Genetic Programming Theory and Practice*, chapter 17, pages 271–290. Kluwer, 2003.
42. M. Khoury. pystep or python strongly typed genetic programming. Available online at: http://pystep.sourceforge.net/, 2009.
43. R. Jensen and Q. Shen. New approaches to fuzzy-rough feature selection. *IEEE Transactions on Fuzzy Systems*, 17(4):824–838, 2009.
44. H. Liu. A fuzzy qualitative framework for connecting robot qualitative and quantitative representations. *IEEE Transactions on Fuzzy Systems*, 16(6):1522–1530, 2008.
45. N.D. Lawrence. Mocap toolbox for matlab. Available on-line at http://www.cs.man.ac.uk/~neill/dmocap/, 2007.
46. C. Cokes. *The Complete Book of Boxing for Fighters and Fight Fans*. Etc Publications, 1980.
47. M. Hall, E. Frank, G. Holmes, B. Pfahringer, P. Reutemann, and I.H. Witten. The weka data mining software: An update. *ACM SIGKDD Explorations Newsletter*, 11(1):10–18, 2009.

Chapter 11
Study of Human Action Recognition Based on Improved Spatio-Temporal Features

11.1 Introduction

In recent years, visual based human action recognition has gradually become a very active research topic. Analysis of human actions in videos is considered a very important problem in computer vision because of such applications as human-computer interaction, content-based video retrieval, visual surveillance, analysis of sports events, and more [1]. Due to the complexity of the action, such as different body wearing and habits leading to different observation of the same action, the camera movement in the external environment, illumination change, shadows, viewpoint and so on, these influence of factors make action recognition still a challenging project [2, 3].

The representation of human motion in video sequences is crucial in action recognition. Other than having enough discrimination between different categories, reliable motion features are also required to deal with rotation, scale transform, camera movement, complex background, shade and so on. At present, most commonly used features in action recognition are based on motion, such as optical flow [4, 5], motion trajectory [6–8] etc., or based on the appearance shape, for example silhouette contour [9, 10] etc. The former features are greatly influenced by illumination, shadow. The latter features rely on accurate localisation, background subtraction or tracking and they are more sensitive to noise, partial occlusions and variations in viewpoint. Compared with the former two kinds of features, local spatio-temporal features are somewhat invariant to changes in viewpoint, person appearance and partial occlusions [11]. Due to their advantage, local spatio-temporal features based on interest points are more and more popular in action recognition [12–15].

Spatio-temporal interest points are those points where the local neighborhood has a significant variation in both the spatial and the temporal domain. Most of local spatio-temporal feature descriptors are the extension of the information extracted form previously 2D space to 3D spatio-temporal based on the interest points detected by Laptev [13] or Dollar [12]. They capture motion variation in space and time dimensions in the neighbourhood of the interest points. Up to now, many efforts have been devoted to the description of the spatio-temporal interest points. The most common descriptions are SIFT [16], SURF [17] and so on, which have advantages of scale, affine, view and rotation invariance. Dollar [12] applied the cuboids and

H. Liu et al., *Human Motion Sensing and Recognition*,
Studies in Computational Intelligence 675, DOI 10.1007/978-3-662-53692-6_11

selected smooth gradient as a descriptor. Later, Scovanner et al. [18] put forward the 3D SIFT to calculate spatio-temporal gradients direction histograms for each pixel within its neighborhood. Another extension to the SIFT was proposed by Klaser et al. [19] based on a histogram of 3D gradient orientations, where gradients are computed using an integral video representation. Williems et al. [20] extended the SURF descriptor to video, by representing each cell as a vector of weighted sums of uniformly sampled responses to Haar wavelets along the three axes.

In the process of recognition, with the extraction of cuboids feature descriptor, Dollar [12] adopted the PCA algorithm to reduce feature dimension and finally made use of nearest neighbour classifier and SVM to recognise human actions on KTH dataset. Niebles [14] considered videos as spatio-temporal bag-of-words by extracting space-time interest points and clustering the features, and then used a probabilistic Latent Semantic Analysis(pLSA) model to localise and categorise human actions. Li [21] got interest points from Harris detector and then extracted 3D SIFT descriptor, in recognition process he made use of SVM with leave-one-out method and also did the experiment on KTH dataset. The above recognition methods have achieved good recognition results, but most studies only stayed on the previous description of the interest points, mainly utilized local spatio-temporal descriptors of single interest point ignoring its overall distribution information in the global space and time. The interest points represent the key position of the human body movement. So the distribution of interest points change according to the human motion. And its implication of sports information also changes accordingly. In addition, it can't simply rely on the local feature of spatio-temporal interest points to represent the target motion when the motion lacks time dimension information. Bregonzio et al. [22] defined a set of features which reflect the interest points distribution based on different temporal scales. In their study, global spatio-temporal distribution of interest points is studied but the excellent performance of local descriptor is also abandoned. When the influence factors interfere body movement, it cannot reliably represent the action only depending on the global information. These above experiments are all conducted on the KTH dataset for the recognition test, but the adaptability of feature applied in different scenarios on the KTH is not studied and discussed. Moreover, despite Dollar [12] detection method's popularity, it tends to generate spurious detection background area surrounding object boundary due to the shadow and noise, and then the subsequent recognition process is affected too.

Based on the above discussions, a novel feature is proposed in this chapter to represent human motion by combining local and global information. That is a combination of 3D SIFT descriptor and the spatio-temporal distribution information based on interest points. First step, an improved detecting method [22] is used to detect spatio-temporal interest points, different from Dollar's method, effectively avoiding the error detecting in the background. Then these interest points are represented by 3D SIFT descriptor and the positional distribution information of these interest points is calculated at the same time. In order to achieve perfect combination, the dimension reduction issue is performed on descriptor twice, which is based on single-frame and multi-frame. Then the processed descriptor is combined with positional distribution information of interest points. Finally, the combined features are quantified and selected to obtain more concise feature descriptors. 3D SIFT descriptor contains human body posture information and motion dynamic information, it describes the local feature of action both in spatio and temporal dimension. As a result of the feature extracted in the key points of motion, it is not affected by changes in human body shape and motion directions etc. So the feature has good adaptability and robustness in complex motion scenarios. The positional distribution information of interest points reflects motion global information by using various location and ratio relationship of the two areas of human body movement and interest points distribution. Different from previous methods of describing the appearance and shape in space, this chapter don't directly pick up the shape information. So when the human appearance and shape changes, the location and ratio relationship of the two areas are not directly affected. Therefore, the positional distribution information of interest points proposed by this chapter is more adaptive for motion description in space. Finally the proposed motion descriptor by combining the above mentioned local feature with global information is tested by using SVM and AdaBoost-SVM recognition algorithm on the public dataset of KTH. Furthermore, the adaptability of proposed method is discussed by testing in each different and mixed scenarios of KTH dataset. By comparing with the related and similar research works in recent years, the results verified the proposed method is better with strong robustness and adaptability.

The rest of the chapter is organised as follows. In Sect. 11.2, the detection method for spatio-temporal interest points is introduced in this chapter. Section 11.3 provides a detailed explanation of 3D SIFT descriptor and positional distribution information of interest points as well as the process of feature dimension reduction, quantification, and selection. And Sect. 11.4 gives experimental results and analysis. Finally, Sect. 11.6 concludes the chapter.

11.2 Interest Points Detection

In computer vision, interest points represent the location which has severe changes in space and time dimensions and considered to be salient or descriptive for the action captured in a video. Among various interest points detection methods, the most widely used for action recognition is the one proposed by Dollar [12]. The method calculates function response values based on the combination of Gabor filter and Gaussian filter, and the extreme values of local response can be considered as spatio-temporal interest points in the video.

Dollar's method is effective to detect the interest points of human motion in video, however, it is prone to false detection due to video shadow and noise, and spurious interest points are easy to occur in the background. It is particularly ineffective to camera movement, or camera zooming. Some of the drawbacks are highlighted in the examples as red square slices shown in Fig. 11.1c.

These drawbacks are due to the shortcomings of the Dollar detector, in particular, the Gabor filtering which do the feature extraction only on the time axis, ignoring the dynamic movement in the prospects. To overcome these shortcomings, we utilise a different interest points detector [22] which explores different filters for detecting salient spatio-temporal local areas undergoing complex motion to get a combined filter response. More specifically, our detector facilitates saliency detection and consists of the follow three steps:

1. frame differencing for focus of human action and region of interest detection, as shown in Fig. 11.1b.
2. utilizing 2D Gabor filter for generating five different directions (0, 22, 45, 67, 90°) filter templates, the example presented in Fig. 11.1b.
3. Filtering on the detected regions of interest got in step 1, using 2D Gabor filters with five different orientations obtained by step 2 in both the spatial and temporal domains to give a combined filter response.

Figure 11.1c shows examples of our interest point detection results (circle points) and the Dollar's [12] (square points). It is evident that the detected interest points are much more meaningful and descriptive compared to those detected using the Dollar detector.

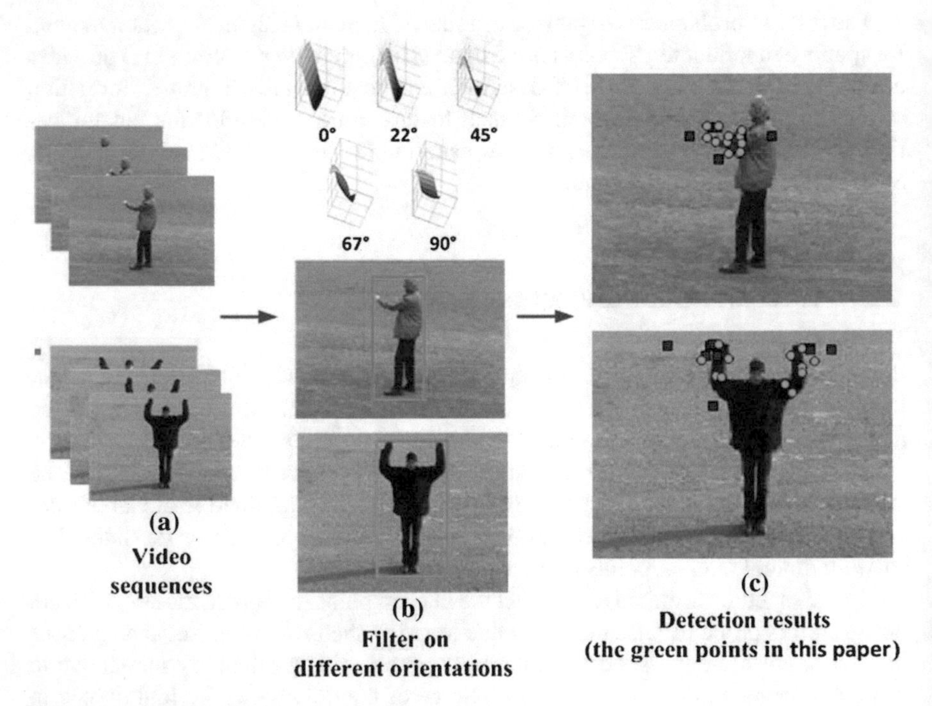

Fig. 11.1 Interest points detection process

11.3 Action Representation

11.3.1 3D SIFT Descriptor

For calculating the SIFT descriptor, firstly spatio-temporal cube is extracted from the interest point as the centre in video sequences and is divided into fixed size unit sub cubes. Then spatio-temporal gradient histogram of each unit cube is calculated by using faceted sphere. Finally the 3D SIFT descriptor is formed by combining all the unit cube histograms [18]. In this chapter, the $12 \times 12 \times 12$ pixel size cube is divided into $2 \times 2 \times 2$ sub cubes, as shown in Fig. 11.2.

3-dimensional gradient magnitude at (x, y, t) is computed by Eq. 11.1.

$$M(x, y, t) = \sqrt{{L^2}_x + {L^2}_y + {L^2}_t} \tag{11.1}$$

The $L_x = L(x+1, y, t) - L(x-1, y, t)$, $L_y = L(x, y+1, t) - L(x, y-1, t)$, $L_t = L(x, y, t+1) - L(x, y, t-1)$ stand for the gradient from x, y, t direction respectively.

According to the amplitude weight in the centre $V(v_{xi}, v_{yi}, v_{ti})$ of each sphere face, three maximums are chosen and added to the corresponding direction of gradient histogram by using the Eq. 11.2.

$$mag = M(x, y, t)G(x', y', t')V(v_{xi}, v_{yi}, v_{ti}) \quad i = 1, 2, \cdots 32 \tag{11.2}$$

(v_{xi}, v_{yi}, v_{ti}) is centre coordinates of each surface on faceted sphere which relied on current pixel as the center. $G(x', y', t') = e^{\frac{x'^2+y'^2+t'^2}{2\sigma^2}}$ serve as the gradient weight, x', y', t' are the difference values between interest point and current pixel in neighbourhood.

This chapter adopts 32 faceted sphere and 32 gradient directions for descriptor, so the feature dimension of each sub cube is 32. The initial whole features of each point are 256 dimensions.

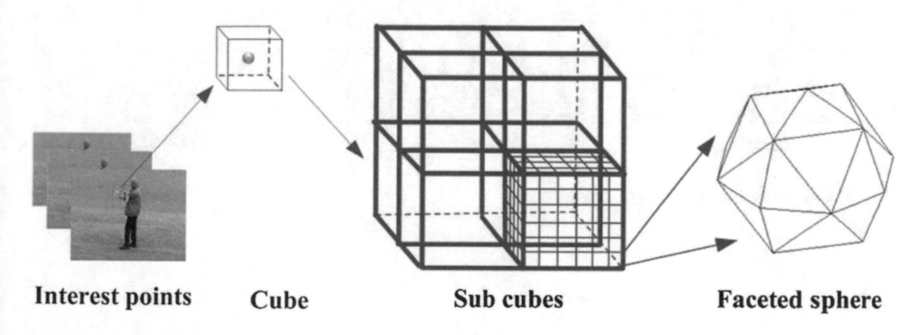

Fig. 11.2 Description process of 3D SIFT

11.3.2 Positional Distribution Information of Interest Points

In terms of interest points in each frame, the positional distribution information is closely related to body motion, it also reflects the amplitude range of action, relevance of human body location and motion parts region. So the positional distribution information of interest points are extracted as another kind of action information. The specific process is described as below.

As shown in Fig. 11.3, we select the example actions from KTH dataset, the distribution region of interest points and body location area are detected in each frame. The region and area are drawn with yellow (Y) and red (R) box respectively. Then the related positional distribution information is calculated with these two areas and expressed by $PDI = [D_{ip}, R_{ip}, R_{ren}, Vertic_{dist}, Orizon_{dist}, W_{ratio}, H_{ratio}, Over_{lap}]$. A calculation method of the features are shown in Table 11.1.

In order to remove the influence of several stray individual points on the overall feature, stray filtering process is utilised on all the interest points before extracting the above positional feature. The points whose distance to the region centroid over a certain threshold value are removed to ensure the validity and reliability of the extracted features. The positional distribution information of interest points is extracted from each frame to represent and reflect the whole attribute of the motion.

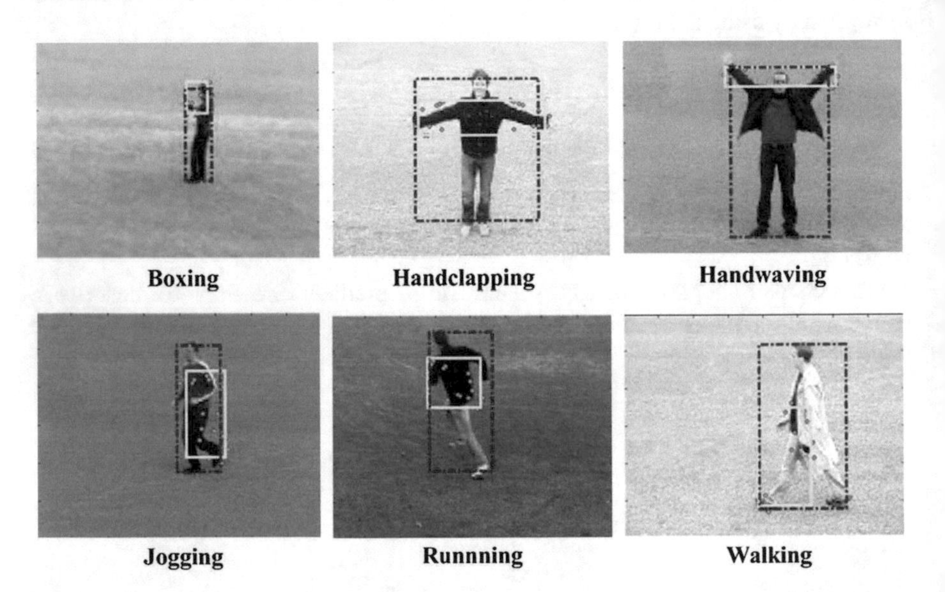

Fig. 11.3 Distribution of interest points

Table 11.1 Calculation method of PDI feature

PDI	Calculation method
D_{ip}	The total number of points normalised by the area Y
R_{ip}	The height and width ratio of Y
R_{ren}	The height and width ratio of R
$Vertic_{dist}$	The vertical distance between the geometrical centre (centroid) of Y and R
$Orizon_{dist}$	The horizontal distance between the geometrical centre (centroid) of Y and R
W_{ratio}	The width ratio between the two areas Y and R
H_{ratio}	The height ratio between the two areas Y and R
$Over_{lap}$	The ratio by the amount of overlap and total width between Y and R

11.3.3 Motion Features

In Sect. 11.3.1, the 3D SIFT descriptor of each interest point is 256 dimensional vector. If the number of interest points in each frame is N, the dimension of the features is $N \times 256$ to represent spatio-temporal information in this frame. The dimension of feature is so high that it cannot reasonable combine with the distribution information PDI obtained by Sect. 11.3.2, therefore, dimension reduction is performed on this part information, as shown in Fig. 11.4. Furthermore, in order to remove redundant

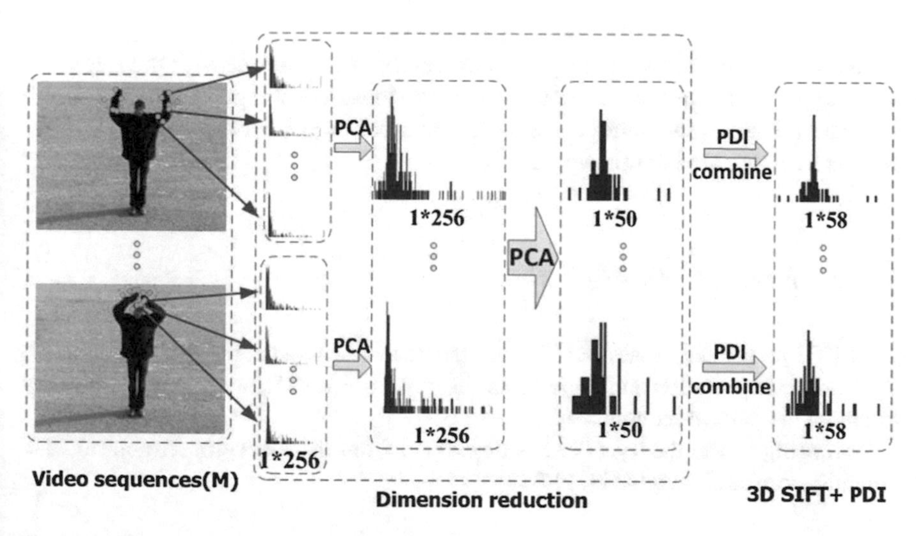

Fig. 11.4 The description of motion feature

data and make features more concise, above combined feature are processed by using quantization and selection. The following five steps are listed as:

1. *Single frame dimension reduction*: Principle component analysis(PCA) is used to perform longitudinal dimension reduction for 3D SIFT descriptor extracted from interest points in the same frame. It means that $N \times 256$ features can be reduced (N is the number of interest points in this frame) to 1×256 by gathering principal component of all descriptors for each frame. The single-frame dimension reduction is helpful to achieve a whole description of the motion in the frame. Although it will lose part of information, the loss of the information is acceptable when the N is chosen as 20 in our experiment.
2. *Multi-frame dimension reduction*: Horizontal dimension reduction is done on the preprocessed descriptors got by step 1. The dimension reduction is used again on all frames to set $M \times 256$ (M for total number of this video frames) to $M \times 50$.
3. *Feature combination*: Make the combination of 1×50 spatio-temporal features and the corresponding positional distribution information (PDI) of interest points in each frame, finally gets 58 dimension features for each frame (3D SIFT+PDI).
4. *Feature quantization*: The linear quantisation is utilised on $M \times 58$ dimension feature of each video to a histogram containing $n(n < M)$ bins, thus it turned to be $n \times 58$ dimension feature.
5. *Feature selection*: Compute the mean of distances from different people performing the same action and make the arrangement, then select the front location $S(S < n \times 58)$ features as the final features.

11.4 SVM Algorithm and Results Analysis

In this section, experiments are performed on the KTH dataset with the improved spatio-temporal feature. By comparing with the most recent reports associated with the related features and dataset, the outstanding performance of the proposed algorithm is demonstrated in this chapter.

11.4.1 Recognition Algorithm

SVM [23] as the data classification of statistical learning method, it has intuitive geometric interpretation and good generalisation ability, so it have gained popularity within visual pattern recognition.

According to the theory, SVM is developed from the theory of Structural Risk Minimisation, as shown in Eq. 11.3.

$$\min \frac{1}{2}\|w\|^2 + C\sum_{i=1}^{R} \varepsilon_i$$
$$y_i(w, \phi(x_i) - b) \geq 1 - \varepsilon_i, 0 \leq \varepsilon_i \leq 1 \tag{11.3}$$

For a given sample set x_i, $y_i \in \{-1, 1\}$,$i = 1, \cdots n$, where $x_i \in R^N$ is a feature vector and y_i is its class label. $(\phi(x_i), \phi(x_j)) = k(x_i, x_j)$ is the kernel function. k is corresponding to the dot product in the feature space, and transformation ϕ implicitly maps the input vectors into a high-dimensional feature space. Define the hyperplane $(w, \phi(x)) - b = 0$ to make a compromise between class interval and classification errors when the sample is linear inseparable. Here, ε_i is the i-th slack variable and C is the regularisation parameter. This minimisation problem can be solved using Lagrange multiplier and KKT conditions, and the dual function is written as:

$$\max_{\alpha} \sum_{i=1}^{n} \alpha_i - \frac{1}{2}\sum_{i,j=1}^{n} \alpha_i\alpha_j y_i y_j (\phi(x_i), \phi(x_j))$$
$$C \geq \alpha_i \geq 0, i = 1, \cdots, n \ \sum_{i=1}^{n} \alpha_i y_i = 0 \tag{11.4}$$

where $\alpha = (\alpha_1,\alpha_2, \cdots, \alpha_n)^T$ is a Lagrange multiplier which corresponds to $y_i(w, \phi(x_i) - b) \geq 1 - \varepsilon_i, \varepsilon_i \geq 0$. We selected kernel function $k(x_i, x_j)=$ $\exp(-\frac{(x_i - x_j)^2}{2\sigma^2})$ and put it into the Eq 11.4 to get the final decision function:

$$y(x) = \text{sgn}(\sum_{i=1}^{n} a_i y_i k(x_i, x) - b) \tag{11.5}$$

Instead of establishing SVMs between one against the rest types, we adopt the method of one against one to establish a SVM between any two categories. The current sample belongs to which category determined by the decision function, and its final type is decided to the category with highest vote.

11.4.2 Dataset

To test our proposed approach for action recognition, we choose the standard KTH dataset, which is one of the most popular benchmark datasets for evaluating action recognition algorithms. This dataset is challenging because there are large variations in human body shape, view angles, scales and appearance. As shown in Fig. 11.5, the KTH dataset contains six types of different human actions respectively performed by 25 different persons: boxing, hand clapping, hand waving, jogging, running, walking. And the sequences are recorded in four different scenarios: outdoors (SC1), outdoors with scale variations (SC2), outdoors with different clothes (SC3), and indoors with

Fig. 11.5 The description of motion feature

lighting variations (SC4). There are obvious changes of visual sense or view between different scenarios, and the background is homogeneous and static in most sequences with some slight camera movement. The sequences are down-sampled to the spatial resolution of 160 × 120 pixels. The examples of the above four scenarios is shown in Fig. 11.5. Apparently, due to the change of camera zooming situation, the size of human body change a lot in SC2(captured in t1 and t2). Furthermore person in SC3 put on different coat, or wear a hat or a bag leading to larger changes in body appearance.

11.4.3 Testing Results in Portion Scenario

In this part recognition experiments are performed by using combined feature of 3D SIFT and PDI on four scenarios (SC1, SC2, SC3 and SC4) according to KTH dataset. Through the tests in various scenarios with the features proposed in this chapter to check out the reliability for motion description and the adaptability to scenarios. Leave-one-out cross validation method is adopted throughout the process, in turns using six action of each actor as test samples, and the rest of all the actions as the training, circulation continued until all actions are completed testing. The experimental results are shown in Table 11.2.

The experimental results show that the feature of 3D sift has the better discriminative ability than PDI feature. SC1 and SC4 are more stable than the other two

Table 11.2 Testing results in portion scenario

Scenario	3D SIFT	PDI	3D SIFT+PDI
SC1	0.9600	0.8000	0.9600
SC2	0.8867	0.8268	0.9200
SC3	0.8542	0.7569	0.9167
SC4	0.9600	0.9000	0.9600

scenarios. We obtained the same recognition rate (96%) by using 3D SIFT and combined features (3D SIFT+PDI). Although emerging the influences of motion direction changes and indoor lighting, the 3D SIFT (after the process of dimension reduction and quantisation) still is a good feature description for motion. It also shows that 3D sift has good adaptability and robustness to motion direction, position, speed and so on.

In SC2 and SC3, scenarios become more complex. Not only the human body exists scale variations with camera zooming in SC2, but also there are 45 degree view changes in jogging, running, walking. In SC3, human body shape changes with different wearing, and the phenomenon of inhomogeneous background even emerges. These above situations make the motion area and position distribution of interest points change obviously as well. So the combined features (3D SIFT+PDI) have certain advantages than 3D SIFT to describe motion in these scenarios, the recognition rate greatly increased by using combined features (3D SIFT+PDI) compared to 3D SIFT. The recognition effect and results confusion matrixes by using combined features are shown in Fig. 11.6.

Observed from the matrixes, the motion jogging is easily confused with running and walking. That is because the similarity between these three actions leads to the error classification, this also accords with our visual observation. Figure 11.7 shows the confusion matrixes by using only 3D SIFT feature and broken line graph with both features in SC2 and SC3 (broken line don't represent rate trend, while it can clearly contrast recognition rate high or low). Compared with the corresponding confusion matrixes of combined features (3D SIFT+PDI) in Fig. 11.6, it is note that the identifiability of 3D SIFT to the confusing actions is improved by combining PDI, and the average recognition rate respectively raised 3% (SC2) and 6% (SC3). It also verified that the proposed combined features is stable and adaptable.

11.4.4 *Testing Results in Mixed Scenarios*

From the analysis of experiments in portion scenarios in Sect. 11.4.3, 3D SIFT in combination with PDI has more advantages. Therefore, in this section, we test combined features in mixed scenarios to validate the feasibility of our approach. The testing process still uses leave-one-out cross validation method.

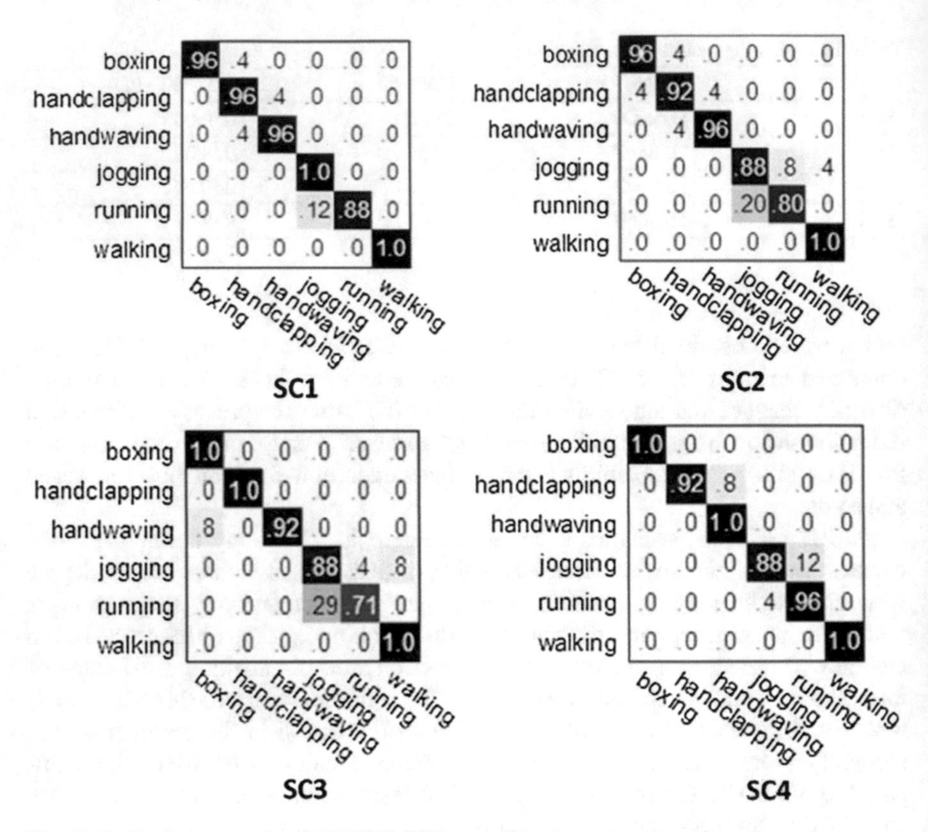

Fig. 11.6 Confusion matrix of 3D SIFT+PDI recognition in portion scenario

As shown in Fig. 11.8, first of all, mix two scenarios, such as stable outdoor SC1 respectively mixed with scale variations SC2, different clothes wearing SC3 mixed with lighting SC4. Through different complexity environment of two mixed scenarios to test our approach. As shown in Fig. 11.8. Due to SC1 and SC4 are relatively simple and stable than other mixed scenarios, so it obtained the best recognition rate 97%. The easily confused actions in above mixed scenarios are still between jogging and running. Summarising all the mixed two scenarios, the average recognition rate reaches 94.10%.

In order to make diversity of scenarios, we extend the number of mixed scenarios, for example SC123(SC1 + SC2 + SC3), SC134(SC1 + SC3 + SC4)mixed three scenarios, finally used all scenarios SC1234 (SC1 and SC2 + SC3 + SC3) to test our approach. The results are shown in Fig. 11.9. The confusion matrix of action recognition from the mixed multiple scenarios remained at about 94%. In order to more clearly verify the adaptability of our approach to mixed scenarios and make a comparison, we drew the results in the form of broken line graph (Confusion SC in Fig. 11.9). The proposed approach in this chapter can get better recognition rate even

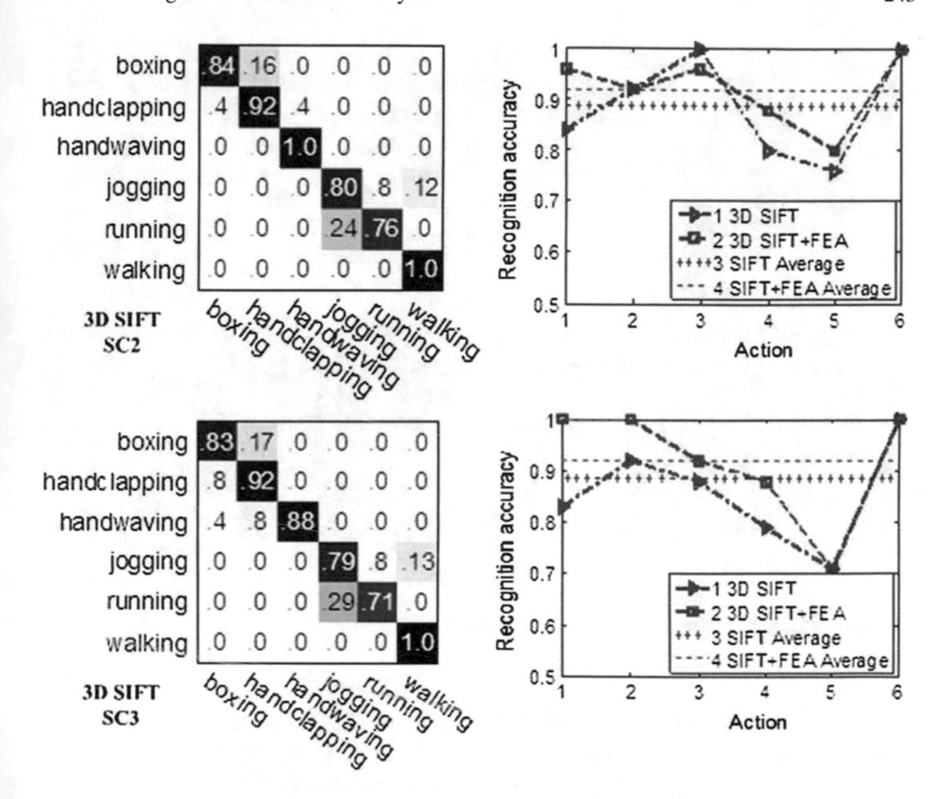

Fig. 11.7 Contrast graph and the confusion matrixes with 3D SIFT

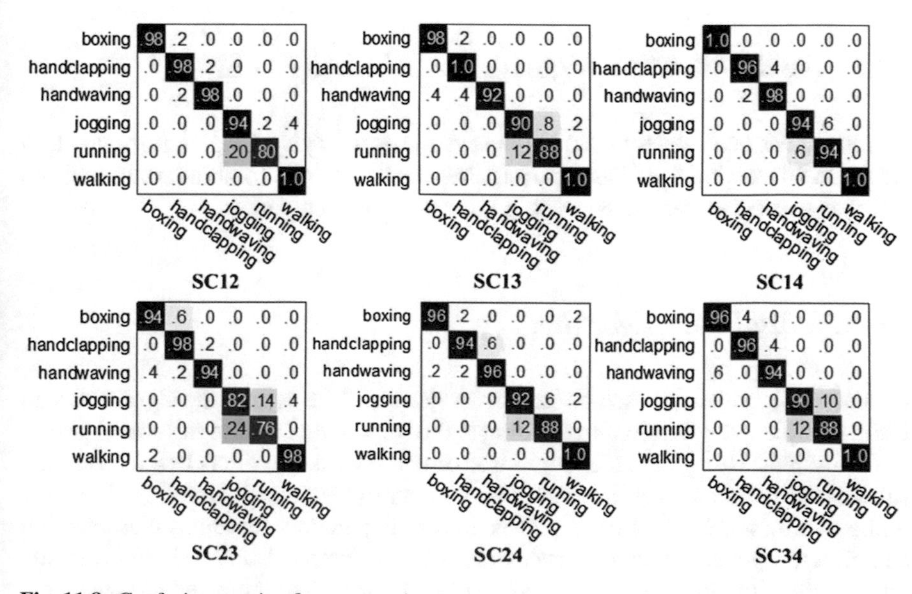

Fig. 11.8 Confusion matrix of two scenarios

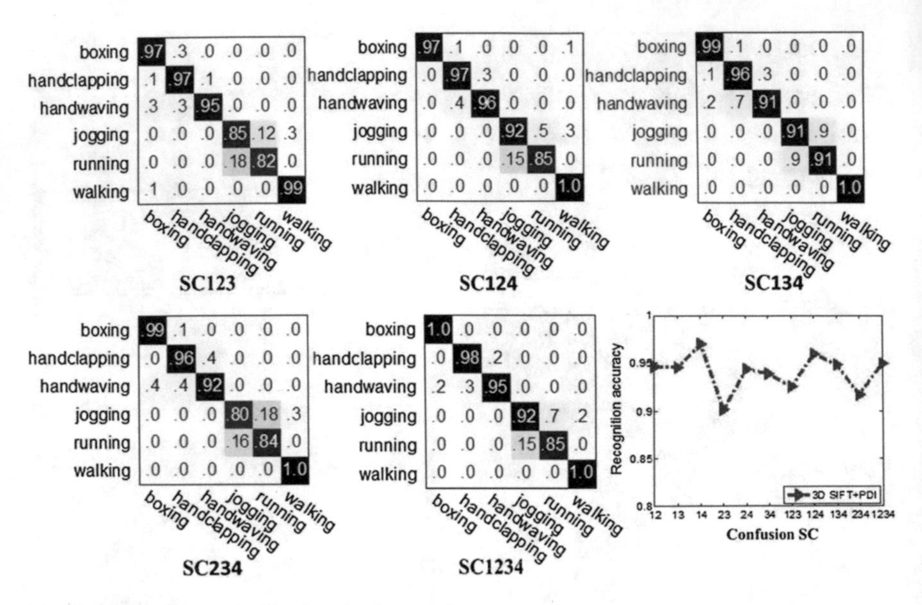

Fig. 11.9 Confusion matrixes in mixed scenarios

in a complex confusion of scenarios. Furthermore it is also found that the actions can be captured in a stable scenario as action training samples, then those samples can be utilised for action recognition in unstable environment.

11.5 AdaBoost-SVM Algorithm and Results Analysis

AdaBoost has been widely used to improve the accuracy of any give learning algorithm. In this section, we focus on designing an algorithm to combine AdaBoost and SVM as weak classifiers to be used in human action recognition.

11.5.1 AdaBoost Algorithm

AdaBoost algorithm is based on iterative algorithm. The core of the algorithm is to train multiple weak classifiers according to the sub training data. AdaBoost classifier is an ensemble strong classifier composed of many weak classifiers based on certain rules. AdaBoost constructs a composite classifier by sequentially training classifiers while putting more and more emphasis on certain patterns. AdaBoost classification can effectively exclude some unnecessary training sample data, and then locate the most critical training sample to improve the accuracy of training and save resource consumption.

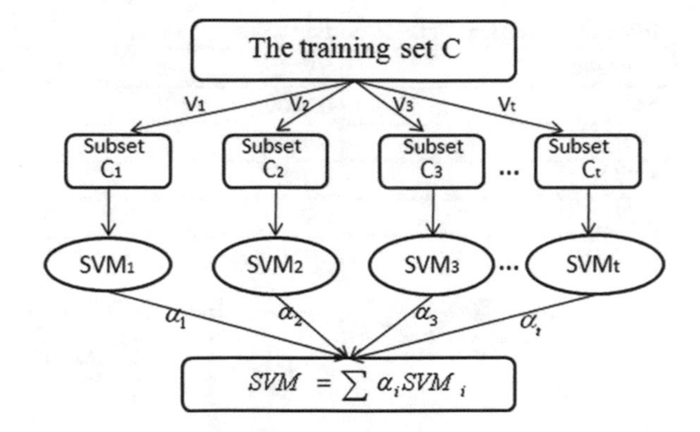

Fig. 11.10 Schematic diagram of AdaBoost Algorithm

The schematic diagram of AdaBoost algorithm is shown in Fig. 11.10. AdaBoost algorithm achieve new training subset by calculating and adjusting the weight of each sample in the iteration process. Given a set of training sample C, AdaBoost maintains a probability distribution V_1. This distribution is initially uniform. AdaBoost algorithm calls Weak Learn algorithm repeatedly in a series of cycle. After the first iteration, new training subset C_1 is obtained. The initial weak classifier $f_1(x)$ is achieved by learning the corresponding training subset C_1. In boosting learning, each example is associated with a weight, and the weights are updated dynamically using a multiplicative rule according to the errors in previous learning so that more emphasis is placed on those examples which are erroneously classified by the weak classifiers learned previously. After t iterations, we can obtain t weak classifiers. The stronger classifier is a linear combination of t weak classifiers. In our method, we choose SVM as weak classifiers.

11.5.2 Experimental Results

In this part the action recognition performance of the AdaBoost-SVM is assessed by using combined feature of 3D SIFT and PDI on four scenarios (SC1, SC2, SC3 and SC4) according to KTH dataset. Leave-one-out cross validation method is adopted throughout the process, in turns using six action of each actor as test samples, and the rest of all the actions as the training, circulation continued until all actions are completed testing. Comparison is also made with the performance of the single SVM classifiers, as shown in Table 11.3 and Fig. 11.11.

Compared with the recognition rate in Table 11.3, the recognition accuracy is greatly improved with the application of AdaBoost-SVM classifiers. SC1 and SC4 are more stable than the other two scenarios. We obtained almost 100% correct recog-

Table 11.3 Testing results in portion scenario

Scenario	SVM	AdaBoost-SVM
SC1	0.9600	0.9933
SC2	0.9200	0.9667
SC3	0.9167	0.9792
SC4	0.9600	1.000

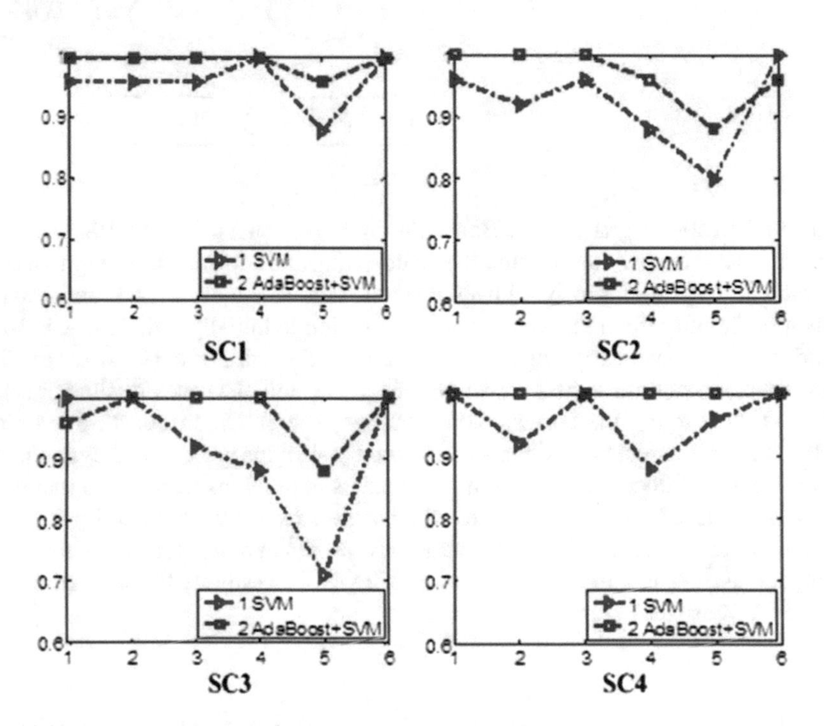

Fig. 11.11 The comparison graph of action recognition rate

nition rate in those two scenarios. Although SC2 and SC3 are relatively complex, the proposed AdaBoost-SVM method can respectively increase the recognition rate by 4 and 6% than traditional SVM method. Compared with the recognition rate in Fig. 11.11, for any action class, the performance of the proposed method is significantly better than the traditional SVM algorithm.

The comparisons of performance between the proposed method and the recent related works based on KTH dataset are shown in Table 11.4. These works are all related to the local spatio-temporal feature. It is worth noting that our method outperforms all of other state of art methods.

Table 11.4 Comparison with related work in recent years

Literature	Method	Accuracy
Niebles [14]	3D SIFT BOW+pLSA	83.33%
Klaser [19]	3D Gradients+SVM	91.4%
Bregonzie [22]	Interest point clouds+NNC	93.17%
Umakanthan [24]	HOG3D+ SVM	92.7%
Our approach	3D SIFT+PDI + SVM	94.92%
Our approach	3D SIFT+PDI + Adaboost-SVM	98.48%

11.6 Conclusion

This chapter proposed a novel video descriptor by combining local spatio-temporal feature and global positional distribution information (PDI) of interest points. Considering that the distribution of interest points contains rich motion information and also reflects the key position in human action, so we combine the 3D SIFT with PDI to achieve a more complete representation of the human action. In order to obtain compact description and efficient computation, the combined features are processed by dimension reduction, feature quantisation and feature selection. Eventually, compared with previous works of 3D SIFT descriptor, the proposed approach further improved the recognition rate. In future work, our approach will be applied to more complex datasets and applications and provide an additional performance improvement [25].

References

1. H.J. Seo and P. Milanfar. Action recognition from one example. *IEEE Transactions on Pattern Analysis and Machine Intelligence*, 33(5):867–882, 2011.
2. D. Weinland, R. Ronfard, and E. Boyer. A survey of vision-based methods for action representation, segmentation and recognition. *Computer Vision and Image Understanding*, 115(2):24–241, 2011.
3. X. Ji and H. Liu. Advances in View-Invariant Human Motion Analysis: A Review. *IEEE Transactions on Systems, Man and Cybernetics Part C*, 40(1):13–24, 2010.
4. X. Li. Hmm based action recognition using oriented histograms of optical flow field. *Electronics Letters*, 43(10):560–561, 2007.
5. S. Ali and M. Shah. Human action recognition in videos using kinematic features and multiple instance learning. *IEEE Transactions on Pattern Analysis and Machine Intelligence*, 32(2):288–303, 2007.
6. M. Hahn, L. Krüger, and C. Wöhler. 3d action recognition and long-term prediction of human motion. *Computer Vision Systems*, pages 23–32, 2008.
7. F. Jiang, Y. Wu, and A. K. Katsaggelos. A dynamic hierarchical clustering method for trajectory-based unusual video event detection. *IEEE Transactions on Image Processing*, 18(4):907–913, 2009.

8. H. Zhou, H. Hu, H. Liu, and J. Tang. Classification of upper limb motion trajectories using shape features. *IEEE Transactions on Systems, Man, and Cybernetics, Part C: Applications and Reviews*, 42(6):970–982, 2012.
9. X. Cao, B. Ning, P. Yan, and X. Li. Selecting key poses on manifold for pairwise action recognition. *IEEE Transactions on Industrial Informatics*, 8(1):168–177, 2012.
10. A.A. Chaaraoui, P. Climent-Pérez, and F. Flórez-Revuelta. Silhouette-based human action recognition using sequences of key poses. *Pattern Recognition Letters*, 34(15):1799–1807, 2013.
11. R. Poppe. A survey on vision-based human action recognition. *Image and Vision computing*, 28(6):976–990, 2010.
12. P. Dollár, V. Rabaud, G. Cottrell, and S. Belongie. Behavior recognition via sparse spatio-temporal features. In *Proceeding of Joint IEEE International Workshop on Visual Surveillance and Performance Evaluation of Tracking and Surveillance*, pages 65–72. Beijing, 2005.
13. I. Laptev and T. Lindeberg. Local descriptors for spatio-temporal recognition. *In Proceeding of First Workshop on Spatial Coherence for Visual Motion Analysis, Springer*, pages 91–103, 2006.
14. J.C. Niebles, H. Wang, and L. Fei-Fei. Unsupervised learning of human action categories using spatial-temporal words. *International Journal of Computer Vision*, 79(3):299–318, 2008.
15. J. Zhu, J. Qi, and X. Kong. An improved method of action recognition based on sparse spatio-temporal features. *Artificial Intelligence: Methodology, Systems, and Applications*, pages 240–245, 2012.
16. P. Liu, J. Wang, M. She, and H. Liu. Human action recognition based on 3d sift and lda model. In *Proceeding of IEEE Workshop on Robotic Intelligence In Informationally Structured Space (RiiSS)*, pages 12–17, 2011.
17. X. Jiang, T. Sun, B. Feng, and C. Jiang. A space-time surf descriptor and its application to action recognition with video words. In *Proceeding of Eighth International Conference on Fuzzy Systems and Knowledge Discovery*, pages 1911–1915. vol.3, 2011.
18. P. Scovanner, S. Ali, and M. Shah. A 3-dimensional sift descriptor and its application to action recognition. In *Proceeding of the 15th international conference on Multimedia*, pages 357–360. ACM, 2007.
19. A. Kläser, M. Marszałek, C. Schmid, and L. Lear. A spatio-temporal descriptor based on 3d-gradients. *In Proceeding of British Machine Vision Conference, UK*, pages 1–10, 2008.
20. G. Willems, T. Tuytelaars, and L. Van Gool. An efficient dense and scale-invariant spatio-temporal interest point detector. In *Proceeding of European Conference on Computer Vision, Springer,France*, pages 650–663, 2008.
21. F. Li, C. Xiamen, and J. Du. Local spatio-temporal interest point detection for human action recognition. In *IEEE 5th International Conference on Advanced Computational Intelligence*, pages 1–10, 2012.
22. M. Bregonzio, S. Gong, and T. Xiang. Recognising action as clouds of space-time interest points. In *Proceeding of IEEE Conference on Computer Vision and Pattern Recognition*, pages 1948–1955, Florida, USA, 2009. IEEE.
23. C.C. Chang and C.J. Lin. Libsvm: A library for support vector machines. *ACM Transactions on Intelligent Systems and Technology*, 2:27:1–27:27, May 2011.
24. S. Umakanthan, S. Denman, S. Sridharan, C. Fookes, and T. Wark. Spatio temporal feature evaluation for action recognition. In *Proceeding of International Conference on Digital Image Computing Techniques and Applications*, pages 1–8, 2012.
25. Jose M Chaquet, Enrique J Carmona, and Antonio Fernández-Caballero. A survey of video datasets for human action and activity recognition. *Computer Vision and Image Understanding*, 117(6):633–659, 2013.

Chapter 12
A View-Invariant Action Recognition Based on Multi-view Space Hidden Markov Models

12.1 Introduction

In recent years, visual-based action recognition has already been widely used in the human-machine interfaces [1]. More natural and intelligent communication between human and machines can be achieved by the human action recognition and behaviour interpretation [2–4]. A challenging issue in this research field is the diversity of action information which originates from different camera viewpoints [5]. A reliable and generic action recognition system should be robust to camera parameters and viewpoints while observing an action sequence [6–8]. That is to say, the recognition methods should exhibit some view-invariance.

Recent research on view-invariant human action recognition can be characterised by two classes of methods: template matching methods and state-space approaches [5]. Template based approached focus on extracting low-level image features which are then compared it to the pre-stored action prototypes during recognition. The view constraint is mostly removed during feature extraction in these recognition methods. A kind of template-based methods directly choose the view-independent motion features [6, 9–11]. For example, Weinland et al. [9] extracted the view-invariant features from Fourier space of Motion History Volumes (MHVs) for action recognition. The advantages of this kind of method are low computational cost and simple implementation. However, they are usually more sensitive to noise and variance of the time interval of the movements. Another kind of template-based methods estimates the parameters of camera viewpoint according to the human body movement direction. Then the observation information from every frame is projected into specific orthogonal space for regularising so that the human action viewpoint can be normalised. Rogez et al. [12] estimated the 3D principal directions of man-made environments and the direction of motion, then transformed both 2D-Model and input images to a common frontal view before the fitting process. Though this kind of approaches can remove the viewpoint effect directly, the recognition results completely depend on the robustness of the body orientation estimation. Furthermore, the computational cost is significantly high.

H. Liu et al., *Human Motion Sensing and Recognition*,
Studies in Computational Intelligence 675, DOI 10.1007/978-3-662-53692-6_12

State-space methods, e.g., Hidden Markov Model(HMM) and its variants have been the dominant tools in view-invariant human motion modelling. Nowadays, there are usually two kinds of state-space recognition methods to solve the viewpoint issue. The former approaches recover the 3D human pose from the motion sequence by directly using the 3D tracking technology. And then these 3D pose representations are utilised as the state information of the graph model for action training and recognising [13–17]. Lv and Nevatia [13] exploited a large number of HMMs to model 3D human joints. And each HMM corresponds to the motion of a single joint or combination of related multiple joints. In the recognition process, multiple weak classifiers based on HMMs are combined by the AdaBoost algorithm to recognise actions.The proposed algorithm is effective in that the results are convincing in recognising 22 actions. Peursum et al. [14] propose a variant of the hierarchical HMMs to achieve human action modelling and human body tracking simultaneously. This model can be broken down into two levels: lower-level corresponds to the sub-actions modelling based on 3D human pose; higher-level corresponds to actions modelling within each sub-actions. This kind of method can fundamentally remove the viewpoint constraint by using 3D pose information as action feature, however the accuracy of the recognition is greatly affected by the performance of 3D tracking algorithm. Furthermore 3D tracking methods always involve huge computational cost, which limits the realtime application.

The latter approach based on state-space usually represents the human static posture by using multi-view posture images of the human action. The constraints on transition of synthetic poses and different viewpoints are represented by a graphical model [18–20]. For example, Lv et al. [18] capture the key poses by clustering 3D motion capture data of the different human actions. Then the key poses are rendered from a variety of viewpoints. The silhouette matching between the input frames and the key poses is performed by an enhanced Pyramid Match Kernel (PMK) algorithm. An action net is used to model the contextual constraints imposed by actions. The best-matched sequence of actions is then tracked by using the Viterbi algorithm. This approach was tested on a challenging video sets consisting of 15 complex action classes. However the accuracy of the result strongly depends on the viewpoint similarity between the input images and training set. So a large number of training data is needed to cover the possible camera viewpoints.

Above methods are based on traditional cameras as sensors, some researchers adopt depth cameras to develop solution to view-invariant human action recognition.The release of the Microsoft Kinect can provide both an RGB image and depth image streams. One kind of methods directly extract the 3D skeletal joint location from the action depth sequences to obtain view-invariant human action recognition. Shotton et al. [21] proposed a method to quickly and accurately estimate 3D positions of skeletal joints from a single depth image from Kinect. They provide accurate estimates of 3D skeletal joint locations at 200 frames per second on the Xbox 360 GPU. Xia et al. [22] extracted the 3D skeletal joint locations from Kinect depth maps by using Shotton et al.'s method, then use the histograms of 3D joint locations (HOG3J) as a compact representation of posture. Finally model the human action by using discrete hidden Markov models. The other kind of methods directly extract view-invariant action feature descriptors from RGBD sequence to achieve view-invariant human action recognition [23, 24]. For example, Yan et al. [25] proposed the unique robust Self-Similarity Matrix (SSM) descriptors by combining the HOG descriptor for RGB data with MHIs descriptor of the variations for depth frames as the view-invariant feature representation. Then utilised a novel Multi-Task Linear Discriminant Analysis (MT-LDA) algorithm to learn the dependencies between different views. The experimental results on the publicly available dataset demonstrate the effectiveness of the proposed method.

There are a large number of variation should be considered in view-invariant action recognition, such as different actions, different actors, viewpoint changing etc. It involves a large parameter space, requiring substantial amounts of training data and with not high classification accuracies for real application. In this chapter, a novel multi-view space HMMs algorithm for view-invariant action recognition based on view space partitioning is proposed. A view-insensitive feature representation and a typical state-space method are effectively combined. And the viewpoint issue is solved from two levels: feature extraction level and action recognition level. The feature representation is combination of the bag-of-words of interest point in shot length-based video and the grid-based amplitude histogram of optical flow. The combined feature has been verified to be insensitive for viewpoint changing [26]. During action training process, the view space is partitioned into finite sub-view spaces according to the camera rotation viewpoint. The sub-HMMs corresponding to each human action model are trained in the sub-view space. The model training process can be greatly simplified by decomposing the parameter space into multiple sub-view spaces. During the recognition process, the sub-view space information is obtained by calculating the probabilities between the test sequence and the each sub-HMM. Then the actions from arbitrary viewpoint can be effectively recognized by fusing the sub-view space information.

In order to assess the discriminative power of the proposed method, it was tested on a standard dataset, i.e., the multiple view IXMAS dataset. The experimental results demonstrated the performance of the proposed method is better than the other methods.

12.2 Framework of the Proposed Method

The framework of the approach is sketched in Fig. 12.1, which includes the following modules:

1. View Space Partitioning: In order to improve the recognition accuracy of view-independent action recognition, the parameter space is decomposed according to viewpoint. The 360° pan angle of human body around the camera is equally partitioned into V pieces. Each piece will be regarded as a sub-view space. The training data for each action class (H classes in total) can be collected from

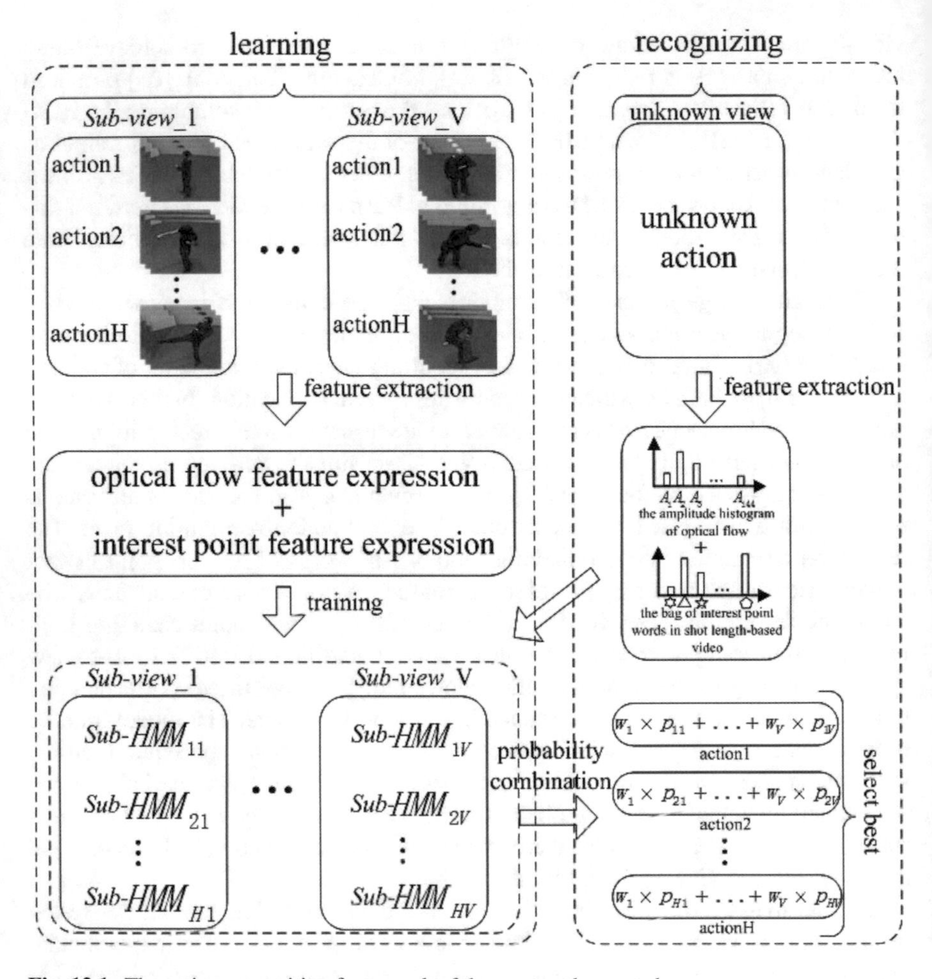

Fig. 12.1 The action recognition framework of the proposed approach

different actors in the sub-view space. In other words, it is not necessary to synchronously collect the training data by using multiple cameras.

2. Feature Extraction: The combined features of the bag-of-words of interest point in shot length-based video and the grid-based amplitude histogram of optical flow are selected to represent the human action sequences. The combined features are robust to the viewpoint change in the sub-view space. In the process of feature extraction, the frames with less dynamic information are removed by using the interest point detection method.
3. Model Training: The sub-models for each action are built by using the combined features which have been extracted from each sub-view space (each sub-view space contains H action models). The HMM is chosen as the action model, because

it could capture the dynamics of the human action and has the robustness to noise, variance of the time interval of the movements and small viewpoint change.

4. Action Recognition: When the test action sequence with unknown viewpoint is given, feature extraction is performed. Then the probabilities of the test sequence for the sub-HMMs are computed. By analysing the discrimination among the action models in the sub-view space and the similarity between the sub-view space and the test sequence viewpoint, the probabilities of the test sequence for each sub-HMM corresponding to the same action are weighted and combined. At last, the test action is recognised by using the maximum likelihood criterion.

12.3 Feature Extraction

In order to improve the robustness of the method to viewpoint changing, a view-insensitive motion feature is chosen as the motion feature. It is the combination of the bag-of-words of interest point and the grid-based amplitude histogram of optical flow. The combined feature has been verified to be insensitive for viewpoint changing in our previous work [26]. The detail of the feature is shown as following.

12.3.1 Interest Point Feature Extraction

The conventional bag-of-words is usually extracted throughout the whole video. So it usually can't associate with HMMs due to the lack of the dynamic information in temporal domain. To address this problem, the bag-of-words in shot length-based video is proposed to improve the temporal characteristic and the contextual semantic. The detailed process is outlined in the following (see Fig. 12.2): firstly, the frames which contain rich motion information are found by interest point detection. Then the 3D Scale-invariant Feature Transform (3D SIFT) descriptor is utilised to represent the interest points in these frames. The 3D SIFT feature is verified it can capture rich local motion features and is robust to minor variations of viewpoint [26]. The 3D SIFT descriptors are clustered into a C-dimension vocabulary. When the new frame of test sequence is input, the bag-of-word of interest point (C-dimension) in the neighbour F frames is used to represent the current frame feature. The descriptors are projected into the C-dimension vocabulary by minimising the Euclidean distance between the descriptors of current neighbour frames and the vocabulary words. Finally, the occurrence frequencies of the vocabulary words are counted as the bag-of-words representation of interest point for the current frame. In order to investigate the effect of the different values of C and F to the system performance, the algorithm was tested with $C = 30$, 45, 75, 100, 144 and $F = 2$, 3, 4, 5. The results have shown that the combined feature reached the best performance with $C = 60$ and $F = 4$. Consequently only the results for these values are listed.

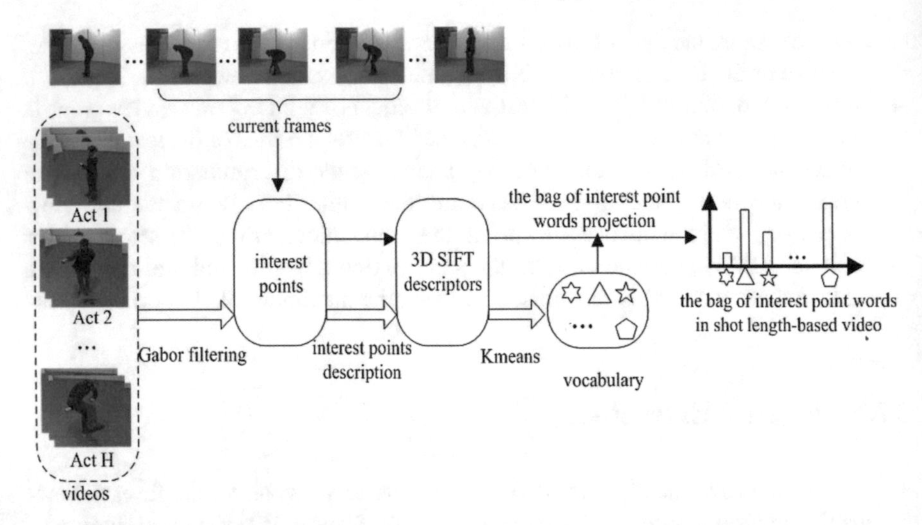

Fig. 12.2 Bag-of-words of interest point feature extraction in shot length-based video

The interest point is referred to the changing point in spatio-temporal domain which can be distinguished from other points within a neighbourhood. Thus the interest points can be used as a feature to represent the action sequence. Dollar detector [27] is tried to be improved inspired by the work of Bregonzio [28]. The region of interest detected by the frame differencing algorithm is filtered using 2D Gabor filters from different orientation (0°, 22°, 45°, 67° and 90° orientations selected). Then combination of different orientation filtering responses is used for the final detection.

To ensure the accuracy of interest point detection, the detected interest points are selected for the second time. The coordinates of the centre of the interest points in the current frame is obtained by calculating the mean position information of the interest points distributed in the n neighbour frames. Then the Euclidean distance between the position of detected interest point and the centre position is used as the dynamic criterion (the point will be removed, if the distance is greater than the fourfold average distance) to check the point for 'true or false'.

12.3.2 Optical Flow Feature Extraction

The global motion information could not be effectively represented by the bag-of-words of interest point in shot length-based video. Since the optical flow contains more motion information and is more robust to the complex environments. The optical flow feature is used to combine with interest point feature. The optical flow images are also extracted from the frames with rich motion information found by the interest point detection. Then the computation complexity of optical flow feature and the redundancy can be reduced in this way.

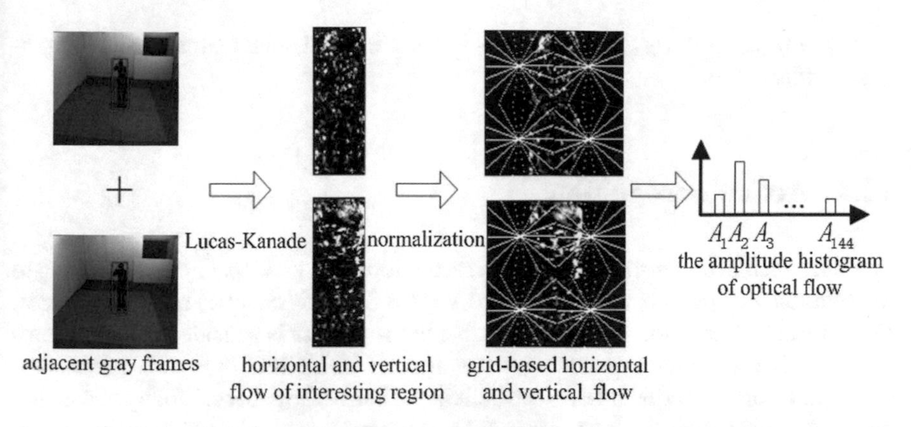

Fig. 12.3 Grid-based amplitude histogram of optical flow feature extraction

Firstly, optical flow images for horizontal and vertical channels are calculated in the interest regions of the adjacent grey frames by using Lucas-Kanade algorithm. Secondly, considering the different sizes of the interest region in every frame, normalisation is utilised to scale the interest region into a fixed $p \times q$ ($p = q = 120$, in this chapter) dimension by employing the bilinear interpolation. In order to increase the anti-disturbance ability, the grid based method [29] is adopted to divide each normalised optical flow image into 2×2 sub-windows. Then a 18-dimension radial histogram of the optical flow amplitude is computed by calculating optical flow amplitude in 18 pie slices. Finally, each frame can be represented by a 144-dimension ($2 \times 2 \times 2 \times 18$) optical flow feature. The process of grid-based optical flow feature extraction is shown in Fig. 12.3.

12.3.3 Mixed Feature

To make the proposed approach more robust to complex environments and viewpoint changing, the bag-of-words of interest point in shot length-based video are combined with the grid-based amplitude histogram of optical flow. The combination details are shown as Eq. 12.1. The combined features can not only capture the global motion information, but also achieve robustness to occlusions and viewpoint changes. Finally each frame will be represented by a 204-dimension ($144 + 60$) mixed feature.

$$O_t = [O_{Fx}, O_{Fy}, O_H] \tag{12.1}$$

where $O_{Fx} = [o_{Fx_{1,1}}, o_{Fx_{1,2}}, \ldots, o_{Fx_{1,18}}, \ldots, o_{Fx_{4,1}}, o_{Fx_{4,2}}, \ldots, o_{Fx_{4,18}}]$ is the horizontal optical flow feature;

$O_{Fy} = [o_{Fy_{1,1}}, o_{Fy_{1,2}}, \ldots, o_{Fy_{1,18}}, \ldots, o_{Fy_{4,1}}, o_{Fy_{4,2}}, \ldots, o_{Fy_{4,18}}]$ is the vertical optical flow feature;

$O_H = [o_{H_1}, o_{H_2,...,} o_{H_{60}}]$ is the bag-of-words of interest point feature in shot length-based video.

12.4 Action Recognition

In order to achieve view-invariant action recognition, the view space corresponding to the rotation viewpoint is partitioned into V (V = 8, in this chapter) sub-view spaces. Only the actor's rotation viewpoint relative to the camera is considered. This means that the view-variation in each sub-view space will be $360°/8 = 45°$. After view space partitioning, action models are built in each sub-view space. Finally, the action with unknown viewpoint is recognised by combining the weighted probabilities of the test sequence for each sub-HMM in the multi-view spaces.

12.4.1 *Sub-view Space HMM Building*

The recognition algorithm of HMM considers the dynamic process of human motion and takes the probability method to model the changes on temporal scale and spatial scale. Meanwhile, the temporal characteristic and contextual semantic information of our features are considered in this chapter.

Supposing that $\lambda^{(1,v)}, \lambda^{(2,v)}, \ldots, \lambda^{(H,v)}$are the trained HMMs of action1, action2,. . ., actionH in the current v sub-view space. When a test observation sequence $O = \{o_1^{test}, o_2^{test}, \ldots o_T^{test}\}$ is given, the probability of the test sequence (i.e., observation probability) for the given sub-HMM is computed by using Forward-Backward algorithm.

$$p\left(O^{test}|\lambda^{(1,v)}\right), p\left(O^{test}|\lambda^{(2,v)}\right), \ldots, p\left(O^{test}|\lambda^{(H,v)}\right) \quad (12.2)$$

where v is the order number of the current sub-view space; H is the total class number of action model in the current v sub-view space.

The essence of HMMs training problem with the given structure (5 states and full-connected HMMs in this chapter) is to maximise the observation probability by adjusting the model parameter for the observation sequence. Baum-Welch algorithm is commonly used to obtain the optimal model parameters. However, it's performance is depended on the choice of initial parameter. If the improper initial parameters are chosen, it can lead procedure to the local minimum so that the action model can't be optimal. Thus, the result of kmeans algorithm is taken as initial input of the Baum-Welch algorithm. Then the Gaussian probability density function is used for computing the probability of generating observation symbol from each state.

12.4.2 Multi-view Space HMMs Probability Fusion

According to the experiments, any action sequence in the dataset has a higher observation probability with itself action model no matter in whatever sub-view space. It demonstrates that the observation probabilities of the sub-HMMs corresponding to the test action class have great contribution to the recognition. The weighted fusion method is used for effectively combining the sub-HMMs information of the given action in each sub-view space. The likelihoods between test sequence and each action class in the whole multi-view space are represented by the summations of the weighted observation probabilities of the corresponding action sub-models in the sub-view space. The test action is recognised as the action class with the highest likelihood, as shown in Eq. 12.3,

$$action\_class = \arg\max_{1\leq h\leq H}\left(\sum_{1\leq v\leq V}^{v} w_v \frac{p(O^{test}|\lambda^{(h,v)})}{\sum_{1\leq h\leq H}^{h} p(O^{test}|\lambda^{(h,v)})}\right) \tag{12.3}$$

where w_v is the sub-view space weight (e.g., $inter - w_v$, $intra - w_v$, $mean - w_v$).

In the recognition process, larger weight usually will be assigned to the sub-view space with better recognition performance. There are two reasons to be considered for selecting the sub-view space weight: (1) The similarity between the sub-view space and the test sequence viewpoint. If the current sub-view space has a higher similarity with the test sequence viewpoint, then the current sub-view space weight will be increased. (2) The discrimination among action models in the sub-view space. If the discrimination among the action models in the current sub-view space is better, then the current sub-view space weight will be improved. Four different weight selecting algorithms are used based on above reasons.

(1) No sub-view space weight: The likelihood between test sequence and each action class in the whole multi-view space is computed by directly summing the observation probability of the corresponding action sub-model in each sub-view space. The action class with the highest likelihood is chosen as the test action class, as shown in Eq. 12.4,

$$action\_class = \arg\max_{1\leq h\leq H}\left(\sum_{1\leq v\leq V}^{v} \frac{p(O^{test}|\lambda^{(h,v)})}{\sum_{1\leq h\leq H}^{h} p(O^{test}|\lambda^{(h,v)})}\right) \tag{12.4}$$

(2) The sub-view space inter-weight: The sub-view space inter-weight is measured by the similarity between the sub-view space and the test sequence viewpoint. If the summation of observation probabilities for each action sub-model is large in the current sub-view space, the similarity between the current sub-view space and the test sequence viewpoint is usually high. Therefore, the summation of observation probabilities in the sub-view space is chosen as the criterion function of the sub-view space inter-weight,as shown in Eq. 12.5,

$$inter\_w_v = \frac{\sum_{1\le h\le H}^{h} p(O^{test}|\lambda^{(h,v)})}{\sum_{1\le v\le V}^{v}\sum_{1\le h\le H}^{h} p(O^{test}|\lambda^{(h,v)})} \tag{12.5}$$

(3) The sub-view space intra-weight: The sub-view space intra-weight is measured by the discrimination among the action models in certain sub-view space. The actions with similarly observation probabilities are usually mis-recognised each other, such as wave and scratch head; walk and turn around, etc. This phenomenon suggests that we should pay more attention to the discrimination among the action models in the sub-view space. Thus, the difference of observation probabilities corresponding to the similar action models in the sub-view space is selected as the criterion function of the sub-view space intra-weight, as shown in Eq. 12.6,

$$intra\_w_v = \frac{\max\limits_{1\le i\le H}(p(O^{test}|\lambda^{(i,v)})) - \frac{1}{N}\sum_{1\le n\le N}^{n} p(O^{test}|\lambda^{(n,v)})}{\max\limits_{1\le i\le H}(p(O^{test}|\lambda^{(i,v)})) + \frac{1}{N}\sum_{1\le n\le N}^{n} p(O^{test}|\lambda^{(n,v)})}$$

$$n \in A = \{j|\max_{1\le i\le H}(p(O^{test}|\lambda^{(i,v)})) - p(O^{test}|\lambda^{(j,v)}) \le th, \tag{12.6}$$

$$1 \le j \le H, j \ne \arg\max_{1\le i\le H}(p(O^{test}|\lambda^{(i,v)}))\}$$

where th is the observation probability threshold among the similar action models. In our experimental setup, action classes with the two highest probabilities are directly chosen as the similar classes in the current sub-view space. The value of the th is not fixed.

(4) The sub-view space mixed-weight: In order to recognise the action with unknown viewpoint more effectively, the inter-weight and the intra-weight methods are combined in this section. The mixed-weight could simultaneously give consideration to the similarity between the sub-view space and the test sequence viewpoint and the discrimination among action models. The detail of selecting the sub-view space mixed-weight is given by the following formulas.

$$mean\_w_v = \frac{\max\limits_{1\le i\le H}(p(O^{test}|\lambda^{(i,v)})) - \frac{1}{H-1}\sum_{1\le j\le H}^{j} p(O^{test}|\lambda^{(j,v)})}{\sum_{1\le v\le V}^{v}\sum_{1\le h\le H}^{h} p(O^{test}|\lambda^{(h,v)})}$$

$$j \ne \arg\max_{1\le i\le H}(p(O^{test}|\lambda^{(i,v)})) \tag{12.7}$$

12.5 Experiments

The proposed method is tested on the Inria Xmas Motion Acquisition Sequences (IXMAS) multi-view action dataset, which contains twelve daily-live actions. Each action is performed three times by twelve actors and recorded simultaneously from five different cameras: four side cameras and one top camera. During the action

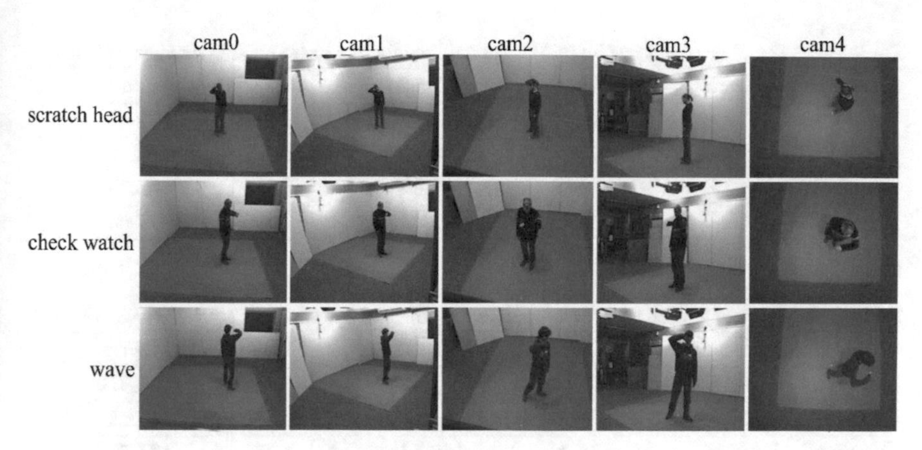

Fig. 12.4 Exemplar frames from IXMAS multi-view dataset

performing, the viewpoint of human body relative to each camera is not restricted. Therefore, it is widely used to test the performance of view-invariant action recognition algorithms. Eleven actions are chosen for verifying our approach during the experiment, namely, check-watch, cross-arms, scratch-head, sit-down, get-up, turn-around, walk, wave, punch, kick and pick-up. So there are $5 * 11 * 3 * 12$ action samples for the experiments. The test method of leave one actor out (LOAO) is used in every experiment. It involves employing a group of videos from a single actor as the testing data and the remaining videos as the training data. This is repeated so that each group of videos in the dataset is used once as the training data. View clusters on action data under every camera are first applied. Then data under each camera are clustered into three sub-view space data to verify the proposed method. Each column in Fig. 12.4 is the exemplar frames under one camera. In each camera, there are three row exemplar frames which are corresponding to the three sub-view space data clustered previously.

Experiment 1 view-insensitive performance of the proposed feature: The robustness of the proposed feature to viewpoint changing is helpful to improve the performance of recognition method. Firstly the view-insensitive performance of the proposed feature is tested in the sub-view space. In this experiment, the training data and the testing data are taken from the same sub-view space data of the same camera which are clustered previously. There are $360°/8 = 45°$ rotation viewpoint changing in every sub-view space. So it is suitable to verify the view-insensitive performance of the proposed feature. The optical flow based feature, the interest point based feature and the mixed feature are respectively tested with the HMMs. The recognition results of optical flow feature based on HMMs in each sub-view space taken from 5 cameras are shown in Table 12.1.

The recognition results of interest point feature based on HMMs in each sub-view space taken from 5 cameras are shown in Table 12.2.

Table 12.1 The recognition results (%) of optical flow feature based on sub-view space HMMs

Subview	Cam0	Cam1	Cam2	Cam3	Cam4
Sub-view1	77.27	80.31	84.85	77.28	72.72
Sub-view2	90.91	90.91	90.91	89.09	84.55
Sub-view3	90.91	88.89	88.89	83.84	75.76
Average	86.36	86.70	88.22	83.40	77.68

Table 12.2 The recognition results (%) of interest point feature based on sub-view space HMMs

Subview	Cam0	Cam1	Cam2	Cam3	Cam4
Sub-view1	60.61	65.15	43.94	68.18	40.91
Sub-view2	79.09	79.09	65.45	84.55	44.55
Sub-view3	66.67	75.76	52.53	76.77	51.52
Average	68.79	73.33	53.97	76.50	45.66

Table 12.3 The recognition results (%) of mixed feature based on sub-view space HMMs

Subview	Cam0	Cam1	Cam2	Cam3	Cam4
Sub-view1	84.84	78.79	86.36	83.33	78.79
Sub-view2	93.64	91.82	95.45	89.09	90.91
Sub-view3	95.96	93.94	86.87	87.88	84.85
Average	91.48	88.18	89.56	86.77	84.85

The recognition results of mixed feature based on HMMs in each sub-view space taken from 5 cameras are shown in Table 12.3.

The above experiment results show that the optical flow based features have better view-insensitive performance than the interest point based features. The combined features using both interest point based features and optical flow based features provide a richer representation of human actions, so the accuracy of the action recognition method is significantly improved.

Experiment 2 the view-invariant action recognition under particular camera: The viewpoint variation of the human body relative to the camera can be divided into two types: rotation variation and pitch variation. In the dataset, the pitch viewpoint variation of the human body relative to the particular camera is relatively small. In this experiment, the performance of the multi-view space HMMs for view-invariant action recognition under particular camera is tested. Moreover, the contribution of different sub-view space weights into recognition performance are explored. The experimental data respectively used for model training and testing are taken from the same camera data(the testing actor sequences are not included in the training data). That means the pitch viewpoint is fixed. However, the rotation viewpoint of human body relative to the camera is not fixed and exists certain change as shown in Fig. 12.4. Hence, the performance of the proposed method to the rotation viewpoint changing

Table 12.4 The recognition results (%) of different sub-view space weights under particular camera

Parameter	Cam0	Cam1	Cam2	Cam3	Cam4	Average
$without - w_v$	89.26	87.68	86.84	80.61	78.99	84.68
$inter - w_v$	89.26	88.01	87.34	80.61	78.99	84.84
$intra - w_v$	89.56	87.51	89.93	86.43	81.55	87.00
$mean - w_v$	90.23	88.65	91.15	87.54	82.76	88.07

can be verified in this experiment. The recognition rates for different sub-view space weights and each camera are listed in Table 12.4.

The recognition rates of the $inter - w_v$ based algorithm and the $intra - w_v$ based algorithm are increased on the basis of directly combination. It is enough to prove that the probability of the test sequence in different sub-view space provides effective information for view-invariant recognition by the properly operation. In addition, the results of the $mean - w_v$ based recognition algorithm are further improved due to introducing the similarity of viewpoints and the discrimination of different actions simultaneously.

Experiment 3 the view-invariant action recognition under different cameras: On the basis of the above experiment, the performance of the proposed method for the real view-invariant recognition (rotation-invariant and pitch-invariant) is considered in this experiment. The data from different cameras (four side cameras, not include the top camera) are used to train the multi-view space HMMs. The view space is partitioned as usual according to the rotation viewpoint of human body to the camera. That is to say the data with the same rotation viewpoint will be clustered into the same sub-view space, no matter which camera the data belong to. A test is carried out by the leave one actor out from all cameras shown in Fig. 12.5. At each iteration, one sample corresponding to one action from one camera is chosen as testing sample, and new multi-view space HMMs are learned by using the remaining training samples.

Figure 12.6 shows the view-independent recognition confusion matrixes of the multi-view space HMMs with different sub-view space weights in experiment 3, and the average recognition rate is given under each confusion matrix.

The above confusion matrixes illustrate that most of actions with unknown viewpoint can be correctly recognised by the proposed approach. The average recognition rate reaches more than 85%, and the rate can be further improved by introducing the different sub-view space weight. Moreover, it is shown that the mis-recognition rates are mainly focus on "scratch head" and "wave". The main reason is the high similar appearances of the two actions in certain sub-view space.

Performance comparison: The proposed approach is compared with the state-of-the-art view-invariant methods. To make the comparison, our experiment is performed under the same test conditions with them. All of the videos in the dataset are used as experimental data. The leave one actor out (LOAO) is used as standard

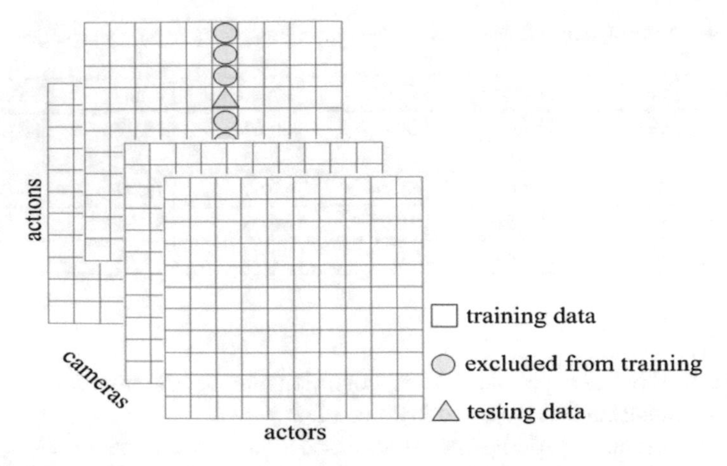

Fig. 12.5 Leave one actor out from all cameras

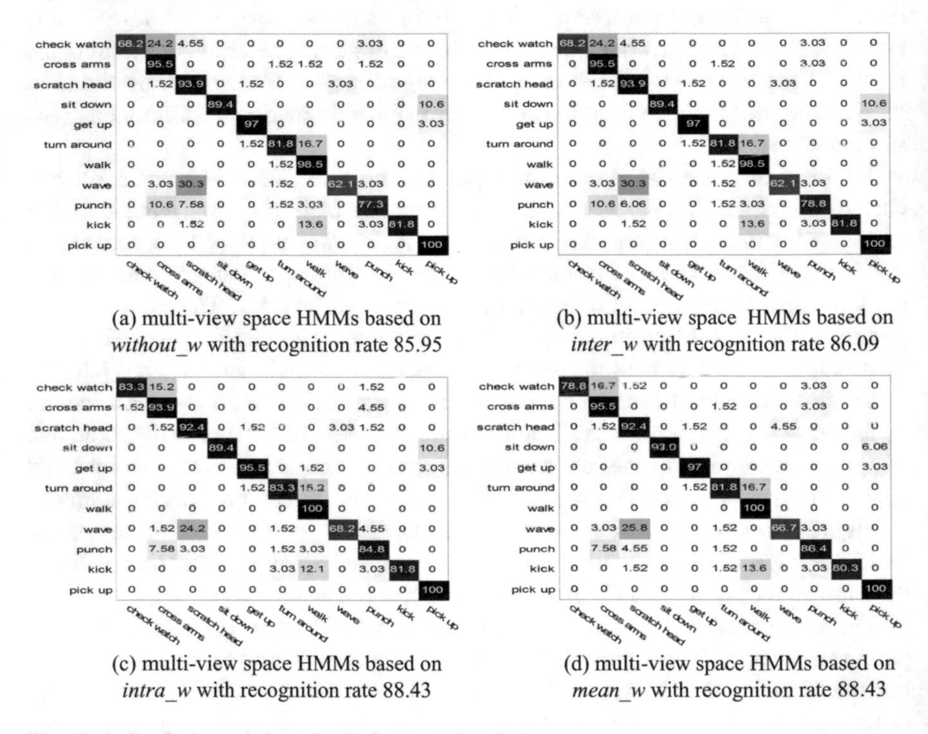

Fig. 12.6 Confusion matrixes (in %) for the IXMAS dataset

testing method. And the performance for different recognition methods is illustrated as Table 12.5.

From the Table 12.5, the results verified that the proposed approach outperforms other view-invariant methods. During feature extraction process, neither the key

Table 12.5 Comparison with related work in recent years

Literature	Method	Accuracy (%)
Lv et al. [18]	Shape context of silhouettes + PMK-NUP	80.6
Junejo et al. [30]	SSM desciptor + SVM	72.7
Weiland et al. [31]	3D HOG descriptor + Local SVM	83.4
Liu et al. [32]	Silhouette -optical flow -interest point + LWE	82.8
Our approach	Optical flow-SIFT + HMM based on $mean - w_v$	88.4

poses [18] are extracted from motion capture sequences nor the camera information [32] are required in advance. The proposed feature could be conveniently extracted and robust to the noise. During the recognition process, the contextual semantic information is used to combine with the HMMs due to the temporal characteristic of the feature. Then the multi-view space HMMs recognition method is employed. By effectively utilising the information among each sub-view space, the observation probabilities of each sub-HMM for one action are weighted and combined. Thus, the recognition accuracy and the robustness for view changes are improved. The experimental results show that the proposed approach is easy to implement and the performance for view-invariant action recognition is satisfactory. Our system runs at 3.82 sec/frames with a 2.94 GHz CPU. Most of the CPU time is spent on feature extracting process. The computational cost of recognition process only takes 0.03 sec/frames.

12.6 Conclusion

A novel multi-view space HMMs algorithm for view-invariant action recognition based on view space partitioning is proposed. The interest point feature based on local information and the optical flow feature based on global motion information are effectively combined. The combined feature is view-insensitive in the sub-view space. The multi-view space HMMs are built in the recognition process. Then the observation probabilities of each sub-HMM for one action are weighted and combined according to the similarity between the sub-view space and test sequence viewpoint, and the discrimination among the action models in the sub-view space. Finally, the action with unknown viewpoint is recognised by the combination results. Since there is no viewpoint transition is unconsidered in the proposed approach, the actions shared with the similar motions are usually mis-recognised in our approach. Therefore, study of a novel hierarchical model with view transition is the future work.

References

1. O.C. Jenkins, G.G. Serrano, and M.M. Loper. Interactive human pose and action recognition using dynamical motion primitives. *International Journal of Humanoid Robotics*, 4(02):365–385, 2007.
2. L. Wang, W. Hu, and T. Tan. Recent developments in human motion analysis. *IEEE Transactions on Systems, Man, and Cybernetics, Part C*, 34(3):334–352, 2004.
3. S. Lin H. Liu, Y. Zhang, and K. Tao. Automatic video-based analysis of athlete action. in. In *International Conference Image Analysis and Processing*, pages 205–210, Italy, 2007.
4. Xiao-Fei Ji, Qian-Qian Wu, Zhao-Jie Ju, and Yang-Yang Wang. Study of human action recognition based on improved spatio-temporal features. *International Journal of Automation and Computing*, 11(5):500–509, 2014.
5. X. Ji and H. Liu. Advances in View-Invariant Human Motion Analysis: A Review. *IEEE Transactions on Systems, Man and Cybernetics Part C*, 40(1):13–24, 2010.
6. A. Yilmaz and M. Shah. Actions sketch: a novel action representation. *Proc. IEEE Conf. Computer Vision and Pattern Recognition*, 1:984–989, 2005.
7. D. Weinland, R. Ronfard, and E. Boyer. Free viewpoint action recognition using motion history volumes. *Computer Vision and Image Understanding*, 104(2):249–257, 2006.
8. J. J. Guerrero G. Rogez, J. Martinez, and C. Orrite. Viewpoint independent human motion analysis in man-made environments. In *the 17th British Machine Vision Conference(BWVC)*, pages 659–668, UK, 2006.
9. Daniel Weinland, Remi Ronfard, and Edmond Boyer. Free viewpoint action recognition using motion history volumes. *Computer vision and image understanding*, 104(2):249–257, 2006.
10. Xinxiao Wu, Dong Xu, Lixin Duan, and Jiebo Luo. Action recognition using context and appearance distribution features. In *Computer Vision and Pattern Recognition (CVPR), 2011 IEEE Conference on*, pages 489–496. IEEE, 2011.
11. Yan Yan, Elisa Ricci, Ramanathan Subramanian, Gaowen Liu, and Nicu Sebe. Multitask linear discriminant analysis for view invariant action recognition. *IEEE Transactions on Image Processing*, 23(12):5599–5611, 2014.
12. Grégory Rogez, José Jesús Guerrero, Jesús Martínez, and Carlos Orrite-Urunuela. Viewpoint independent human motion analysis in man-made environments. In *BMVC*, volume 6, page 659, 2006.
13. F. Lv and R. Nevatia. Recognition and segmentation of 3-D human action using HMM and multi-class AdaBoost. *Proceedings of European Conference on Computer Vision*, 4:359–372, 2006.
14. P. Peursum, S. Venkatesh, and G. West. Tracking as recognition for articulated full body human motion analysis. In *IEEE International Conference on Computer Vision and Pattern Recognition*, pages 1–8, USA, 2007.
15. V. K. Singh and R. Nevatia. Simultaneous tracking and action recognition for single actor human actions. *IEEE Transactions on Visualization and Computer Graphics*, 27(12):1115–1123, 2011.
16. Xiaofei Ji, Ce Wang, and Yibo Li. A view-invariant action recognition based on multi-view space hidden markov models. *International Journal of Humanoid Robotics*, 11(01):1450011, 2014.
17. Xiaofei Ji, Zhaojie Ju, Ce Wang, and Changhui Wang. Multi-view transition hmms based view-invariant human action recognition method. *Multimedia Tools and Applications*, pages 1–18, 2015.
18. F. Lv and R. Nevatia. Single view human action recognition using key pose matching and viterbi path searching. In *IEEE International Conference on Computer Vision and Pattern Recognition*, USA, 2008.
19. R. Natarajan and View and P. Nevatia. View and scale invariant action recognition using multiview shape-flow models. *IEEE International Conference on Computer Vision and Pattern Recognition*, 2008:1–8, 2008.

20. B. Peng, G. Qian, and S. Rajko. View-invariant full-body gesture recognition from video. In *International Conference on Pattern Recognition*, pages 1–5, USA, 2008.
21. J. Shotton, A. Fitzgibbon, M. Cook, T. Sharp, M. Finocchio, R. Moore, A. Kipman, and A. Blake. Real-time human pose recognition in parts from single depth images. In *IEEE Conference on Computer Vision and Pattern Recognition*, pages 1297–1304. IEEE, 2011.
22. Lu Xia, Chia-Chih Chen, and JK Aggarwal. View invariant human action recognition using histograms of 3d joints. In *IEEE Computer Society Conference on Computer Vision and Pattern Recognition Workshops*, pages 20–27. IEEE, 2012.
23. Z. Cheng, L. Qin, Y. Ye, Q. Huang, and Q. Tian. Human daily action analysis with multi-view and color-depth data. In *Computer Vision–ECCV 2012. Workshops and Demonstrations*, pages 52–61. Springer, 2012.
24. A.R. Lee, H. Suk, and S.W. Lee. View-invariant 3d action recognition using spatiotemporal self-similarities from depth camera. In *International Conference on Pattern Recognition*, pages 501–505, Aug 2014.
25. Y. Yan, E. Ricci, G. Liu, R. Subramanian, and N. Sebe. Clustered multi-task linear discriminant analysis for view invariant color-depth action recognition. In *International Conference on Pattern Recognition*, pages 3493–3498, Aug 2014.
26. X. Ji, C. Wang, and Y. Li. View insensitive representation based recognition algorithm. In *the Chinese Control Conference*, pages 3877–3881, 2013.
27. V. Rabaud P. Dollar, G. Cottrell, and S. Belongie. Behavior recognition via sparse spatio-temporal features. In *Joint IEEE International Workshop on Visual Surveillance and Performance Evaluation of Tracking and Surveillance*, pages 65–72, 2005.
28. M. Bregonzio, S. Gong, and T. Xiang. Recognising action as clouds of space-time interest points. In *Proceeding of IEEE Conference on Computer Vision and Pattern Recognition*, pages 1948–1955, Florida, USA, 2009. IEEE.
29. D. Tran and A. Sorokin. Human activity recognition with metric learning. In *the 10th European Conf Computer Vision*, pages 548–561, France, 2008.
30. I. Junejo, E. Dexter, I. Laptev, and P. Perez. Cross-view action recognition from temporal self-similarities. In *the 10th European Conf Computer Vision*, pages 293–306, France, 2008.
31. D. Weinland, M. Ozuysal, and P. Fua. Making action recognition robust to occlusions and viewpoint changes. In *European Conference on Computer Vision*, pages 635–648, Greece, 2010.
32. J. Liu, B. Kuipers M. Shah, and S. Savarese. Cross-view action recognition via view knowledge transfer. In *IEEE International Conference on Computer Vision and Pattern Recognition*, pages 3209–3216, USA, 2011. IEEE.

Appendix A
Arithmetic Operations of Chap. 2

Arithmetic operations with four-tuple fuzzy numbers

Operation	Result	Conditions
−n	$[-d, -c, \delta, \gamma]$	all n
$\sqrt{n}$	$[\sqrt{c}, \sqrt{d}, \sqrt{c} - \sqrt{c-\gamma}, -\sqrt{d} + \sqrt{d+\delta}]$	$n >_0 0$
$1/n$	$\left[1/d, 1/c, \delta/(d(d+\delta)), \gamma/(c(c-\gamma))\right]$	$n >_0 0, n <_0 0$
m + n	$[a+c, b+d, \tau+\gamma, \beta+\delta]$	all m, n
m − n	$[a-d, b-c, \tau+\delta, \beta+\gamma]$	all m, n
$m \times n$	$[ac, bd, a\gamma + c\tau - \tau\gamma, b\delta + d\beta + \beta\delta]$	$m >_0 0, n >_0 0$
	$[ad, bc, d\tau - a\delta + \tau\delta, c\beta - b\gamma + \beta\gamma]$	$m <_0 0, n >_0 0$
	$[bc, ad, b\gamma - c\beta + \beta\gamma, a\delta - d\tau - \tau\delta]$	$m >_0 0, n <_0 0$
	$[bd, ac, -b\delta - d\beta - \beta\delta, \tau\gamma - a\gamma - c\tau]$	$m <_0 0, n <_0 0$
	m = [a,b,τ,β], n = [c,d,γ,δ]	

H. Liu et al., *Human Motion Sensing and Recognition*,
Studies in Computational Intelligence 675, DOI 10.1007/978-3-662-53692-6

Appendix B
Algorithm of Chap. 4

Algorithm 1 QNT BASED HUMAN MOTION RECOGNITION - TRAINING PHASE

Require: Define a human motion skeleton and generate symbolic QNTs as in Eq. 4.15
Ensure: $t_{tracking} < t_{ct}$ {Tracking computational time threshold t_{ct}}
Ensure: $t_{matching} < t_{tm}$ {Template match computational time threshold t_{tm}}
  SET $m_{tracking}$ {Set particle number for the condensation algorithm}
  $\phi_t = [(\phi_t^{pos})^T, (\phi_t^{vel})^T]^T \Leftarrow$ CONDENSATION TRACKING
  SET FQT_O, FQT_T {Set precision parameters for the fuzzy qualitative unit circle.}
  $\hat{\phi}_t \Leftarrow$ DATA QUANTISATION {Project and normalize ϕ_t as Eq. 4.8}
  TEMPLATE MATCHING {Compare $\hat{\phi}_t$ with the symbolic QNTs}
  return $m_{tracking}$, FQT_O and FQT_T

H. Liu et al., *Human Motion Sensing and Recognition*,
Studies in Computational Intelligence 675, DOI 10.1007/978-3-662-53692-6

Appendix C
Algorithm and Proof of Chap. 5

C.1 Algorithm

The EM algorithm of probability based FGMMs is shown in Algorithm 2.

Algorithm 2 EM algorithm of probability based FGMMs.

Require: Fix k, m and ε {k is the number of components $2 < k < n$; m is the degree of fuzziness $m > 1$; ε is a small preset real positive number}.

1: $U \leftarrow fcm(data, k)$ {Use FCM to generate matrix U satisfying Eq. 5.25}
2: **repeat**
3:   **for all** i such that $1 \leq i \leq k$ **do**
4:     $\alpha_i^{new} \leftarrow Eq.\ 5.32$ {Compute the k fuzzy mixture weights α_i^{new} using Eq. 5.32}
5:     $a_i \leftarrow Eq.\ 5.20$
6:     **if** $a_i < \varepsilon$ **then**
7:       $\mu_i^{new} \leftarrow Eq.\ 5.29$ {Compute the k fuzzy mean vectors using Eq. 5.29 }
8:       $\Sigma_i^{new} \leftarrow Eq.\ 5.33$ {Compute the k fuzzy covariance matrices using Eq. 5.33 }
9:     **else**
10:       $\{C_i^{new}, T_i^{new}, Q_i^{new}\} \leftarrow Eq.\ 5.20$ {Compute the k fuzzy control parameters of standard curve, translation and rotation matrices using Eq. 5.20 }
11:       $\mu_i^{new} \leftarrow Eq.\ 5.34$ {Compute the k fuzzy centers using Eq. 5.34}
12:       $\Sigma_i^{new} \leftarrow Eqs.\ 5.35\text{–}5.36$ {Compute the k fuzzy covariance matrices using Eqs. 5.35–5.36 }
13:     **end if**
14:   **end for**
15:   $U^{new} \leftarrow Eq.\ 5.28$ {Upgrade the fuzzy membership using Eq. 5.28 }
16:   $\log(\mathscr{L}(\Theta|\mathscr{X}))^{new} \leftarrow Eq.\ 5.5$ {Compute the log-likelihood using Eq. 5.5}
17: **until** $\frac{\log(\mathscr{L}(\Theta|\mathscr{X}))^{new}}{\log(\mathscr{L}(\Theta|\mathscr{X}))^{old}} - 1 \leq threshold$ {Stop if the relative difference of the log-likelihood between two adjacent iterations is blow the preset threshold}

H. Liu et al., *Human Motion Sensing and Recognition*,
Studies in Computational Intelligence 675, DOI 10.1007/978-3-662-53692-6

C.2 Proof

The proposed updating algorithm not only minimises the objective function 5.27 of FCM but also maximises the log-likelihood function 5.5 of GMMs. The theoretical justification is presented as below.

- **Maximising the log-likelihood function** 5.5 **of GMMs**
  The EM algorithm of normal GMMs guarantees the maximum likelihood function 5.5. In practice, the objective Q function 5.6 is taken instead.
  Comparing the two EM algorithms of normal GMMs (Eqs. 5.7, 5.9, 5.10, 5.11) and Fuzzy GMMs ($|a| < \varepsilon$) (Eqs. 5.28, 5.29, 5.32, 5.33), the only difference is that u_{it}^m is used as w_{it} in Fuzzy GMMs. Therefore, the EM algorithm of Fuzzy GMMs ($|a| < \varepsilon$) would guarantee the objective Q^* function:

$$Q^* = \sum_{t=1}^{n} \sum_{i=1}^{k} u_{it}^m \log[\alpha_i p_i(x_t|\theta_i)] \tag{C.1}$$

  In 5.3.4, we have proven that for probability based FGMMs

$$u_{it}^m \propto [\alpha_i p_i(x_t|\theta_i)]^{\frac{m}{m-1}} \tag{C.2}$$

  and for distance based FGMMs

$$u_{it}^m \propto \exp\left(\frac{-m}{m-1}\frac{(x_t-\mu_i)^T \Sigma_i^{-1}(x_t-\mu_i)}{2}\right) \tag{C.3}$$

  Because

$$w_{it} = \frac{\alpha_i p_i(x_t|\theta_i)}{\sum_{s=1}^{k} \alpha_s p_s(x_t|\theta_s)} \tag{C.4}$$

  for a fixed m, there exists a function f that $u_{it}^m = f(w_{it})$, which is a monotonic increasing function. So that there exists a function g that $Q^* = g(Q)$, which is also a monotonic increasing function. When Q^* reaches to its local maximum, Q also gets local maximum. Therefore, the EM algorithm of FGMM ($|a| < \varepsilon$) guarantees a local optimal search.
  The same result can be achieved for the EM algorithm of FGMM ($|a| \geq \varepsilon$) by the similar justification.
- **Minimising the objection function** 5.27 **of FCM**
  Because the degree of membership u_{it} and the position of centre μ_i in our proposed EM algorithm are achieved by the same way as in FCM, i.e. by the functions 5.28 and 5.29, the objection function 5.27 will be minimised.

Appendix D
Algorithms of Chap. 6

D.1 Fuzzy Empirical Copula algorithm

Algorithm 3 Fuzzy Empirical Copula algorithm

Require: $X = \{x_1, x_2, ..., x_n\}$ {X is a r dimensional dataset with n objects and $r << n$}
Require: $[\rho_{all}, \gamma_{all}]$ {the Spearman's rho and Gini's gamma of the original data X}
1: **for all** i such that $1 \leq i \leq n$ **do**
2:     **for all** j such that $1 \leq j \leq n$ **do**
3:         $d_2(x_i, x_j) \leftarrow Eq.6.8$ {calculate the Euclidean distance of two objects by Eq. 6.8}
4:     **end for**
5: **end for**
6: $K = 0$ {K is the number of nearest neighbours under consideration}
7: **repeat**
8:     $K = K + 1$
9:     **for all** i such that $1 \leq i \leq n$ **do**
10:         $Den_2(x_i) \leftarrow Eq.$ 6.10 {get the density of every object}
11:     **end for**
12:     $c = 0$ {the number of CSOs}
13:     **for all** i such that $1 \leq i \leq n$ **do**
14:         **if** $Den_2(x_i) \geq max(Den_2(y_i))$ where $y_i \in knn(x_i)$ **then**
15:             $c = c + 1$
16:             $CSO_c \leftarrow x_i$ {get CSOs which have the local maximum densities}
17:         **end if**
18:     **end for**
19:     **for all** u such that $1 \leq u \leq r$ **do**
20:         **for all** v such that $(u + 1) \leq v \leq r$ **do**
21:             $[\rho(u, v), \gamma(u, v)] \leftarrow EmpSG(CSO(u), CSO(v))$ {$CSO(i)$ is the ith attribution of CSO, $EmpSG$ is a function to calculate the Spearman's rho and Gini's gamma as showed in algorithm 4, $\rho_{r\times r}$ is the Spearman's rho matrix and $\gamma_{r\times r}$ is the Gini's gamma matrix}
22:         **end for**
23:     **end for**
24:     $error = \|[\rho_{all}, \gamma_{all}] - [\rho_{r\times r}, \gamma_{r\times r}]\|$ {take the Euclidean distance of the ρ and γ as the overall error}
25: **until** $error \geq threshold$ {$threshold$ is for the overall error according to the original dataset}

H. Liu et al., *Human Motion Sensing and Recognition*,
Studies in Computational Intelligence 675, DOI 10.1007/978-3-662-53692-6

D.2 Function of $EmpSG$ for Spearman's rho and Gini's gamma

Algorithm 4 Function of $EmpSG$ for Spearman's rho and Gini's gamma

Require: $CSO(u)$, $CSO(v)$ {two attributes in CSO}
Ensure: SP, GI {Spearman's rho and Gini's gamma of above two attributes}
1: $x = CSO(u)$; $y = CSO(v)$ {x and y are two vectors with c elememts}
2: $x' = sort(x)$; $y' = sort(y)$ {x' and y' are the order statistics of x and y}
3: **for all** i such that $1 \leq i \leq c$ **do**
4:   **for all** j such that $1 \leq j \leq c$ **do**
5:     $num \leftarrow 0$ {initialisation}
6:     **for all** t such that $1 \leq t \leq c$ **do**
7:       **if** $x(t) \leq x'(i)$ and $y(t) \leq y'(j)$ **then**
8:         $num \leftarrow num + 1$
9:       **end if**
10:     **end for**
11:     $EC(i, j) \leftarrow num/c$
12:   **end for**
13: **end for**
14: $SP \leftarrow 0$ {the return value of Spearman's rho}
15: $GI \leftarrow 0$ {the return value of Gini's gamma}
16: **for all** i such that $1 \leq i \leq c$ **do**
17:   **for all** j such that $1 \leq j \leq c$ **do**
18:     $SP \leftarrow SP + EC(i, j) - (j * i)/(c * c)$
19:   **end for**
20:   **if** $i \neq b$ **then**
21:     $GI \leftarrow GI + EC(i, c - i) - i/c + EC(i, i)$
22:   **end if**
23: **end for**
24: $SP \leftarrow SP * 12/(c * c - 1)$
25: $GI \leftarrow 2 * c/(c * c/2) * (GI - 1 + EC(c, c))$
26: **return** SP and GI

Appendix E
Algorithms of Chap. 9

E.1 Algorithm for Hand Edge Contour Generation

Algorithm 5 Algorithm for Hand Edge Contour Generation

Require: Pixel[r] {Pixel is a 4 dimensional dataset with n pixels in the RGB-D image and $r = \{1, \ldots, n\}$, n is the total number of the pixels in the image}.
Require: Hand_centre {Hand_centre is the location of the centre of the tracked hand}
Require: initial_depth {initial_depth is the average depth of the tracked hand}
  mini_depth = initial_depth-offset {offset is the user chosen threshold}
  max_depth = initial_depth+offset
1: **for all** i such that $1 \leq i \leq n$ **do**
2:   **if** $mini\_depth \leq depth(i)$ and $|pos(i) - Hand\_centre| \leq 0.5 * body\_length$ {body_length is the preset length of the body in pixels; depth(i) returns the depths of pixcel[i]; pos() gets the position of pixel[i]} **then**
3:     $is\_hand[i] \leftarrow true$ {is_hand[] is an array that indicates hand pixels}
4:   **else**
5:     $is\_hand[i] \leftarrow false$
6:   **end if**
7: **end for**
8: **for all** i such that $1 \leq i \leq n$ **do**
9:   **if** $is\_hand[i] == true$ **then**
10:     $K \leftarrow \{k1, \ldots, k8 || pos(k_r) - pos(i)| < 2\}$ {get the eight nearby pixels around the i_{th} pixel, $0 \leq r \leq n$}
11:     **if** $\prod_{j=1}^{8} is\_hand[k_j] == true$ **then**
12:       $is\_contour[i] \leftarrow true$ {$is\_contour$[] is an array that indicates the hand contour}
13:     **else**
14:       $is\_contour[i] \leftarrow false$
15:     **end if**
16:   **else**
17:     $is\_contour[i] \leftarrow false$
18:   **end if**
19: **end for**
20: **return** $is\_contour$

H. Liu et al., *Human Motion Sensing and Recognition*,
Studies in Computational Intelligence 675, DOI 10.1007/978-3-662-53692-6

E.2 Lasso Algorithm

Algorithm 6 Lasso Algorithm

```
Require: m_signature {structure of fingertip feature, which contains
    int a, b {the coordinate values of the fingers}
    bool R_suspended, L_suspended {indicates if the lasso is suspended}
    double_length {the length of the lasso}}
Require: m_signature_num {the number of fingertip features}
Require: L_cur {int, left cursor of the lasso}
Require: R_cur {int, right cursor of the lasso}
Require: L_cur_move {bool, left cursor should be moved or not}
Require: R_cur_move {bool, right cursor should be moved or not}
Require: distance {int, the dynamic length of the lasso}
Require: pre_distance {int, the former dynamic length of the lasso}
    for all i such that 1 ≤ i ≤ m_signature_num do
2:    while distance < m_signature[i].rope_length {if the length of the lasso is longer than the
      preset value of the width of the finger,that means the lasso has clamped completely, the lasso
      procedure will come to an end} do
        if R_suspended and R_cur_move then
4:        R_cur ← R_cur + 1 {move R_cur right}
          calculate R_suspended
6:        if R_suspended == true {if the right side of the lasso is suspended, then its left side
          will move left} then
            R_cur_move ← false
8:          calculate distance
          end if
10:       if distance > pre_distance and !L_suspended {if the length of the lasso becomes
          longer, then its left side will move left} then
            R_cur_move ← false
12:         pre_distance ← distance
          else if !L_suspended and L_cur_move then
14:         L_cur ← L_cur + 1
            calculate L_suspended
16:       end if
          if L_suspended == true {if the left side of the lasso is suspended, then move the right
          side towards the right next} then
18:         R_cur_move ← true
            calculate distance
20:       end if
          if distance > pre_distance && !R_suspended {if the length of the lasso become
          longer,then move the right side towards the right} then
22:         R_cur_move ← true
            pre_distance ← distance
24:       end if
        end if
26:   end while
      m_signature[i].a ← L_cur
28:   m_signature[i].b ← R_cur
    end for
```

Index

H. Liu et al., *Human Motion Sensing and Recognition*,
Studies in Computational Intelligence 675, DOI 10.1007/978-3-662-53692-6

Printed in the United States
By Bookmasters